BIOLOGY
FOR NGSS

Student Workbook

BIOLOGY
FOR NGSS

Student Workbook

First edition 2014
Third printing with corrections

ISBN 978-1-927173-84-8
Copyright © **2014** Richard Allan
Published by **BIOZONE International Ltd**

Printed by ABC Printing Company, Chicago, USA

Purchases of this workbook may be made direct from the publisher:

www.the**BIOZONE**.com

BIOZONE International Ltd.
P.O. Box 5002, Hamilton 3242, New Zealand
Telephone: +64 7-856 8104
Fax: +64 7-856 9243
Toll FREE phone: 1-855-246-4555 (USA-Canada only)
Toll FREE fax: 1-855-935-3555 (USA-Canada only)
Email: sales@biozone.co.nz
Website: www.the**BIOZONE**.com

Cover photograph

The ring-tailed lemur (*Lemur catta*) is one of the most recognized primates with a distinctive black and white ringed tail. It belongs to Lemuridae, one of five lemur families, and is the only member of the *Lemur* genus. Like all lemurs it is endemic to the island of Madagascar. It is a highly social species, living in groups of up to 30, with a female dominance hierarchy. Although it is listed as near threatened, it breeds readily in captivity and there are more than 2000 in captive breeding programs in zoos worldwide.
PHOTO: © irakite/www.istockphoto.com

Note to the student

BIOZONE's new title for the High School Life Sciences component of NGSS has been designed with today's students in mind. It contains a wealth of engaging, accurate, and up-to-date material, and is fully compliant with the content and aims of the new Next Generation Science Standards.

To use this workbook most effectively, take note of the features outlined in this introduction. Understanding the activity coding system and using the external resources we have provided will help to scaffold your learning so that you understand the principles and processes involved in each topic of study.

Key features include:

▶ Full alignment to the High School Life Sciences component of the NGSS. This workbook is structured on the Disciplinary Core Ideas, with science practices and crosscutting concepts integrated throughout to provide an thorough and balanced treatment of content.
▶ A key idea for every activity that clearly and concisely summarizes the key points of the activity.
▶ High proportion of content relates directly to performance expectations. Activities suitable as assessment tasks are written and organized in a way to help you meet these performance expectations.
▶ A direct questioning style simplifies questions and helps you to craft your answers based on your own knowledge and experience.
▶ The language used is appropriate for grade 9-10 students and the chapter structure scaffolds learning by building on prior knowledge.
▶ Specific, clear CORE IDEAS correspond directly to the DCIs of the NGSS document.
▶ We have provided a strong focus on scientific literacy and learning within relevant contexts.
▶ Engaging graphics and widespread use of annotated diagrams help your understanding of difficult concepts.
▶ WEB and CONNECT tabs help you to locate external resources and provide crosscutting connections between related concepts.

A note to the teacher

This workbook is a student-centered resource, and benefits students by facilitating independent learning and critical thinking. This workbook is just that - a place for the student's answers notes, asides, and corrections. It is **not a textbook** and regular revisions are our commitment to providing a current, flexible, and engaging resource. The low price is a reflection of this commitment. Please **do not photocopy** the activities. If you think it is worth using, then we recommend that students purchase and retain it for their own use. I thank you for your support.
Richard Allan, Founder & CEO

Meet the writing team

Tracey Greenwood
I have been writing resources for students since 1993. I have a Ph.D in biology, specialising in lake ecology and I have taught both graduate and undergraduate biology.

Tracey
Senior Author

Lissa Bainbridge-Smith
I worked in industry in a research and development capacity for eight years before joining BIOZONE in 2006. I have an M.Sc from Waikato University.

Lissa
Author

Kent Pryor
I have a BSc from Massey University majoring in zoology and ecology and taught secondary school biology and chemistry for 9 years before joining BIOZONE as an author in 2009.

Kent
Author

Richard Allan
I have had 11 years experience teaching senior secondary school biology. I have a Masters degree in biology and founded BIOZONE in the 1980s after developing resources for my own students.

Richard
Founder & CEO

Acknowledgements

We thank all those who have contributed towards this edition:
The staff at BIOZONE, including Denise Fort and Gemma Conn for design and graphics support, Paolo Curray for IT support, and Debbie Antoniadis and Tim Lind for office handling and procedures. Gwen Gilbert, Nadège Stoffel, and Rebecca Johnson for sales and marketing.

Contents

CODES: **Activity** is marked: • to be done   ✓ when completed

Contents

CODES: **Activity** is marked: ☐ to be done   ☑ when completed

Contents

CODES: **Activity** is marked: ● to be done ✓ when completed

Using This Workbook

▶ The outline of the chapter structure below will help you to navigate through the material in each chapter.

Introduction
- A check list of core ideas for the chapter
- A list of key terms

Activities
- The KEY IDEA provides your focus for the activity
- Annotated diagrams help you understand the content
- Questions review the content of the page

Review
- Create your own summary for review
- Hints help you to focus on what is important
- Your summary will help you to consolidate your understanding of the chapter

Literacy
- Activities are based on the introductory key terms list
- Several types of activities test your understanding of concepts and biological terms

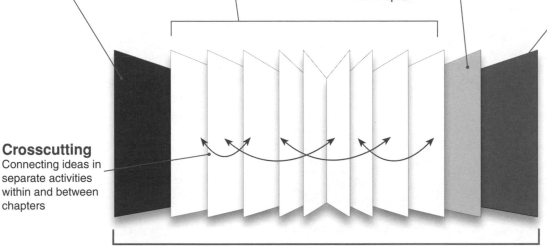

Crosscutting
Connecting ideas in separate activities within and between chapters

Structure of a chapter

▶ The activities make up most of this workbook. Each one has a similar structure and they are organized through the chapter in a way that unpacks the material in a series of steps.

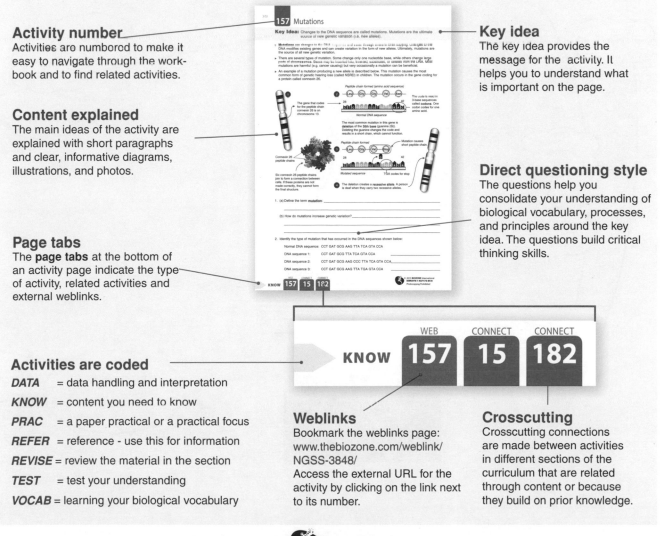

Activity number
Activities are numbered to make it easy to navigate through the workbook and to find related activities.

Content explained
The main ideas of the activity are explained with short paragraphs and clear, informative diagrams, illustrations, and photos.

Page tabs
The **page tabs** at the bottom of an activity page indicate the type of activity, related activities and external weblinks.

Key idea
The key idea provides the message for the activity. It helps you to understand what is important on the page.

Direct questioning style
The questions help you consolidate your understanding of biological vocabulary, processes, and principles around the key idea. The questions build critical thinking skills.

Activities are coded

DATA	= data handling and interpretation
KNOW	= content you need to know
PRAC	= a paper practical or a practical focus
REFER	= reference - use this for information
REVISE	= review the material in the section
TEST	= test your understanding
VOCAB	= learning your biological vocabulary

KNOW **157** CONNECT **15** CONNECT **182** (WEB)

Weblinks
Bookmark the weblinks page: www.thebiozone.com/weblink/NGSS-3848/
Access the external URL for the activity by clicking on the link next to its number.

Crosscutting
Crosscutting connections are made between activities in different sections of the curriculum that are related through content or because they build on prior knowledge.

NGSS: Concepts and Connections

This map shows the structure of the NGSS Life Science program as represented in this workbook. The dark blue boxes indicate the book sections, each of which has its own concept map. The blue ovals are the chapters in each section. We have placed some major connections between topics. You can make more of your own.

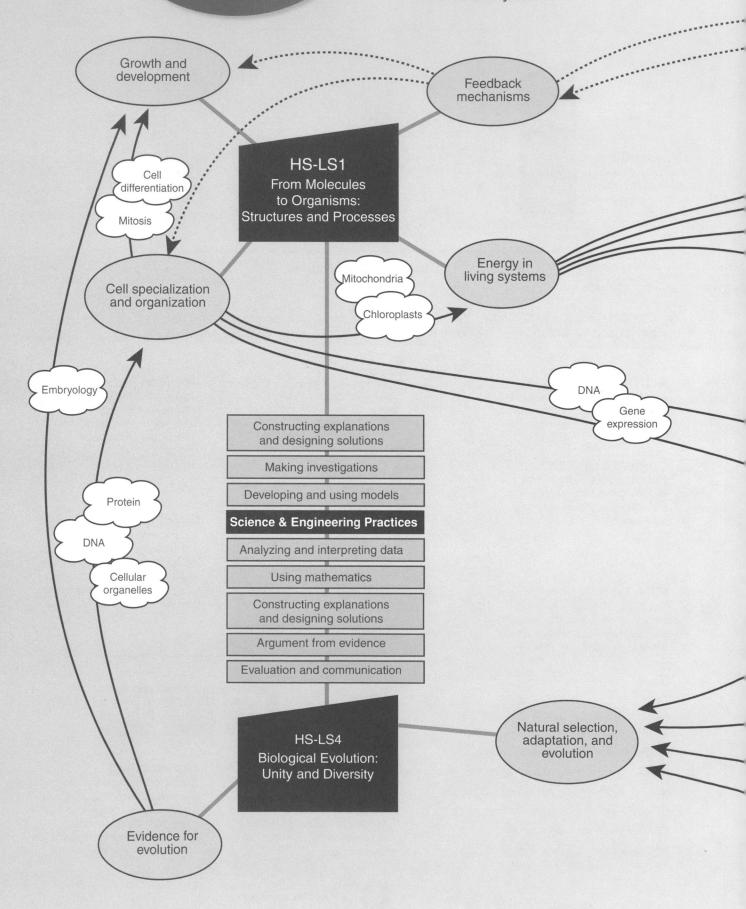

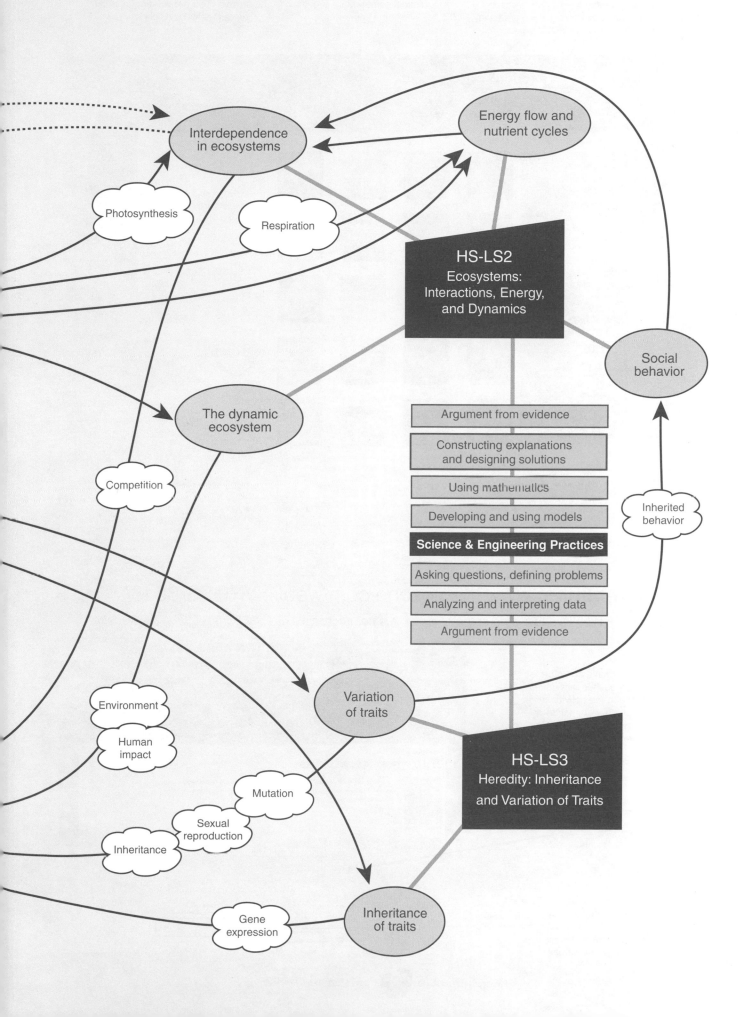

Using BIOZONE's Website

Access the **BIOLINKS** database of web sites directly from the homepage of our new website. The Biolinks page is organized into easy-to-use sub-sections relating to general areas of interest. It's a great way to quickly find out more about specific topics.

Contact us with questions, feedback, ideas, and critical commentary. We welcome your input.

You can search our website for specific content. Use specific key terms if you are wanting to locate resources within Biolinks.

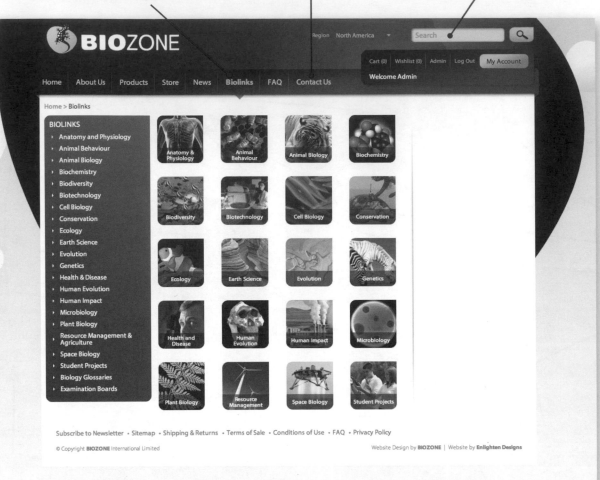

Weblinks: www.thebiozone.com/weblink/NGSS-3848/

Bookmark Weblinks by typing in the address: it is not accessible directly from BIOZONE's website

Throughout this workbook, some pages make reference to websites that have particular relevance to the activity by providing an explanatory animation or video clip. They are easy to use and a very useful supplement to the activity.

Activity reference: The activity on which the weblink is cited.

Weblink: Provides a link to an **external web site** with supporting information for the activity. Additional weblinks linked to the same activity will have the same Weblink number.

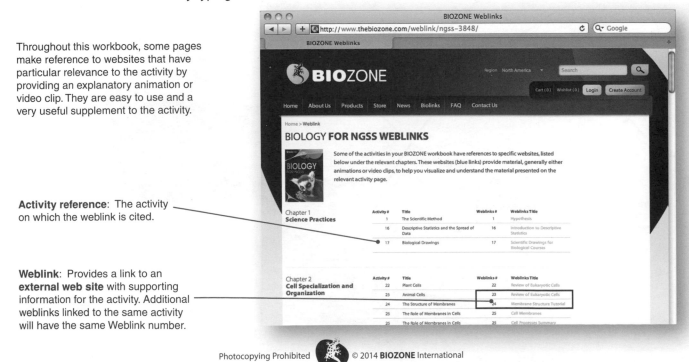

Science Practices

Key terms

accuracy

assumption

biological drawing

control

controlled variable

data

descriptive statistics

dependent variable

graph

hypothesis

independent variable

mean

median

mode

model

observation

precision

prediction

qualitative data

quantitative data

raw data

scientific method

table

variable

Science practices

Activity number

The scientific method

☐ 1. Outline the basis of the **scientific method**, understanding that **observation** is the basis for forming hypotheses and making **predictions** about **systems** or their components.

`1`

☐ 2. Plan and conduct an investigation to provide data to test a **hypothesis** based on observations. Identify any **assumptions** in the design of your investigation.

`3 19`

☐ 3. Develop and use **models** based on evidence to describe systems or their components and how they work.

`2`

Collecting data

☐ 4. In your investigation, consider and evaluate the **accuracy** and **precision** of the **data** that you collect.

`4`

☐ 5. Data may be **quantitative**, **qualitative**, or ranked. Quantitative data can be **continuous** or **discontinuous**.

`5`

☐ 6. **Variables** are factors that can change or be changed in an experiment. Classify the variables in your investigation as **independent**, **dependent**, and **controlled**.

`6`

☐ 7. Understand the purpose of experimental **controls** as a way to insure that the experiment is working properly and the results obtained are due to the variable being tested.

`6`

☐ 8. Use **tables** to accurately and systematically record results and summarize data.

`8 11`

Transforming and graphing data

☐ 9. Demonstrate an ability to transform **raw data** to identify important features. Common transformations include tallies, percentages, and rates.

`9 10`

☐ 10. Use **graphs** to visualize data and identify trends. Use a graph type appropriate to the type of data you have collected.

`12 13 14 15`

☐ 11. Summarize your data set and describe its basic features using **descriptive statistics** (e.g. **mean**, **median**, and **mode**). Use a statistic appropriate to the type of data and its distribution.

`16`

Biological drawings

☐ 12. **Biological drawings** record what a specimen looks like, and provides an opportunity to record its important features. A good biological drawing records what you have actually seen, and includes only as much detail as needed to distinguish different structures or tissues.

`17 18`

1 The Scientific Method

Key Idea: The scientific method involves making and then testing a hypothesis based on observation. A hypothesis is a tentative explanation for an observation.

What is the scientific method?

The scientific method is a way to ask and answer questions about science. A model of how the scientific method might operate is shown below. In reality, the observation, testing, and use of data is fluid and may not proceed in a strict linear direction as the pool of data, techniques, and ideas develops and increases over time.

What is a hypothesis?

A **hypothesis** is a tentative explanation for an observation. It can be tested by experimentation. The hypothesis can be accepted or rejected based on the findings of the investigation.

Features of a sound hypothesis:

- It is based on observations and prior knowledge.
- It offers a possible explanation for an observation.
- It leads to predictions about a system.
- It can be tested by experimentation.

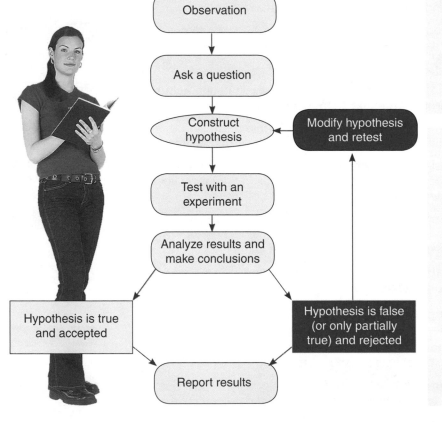

Testing a hypothesis

A hypothesis needs to be tested using sound methodology. Features of a sound method include:

- It is repeatable. Another person should be able to carry out the method and achieve the same results as the initial investigation.
- It tests the validity of the hypothesis.
- It includes a control which does not receive the treatment.
- All variables are controlled where possible.
- The method includes a dependent and independent variable.
- Only the independent variable is changed (manipulated) between treatment groups.
- A hypothesis may be rejected or modified at a later date as new information from later investigations is revealed.

The null hypothesis

Every hypothesis has a corresponding null hypothesis, which is a hypothesis of no difference or no effect.

- A null hypothesis enables a hypothesis to be tested using statistical tests.
- If the results of an experiment are statistically significant, the null hypothesis can be rejected.
- If a hypothesis is accepted, anyone should be able to test the predictions with the same methods and get a similar result each time.

1. What is a hypothesis? _____

_____

2. Why can an accepted hypothesis be rejected at a later date? _____

_____

3. Why is it important that a method being used to test a hypothesis is repeatable? _____

_____

_____

© 2014 **BIOZONE** International
ISBN: 978-1-927173-84-8

2 Systems and System Models

Key Idea: Scientists use models to learn about biological systems. Models usually study one small part of a system, so that the system can be more easily understood.

A **system** is a set of interrelated components that work together. Energy flow in ecosystems (such as the one on the right), gene regulation, interactions between organ systems, and feedback mechanisms are all examples of systems studied in biology.

Scientists often used models to learn about biological systems. A **model** is a representation of a system and is useful for breaking a complex system down into smaller parts that can be studied more easily. Often only part of a system is modelled. As scientists gather more information about a system, more data can be put into the model so that eventually it represents the real system more closely.

Modeling data

There are many different ways to model data. Often seeing data presented in different ways can help to understand it better. Some common examples of models are shown here.

Visual models
Visual models can include drawings, such as these plant cells on the right.

Three dimensional models can be made out of materials such as modeling clay and ice-cream sticks, like this model of a water molecule (below).

Mathematical models
Displaying data in a graph or as a mathematical equation, as shown below for logistic growth, often helps us to see relationships between different parts of a system.

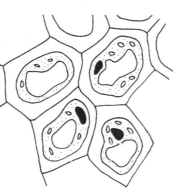

$$N = \frac{N_0 K}{N_0 + (K - N_0)\, e^{-rt}}$$

Population numbers (N) — Carrying capacity (K) — Time

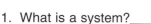
James Hedberg

Analogy
An analogy is a comparison between two things. Sometimes comparing a biological system to an everyday object can help us to understand it better. For example, the heart pumps blood in blood vessels in much the same way a fire truck pumps water from a fire hydrant through a hose. Similarly, ATP is like a fully charged battery in a phone.

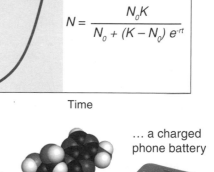
... a charged phone battery

ATP is like...

1. What is a system? _____

_____

2. (a) What is a model? _____

_____

(b) Why do scientists often study one part of a system rather than the whole system? _____

_____

_____

© 2014 **BIOZONE** International
ISBN: 978-1-927173-84-8
Photocopying Prohibited

CONNECT — CONNECT — CONNECT

132 **99** **17** **KNOW**

3 Observations and Assumptions

Key Idea: Observations are the basis for forming hypotheses and making predictions about systems. An assumption is something that is accepted as true but is not tested.

Observations

An observation is watching or noticing what is happening. Observation is the basis for forming hypotheses and making predictions. An observation may generate a number of hypotheses, and each hypothesis will lead to one or more predictions, which can be tested by further investigation.

Observation 1: Some caterpillar species are brightly colored and appear to be highly visible to predators such as insectivorous birds. Predators appear to avoid these caterpillar species. These caterpillars are often found in groups.

Observation 2: Some caterpillar species have excellent camouflage. When alerted to danger they are difficult to see because they blend into the background. These caterpillars are usually found alone.

Assumptions

Any biological investigation requires you to make **assumptions** about the biological system you are working with. Assumptions are features of the system you are studying that you assume to be true but that you do not (or cannot) test. Some assumptions about the two caterpillar systems described above include:

- Insect eating birds have color vision.
- Caterpillars that look bright to us, also appear bright to insectivorous birds.
- Insectivorous birds can learn about the tastiness of prey by eating them.

1. Read the two observations about the caterpillars above and then answer the following questions:

    (a) Generate a hypothesis to explain the observation that some caterpillars are brightly colored and highly visible while others are camouflaged and blend into their surroundings:

    Hypothesis: _____

    _____

    _____

    (b) Describe one of the **assumptions** being made in your hypothesis: _____

    _____

    (c) Generate a **prediction** about the behavior of insect eating birds towards caterpillars: _____

    _____

    _____

 © 2014 **BIOZONE** International
ISBN: 978-1-927173-84-8
Photocopying Prohibited

KNOW

4 Accuracy and Precision

Key Idea: Accuracy refers to the correctness of a measurement (how true it is to the real value). Precision refers to how close the measurements are to each other.

The terms accuracy and precision are two terms that are often used when talking about scientific measurements.

Accuracy refers to how close a measured value is to its true value. Simply put, it is the correctness of the measurement.

Precision refers to the closeness of repeated measurements to each other, i.e. the ability to be exact.

For example, a digital device, such as a pH meter, will give very precise measurements, but its accuracy depends on correct calibration.

Using the analogy of a target, repeated measurements are compared to arrows being shot at a target. This analogy can be useful when thinking about the difference between accuracy and precision.

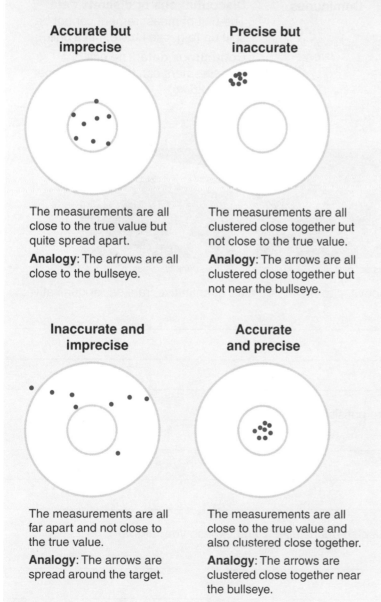

Accurate but imprecise

The measurements are all close to the true value but quite spread apart.

Analogy: The arrows are all close to the bullseye.

Precise but inaccurate

The measurements are all clustered close together but not close to the true value.

Analogy: The arrows are all clustered close together but not near the bullseye.

Inaccurate and imprecise

The measurements are all far apart and not close to the true value.

Analogy: The arrows are spread around the target.

Accurate and precise

The measurements are all close to the true value and also clustered close together.

Analogy: The arrows are clustered close together near the bullseye.

When collecting scientific data, it is important to take measurements that are both accurate and precise. Taking care with your measurements and observations will help to insure that the sample data you collect will be close to the true value of that measured variable in the wider population.

1. What is accuracy?

_____

_____

_____

_____

2. What is precision?

_____

_____

_____

_____

3. Why are precise but inaccurate measurements not helpful in a biological investigation?

_____

_____

_____

_____

_____

_____

_____

KNOW

5 Types of Data

Key Idea: Data is information collected during an investigation. Data may be quantitative, qualitative, or ranked.

Data is information collected during an investigation. Data may be quantitative, qualitative, or ranked. When planning a biological investigation, it is important to consider the type of data that will be collected. It is best to collect quantitative or numerical data, because it is easier to analyze it objectively (without bias).

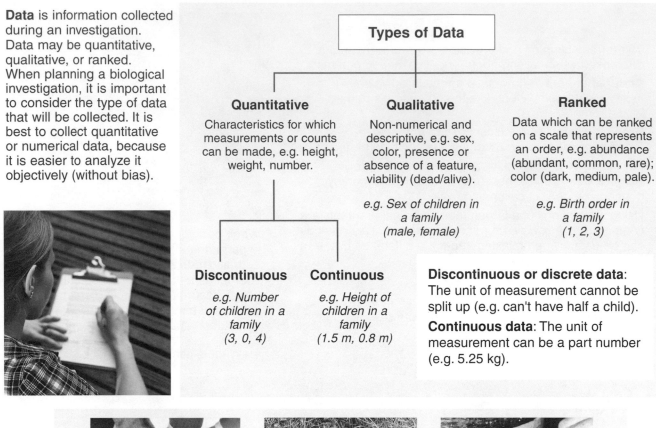

Types of Data

Quantitative
Characteristics for which measurements or counts can be made, e.g. height, weight, number.

Qualitative
Non-numerical and descriptive, e.g. sex, color, presence or absence of a feature, viability (dead/alive).

e.g. Sex of children in a family (male, female)

Ranked
Data which can be ranked on a scale that represents an order, e.g. abundance (abundant, common, rare); color (dark, medium, pale).

e.g. Birth order in a family (1, 2, 3)

Discontinuous
e.g. Number of children in a family (3, 0, 4)

Continuous
e.g. Height of children in a family (1.5 m, 0.8 m)

Discontinuous or discrete data: The unit of measurement cannot be split up (e.g. can't have half a child).
Continuous data: The unit of measurement can be a part number (e.g. 5.25 kg).

A: Skin color

B: Eggs per nest

C: Tree trunk diameter

1. For each of the photographic examples A-C above, classify the data as quantitative, ranked, or qualitative:

   (a) Skin color: _____

   (b) Number of eggs per nest: _____

   (c) Tree trunk diameter: _____

2. Why is it best to collect quantitative data where possible in biological studies? _____

   _____

   _____

   _____

3. Give an example of data that could not be collected quantitatively and explain your answer:

   _____

   _____

   _____

© 2014 **BIOZONE** International
ISBN: 978-1-927173-84-8

6 Variables and Controls

Key Idea: Variables may be dependent, independent, or controlled. A control in an experiment allows you to determine the effect of the independent variable.

Types of Variables

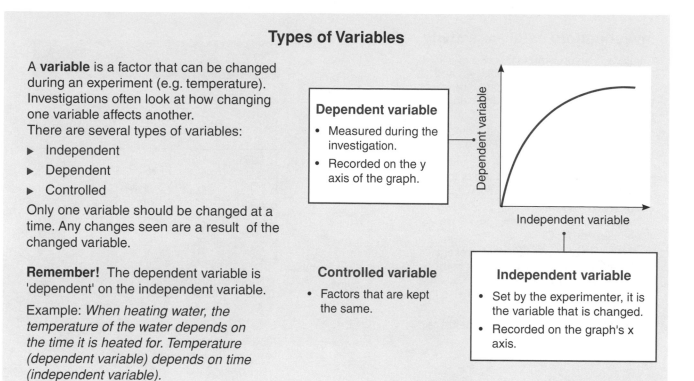

A **variable** is a factor that can be changed during an experiment (e.g. temperature). Investigations often look at how changing one variable affects another.
There are several types of variables:

▶ Independent
▶ Dependent
▶ Controlled

Only one variable should be changed at a time. Any changes seen are a result of the changed variable.

Remember! The dependent variable is 'dependent' on the independent variable.

Example: *When heating water, the temperature of the water depends on the time it is heated for. Temperature (dependent variable) depends on time (independent variable).*

Dependent variable
- Measured during the investigation.
- Recorded on the y axis of the graph.

Controlled variable
- Factors that are kept the same.

Independent variable
- Set by the experimenter, it is the variable that is changed.
- Recorded on the graph's x axis.

Experimental Controls

▶ A **control** is the standard or reference treatment in an experiment. Controls make sure that the results of an experiment are due to the variable being tested (e.g. nutrient level) and not due to another factor (e.g. equipment not working correctly).

▶ A control is identical to the original experiment except it lacks the altered variable. The control undergoes the same preparation, experimental conditions, observations, measurements, and analysis as the test group.

▶ If the control works as expected, it means the experiment has run correctly, and the results are due to the effect of the variable being tested.

Test plant
(nutrient added)

Control plant
(no nutrient added)

An experiment was designed to test the effect of a nutrient on plant growth. The control plant had no nutrient added to it. Its growth sets the baseline for the experiment. Any growth in the test plant above that seen in the control plant is due to the presence of the nutrient.

1. What is the difference between a dependent variable and an independent variable? _____

_____

_____

2. Why should experiments include a control? _____

_____

KNOW

7 A Case Study: Catalase Activity

Key Idea: A simple experiment to test a hypothesis involves manipulating one variable (the independent variable) and recording the response.

Investigation: catalase activity

Catalase is an enzyme that converts hydrogen peroxide (H_2O_2) to oxygen and water. An experiment investigated the effect of temperature on the rate of the catalase reaction.

- 10 cm^3 test tubes were used for the reactions, each tube contained 0.5 cm^3 of catalase enzyme and 4 cm^3 of H_2O_2.
- Reaction rates were measured at four temperatures (10°C, 20°C, 30°C, 60°C).
- For each temperature, there were two reaction tubes (e.g. tubes 1 and 2 were both kept at 10°C).
- The height of oxygen bubbles present after one minute of reaction was used as a measure of the reaction rate. A faster reaction rate produced more bubbles than a slower reaction rate.
- The entire experiment, was repeated on two separate days.

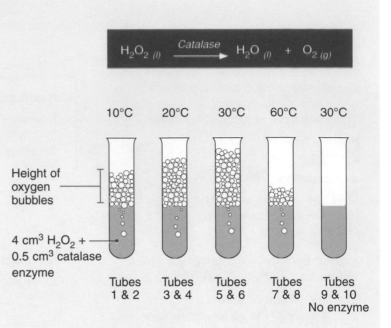

1. Write a suitable aim for this experiment: _____

_____

_____

2. Write an hypothesis for this experiment: _____

_____

_____

3. (a) What is the **independent variable** in this experiment? _____

(b) What is the range of values for the independent variable? _____

(c) Name the unit for the independent variable: _____

(d) List the equipment needed to set the independent variable, and describe how it was used:

_____

_____

4. (a) What is the **dependent variable** in this experiment? _____

(b) Name the unit for the dependent variable: _____

(c) List the equipment needed to measure the dependent variable, and describe how it was used:

_____

_____

5. Which tubes are the control for this experiment? _____

TEST

8 Recording Results

Key Idea: Accurately recording results makes it easier to understand and analyze your data later. A table is a good way to record data.

Recording your results accurately is very important in an experiment. Analyzing and understanding your data is easier when you have recorded your results accurately, and in an organized way.

A table is often the best way to record and present your results. Tables can also be useful for showing calculated values (such as rates and means).

An example of a table for recording results is shown below. It relates to an investigation looking at the growth of plants at three pH levels, but it represents a relatively standardized layout which you can adapt for your own investigations.

Dependent variable and its units

Space for repeats of the experimental design (in this case, three trials).

The labels on the columns and rows are chosen to represent the design features of the investigation.

Space for three plants at each pH

The range of values for the **independent variable** are in this column

Recordings of the dependent variable

Space for calculated means

		Trial 1 (plant mass in grams)						Trial 2 (plant mass in grams)						Trial 3 (plant mass in grams)					
		Day No.						Day No.						Day No.					
		0	2	4	6	8	10	0	2	4	6	8	10	0	2	4	6	8	10
pH 3	1	0.5	1.1																
	2	0.6	1.2																
	3	0.7	1.3																
	Mean	0.6	1.2																
pH 5	1	0.6	1.4																
	2	0.8	1.7																
	3	0.5	1.9																
	Mean	0.6	1.7																
pH 7	1	0.7	1.3																
	2	0.8	1.3																
	3	0.4	1.7																
	Mean	0.6	1.4																

1. On a separate piece of paper page, design a table to record the data you would collect from the case study below. Include room for individual results and means from the three set ups. Staple or paste it into your workbook once you have finished. You can use the table above as a guide to help you.

Case study:
Carbon dioxide levels in a respiration chamber

A datalogger was used to monitor the concentrations of carbon dioxide (CO_2) in respiration chambers containing five green leaves from one plant species. The entire study was performed in conditions of full light and involved three identical set-ups. The CO_2 concentrations were measured every minute, over a period of ten minutes, using a CO_2 sensor. A mean CO_2 concentration (for the three set-ups) was calculated. The study was carried out two more times, two days apart.

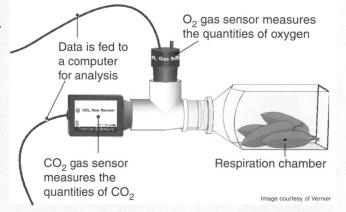

O_2 gas sensor measures the quantities of oxygen

Data is fed to a computer for analysis

CO_2 gas sensor measures the quantities of CO_2

Respiration chamber

Image courtesy of Vernier

9 Transforming Raw Data

Key Idea: Unprocessed data is called raw data. A set of data is often processed or transformed to make it easier to understand and to identify important features.

The data collected by measuring or counting in the field or laboratory is called **raw data**. A set of data often needs to be processed or transformed into a form that makes it easier to identify its important features (e.g. trends). Basic calculations, such as totals (the sum of all data values for a variable), are commonly used to compare treatments. Some common data transformations include tally charts, percentages, and rates. These are explained below.

Tally Chart

Records the number of times a value occurs in a data set

HEIGHT (cm)	TALLY	TOTAL
0–0.99	III	3
1–1.99	++++ I	6
2–2.99	++++ ++++	10
3–3.99	++++ ++++ II	12
4–4.99	III	3
5–5.99	II	2

- A useful first step in analysis; a neatly constructed tally chart doubles as a simple histogram.
- Cross out each value on the list as you tally it to prevent double entries.

Percentages

Expressed as a fraction of 100

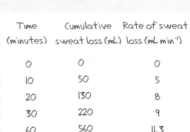

Men	Body mass (kg)	Lean body mass (kg)	% lean body mass
Athlete	70	60	85.7
Lean	68	56	82.3
Normal weight	83	65	78.3
Overweight	96	62	64.6
Obese	125	65	52.0

- Percentages express what proportion of data fall into any one category, e.g. for pie graphs.
- Allows meaningful comparison between different samples.
- Useful to monitor change (e.g. % increase from one year to the next).

Rates

Expressed as a measure per unit time

Time (minutes)	Cumulative sweat loss (mL)	Rate of sweat loss (mL min⁻¹)
0	0	0
10	50	5
20	130	8
30	220	9
60	560	11.3

- Rates show how a variable changes over a standard time period (e.g. one second, one minute, or one hour).
- Rates allow meaningful comparison of data that may have been recorded over different time periods.

Example: Height of 6 day old seedlings

Example: Percentage of lean body mass in men

Example: Rate of sweat loss during exercise in cyclists

1. What is raw data? _____

_____

2. (a) What is data transformation? _____

_____

(b) Why do we transform data? _____

_____

_____

 © 2014 **BIOZONE** International
ISBN: 978-1-927173-84-8
Photocopying Prohibited

10 Practicing Data Transformation

Key Idea: Percentages, rates, and frequencies are commonly used data transformations.

1. Complete the transformations for each of the tables on the right. The first value, and their working, is provided for each example.

   (a) TABLE: Incidence of red clover in different areas:

   Working: 124 ÷ 159 = 0.78 = 78%

   > This is the number of red clover out of the total.

Incidence of red and white clover in different areas

Clover plant type	Frost free area		Frost prone area		Totals
	Number	%	Number	%	
Red	124	78	26		
White	35		115		
Total	159				

   (b) TABLE: Plant water loss using a bubble potometer:

   Working: (9.0 − 8.0) ÷ 5 min = 0.2

   > This is the distance the bubble moved over the first 5 minutes. Note that there is no data entry possible for the first reading (0 min) because no difference can be calculated.

Plant water loss using a bubble potometer

Time (min)	Pipette arm reading (cm^3)	Plant water loss ($cm^3\ min^{-1}$)
0	9.0	–
5	8.0	0.2
10	7.2	
15	6.2	
20	4.9	

   (c) TABLE: Frequency of size classes in a sample of eels:

   Working: (7 ÷ 270) x 100 = 2.6 %

   > This is the number of individuals out of the total that appear in the size class 0-50 mm. The relative frequency is rounded to one decimal place.

Frequency of size classes in a sample of eels

Size class (mm)	Frequency	Relative frequency (%)
0-50	7	2.6
50-99	23	
100-149	59	
150-199	98	
200-249	50	
250-299	30	
300-349	3	
Total	270	

 © 2014 **BIOZONE** International
ISBN: 978-1-927173-84-8
Photocopying Prohibited

DATA

11 Constructing Tables

Key Idea: Tables are used to record and summarize data. Tables allow relationships and trends in data to be more easily recognized.

▶ Tables are used to record data during an investigation. Your log book should present neatly tabulated data (right).

▶ Tables allow a large amount of information to be condensed, and can provide a summary of the results.

▶ Presenting data in tables allows you to organize your data in a way that allows you to more easily see the relationships and trends.

▶ Columns can be provided to display the results of any data transformations such as rates. Basic descriptive statistics (such as mean or standard deviation) may also be included.

▶ Complex data sets tend to be graphed rather than tabulated.

Features of tables

Tables should have an accurate, descriptive title. Number tables consecutively through a report.

Heading and subheadings identify each set of data and show units of measurement.

Independent variable in the left column.

Table 1: Length and growth of the third internode of bean plants receiving three different hormone treatments.

Treatment	Sample size	Mean rate of internode growth (mm day^{-1})	Mean internode length (mm)	Mean mass of tissue added (g day^{-1})
Control	50	0.60	32.3	0.36
Hormone 1	46	1.52	41.6	0.51
Hormone 2	98	0.82	38.4	0.56
Hormone 3	85	2.06	50.2	0.68

Control values should be placed at the beginning of the table.

Each row should show a different experimental treatment, organism, sampling site etc.

Columns for comparison should be placed alongside each other. Show values only to the level of significance allowable by your measuring technique.

Organize the columns so that each category of like numbers or attributes is listed vertically.

1. What are two advantages of using a table format for data presentation?

   (a) _____

   _____

   (b) _____

   _____

2. Why might you tabulate data before you presented it in a graph? _____

   _____

   _____

 © 2014 **BIOZONE** International
ISBN: 978-1-927173-84-8
Photocopying Prohibited

KNOW

12 Drawing Line Graphs

Key Idea: Line graphs are used to plot continuous data when one variable (the independent variable) affects another, the dependent variable.

Graphs provide a way to visually see data trends. Line graphs are used when one variable (the independent variable) affects another, the dependent variable. Important features of line graphs are:

• The data must be continuous for both variables.

• The dependent variable is usually a biological response.

• The independent variable is often time or the experimental treatment.

• Where there is a trend, a line of best fit is usually plotted to show the relationship.

• If fluctuations in the data are likely to be important (e.g. environmental data) the data points are usually connected directly (point to point), as shown top right.

• Line graphs may be drawn with measure of error (bottom right). The data are presented as points (the calculated means), with bars above and below, indicating a measure of variability or spread in the data (e.g. standard error or standard deviation).

• Where no error value has been calculated, the scatter can be shown by plotting the individual data points vertically above and below the mean. Bars are not used to indicate the range of raw values in a data set.

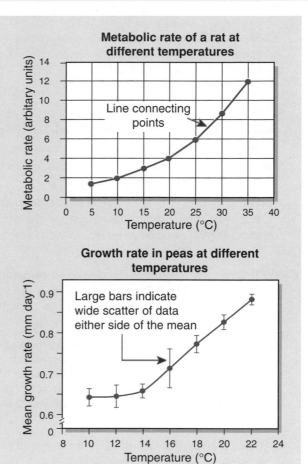

Metabolic rate of a rat at different temperatures

Line connecting points

Growth rate in peas at different temperatures

Large bars indicate wide scatter of data either side of the mean

1. The results (shown right) were collected in a study investigating the effect of temperature on the activity of an enzyme.

   (a) Using the results provided, plot a **line graph** on the grid below:

   (b) Estimate the rate of reaction at 15°C: _____

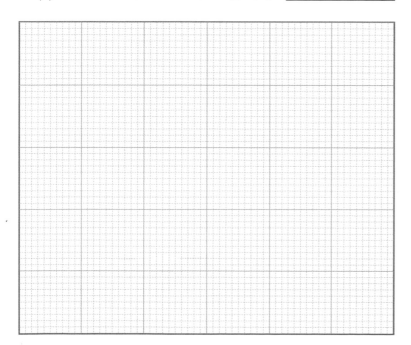

Lab Notebook

An enzyme's activity at different temperatures

Temperature (°C)	Rate of reaction (mg of product formed per minute)
10	1.0
20	2.1
30	3.2
35	3.7
40	4.1
45	3.7
50	2.7
60	0

© 2014 **BIOZONE** International
ISBN: 978-1-927173-84-8
Photocopying Prohibited

13 Drawing Scatter Graphs

Key Idea: Scatter graphs are used to plot continuous data where there is a relationship between two interdependent variables.

Scatter graphs are used to display continuous data where there is a relationship between two interdependent variables.

- The data must be continuous for both variables.
- There is no independent (manipulated) variable, but the variables are often correlated, i.e. they vary together in some predictable way.
- Scatter graphs are useful for determining the relationship between two variables.
- The points on the graph should not be connected, but a line of best fit is often drawn through the points to show the relationship between the variables.

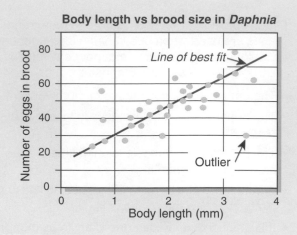

Body length vs brood size in *Daphnia*

1. In the example below, metabolic measurements were taken from seven Antarctic fish *Pagothenia borchgrevinski*. The fish are affected by a gill disease, which increases the thickness of the gas exchange surfaces and affects oxygen uptake. The results of oxygen consumption of fish with varying amounts of affected gill (at rest and swimming) are tabulated below.

   (a) Plot the data on the grid (bottom right) to show the relationship between oxygen consumption and the amount of gill affected by disease. Use different symbols or colors for each set of data (at rest and swimming), and use only one scale for oxygen consumption.

   (b) Draw a line of best fit through each set of points.

2. Describe the relationship between the amount of gill affected and oxygen consumption in the fish:

   (a) For the **at rest** data set:

   _____

   _____

   _____

   _____

   _____

   _____

   (b) For the **swimming** data set:

   _____

   _____

   _____

   _____

   _____

   _____

Oxygen consumption of fish with affected gills

Fish number	Percentage of gill affected	Oxygen consumption (cm^3 g^{-1} h^{-1})	
		At rest	Swimming
1	0	0.05	0.29
2	95	0.04	0.11
3	60	0.04	0.14
4	30	0.05	0.22
5	90	0.05	0.08
6	65	0.04	0.18
7	45	0.04	0.20

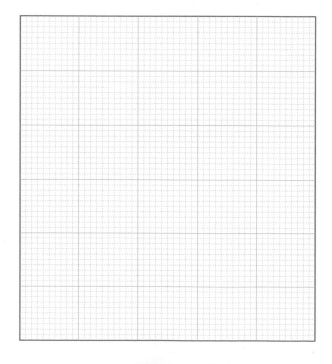

© 2014 **BIOZONE** International
ISBN: 978-1-927173-84-8
Photocopying Prohibited

DATA

14 Drawing Bar Graphs

Key Idea: Bar graphs are used to plot data that is non-numerical or discrete for at least one variable.

Bar graphs are appropriate for data that is non-numerical and discrete for at least one variable.

- There are no dependent or independent variables.
- Data is collected for discontinuous, non-numerical categories (e.g. place, color, and species), so the bars do not touch.
- Multiple sets of data can be displayed side by side for direct comparison.
- Axes may be reversed, i.e. the bars can be vertical or horizontal. When they are vertical, these graphs are called column graphs.

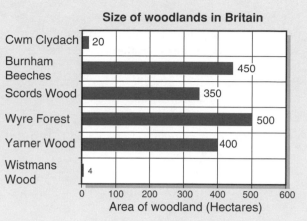

Size of woodlands in Britain

Cwm Clydach 20
Burnham Beeches 450
Scords Wood 350
Wyre Forest 500
Yarner Wood 400
Wistmans Wood 4

Area of woodland (Hectares)
0   100   200   300   400   500   600

1. Counts of eight mollusk species were made from a series of quadrat samples at two sites on a rocky shore. The summary data are presented on the right.

   (a) Tabulate the mean (average) numbers per square meter at each site in the table (below).

   (b) Plot a **bar graph** of the tabulated data on the grid below. For each species, plot the data from both sites side by side using different colors to distinguish the two sites.

Average abundance of 8 mollusk species from two sites along a rocky shore.

Species	Mean (no. m^{-2})	
	Site 1	Site 2

Field data notebook

Total counts at site 1 (11 quadrats) and site 2 (10 quadrats). Quadrats 1 sq m.

Species	Site 1 Total	Site 1 Mean	Site 2 Total	Site 2 Mean
	No (m^{-2})		No (m^{-2})	
Ornate limpet	232	21	299	30
Radiate limpet	68	6	344	34
Limpet sp. A	420	38	0	0
Cats-eye	68	6	16	2
Top shell	16	2	43	4
Limpet sp. B	628	57	389	39
Limpet sp. C	0	0	22	2
Chiton	12	1	30	3

15 Drawing Histograms

Key Idea: Histograms graphically show the frequency distribution of continuous data.

Histograms are plots of continuous data and are often used to represent frequency distributions, where the y-axis shows the number of times a particular measurement or value was obtained. For this reason, they are often called frequency histograms. Important features of histograms include:

- The data are numerical and continuous (e.g. height or weight), so the bars touch.

- The x-axis usually records the class interval. The y-axis usually records the number of individuals in each class interval (frequency).

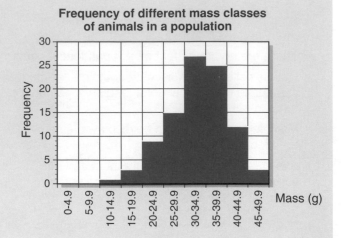

Frequency of different mass classes of animals in a population

1. The weight data provided below were recorded from 95 individuals (male and female), older than 17 years.

   (a) Create a tally chart (frequency table) in the table provided (right). An example of the tally for the weight grouping 55-59.9 kg has been completed for you. Note that the raw data values, once they are recorded as counts on the tally chart, are crossed off the data set in the notebook. It is important to do this in order to prevent data entry errors.

   (b) Plot a **frequency histogram** of the tallied data on the grid below.

Weight (kg)	Tally	Total
45-49.9		
50-54.9		
55-59.9	⊩⊩ ‖	7
60-64.9		
65-69.9		
70-74.9		
75-79.9		
80-84.9		
85-89.9		
90-94.9		
95-99.9		
100-104.9		
105-109.9		

Lab notebook

Weight (in kg) of 95 individuals

63.4	81.2	65
56.5	83.3	75.6
84	95	76.8
81.5	105.5	67.8
73.4	82	68.3
56	73.5	63.5
60.4	75.2	58
83.5	63	59.5
82	70.4	50
61	82.2	92
55.2	87.8	91.5
48	86.5	88.3
53.5	85.5	81
63.8	87	72
69	98	66.5
82.8	71	61.5
68.5	76	66
67.2	72.5	65.5
82.5	61	67.4
83	60.5	73
78.4	67	67
76.5	86	71
83.4	85	70.5
77.5	93.5	65.5
77	62	68
87	62.5	90
89	63	83.5
93.4	60	73
83	71.5	66
80	73.8	57.5
76	77.5	76
56	74	

DATA

16 Descriptive Statistics and the Spread of Data

Key Idea: Descriptive statistics are used to summarize a data set and describe its basic features. The type of statistic calculated depends on the type of data and its distribution.

Descriptive statistics

When we describe a set of data, it is usual to give a measure of **central tendency**. This is a single value identifying the central position within that set of data. **Descriptive statistics**, such as mean, median, and mode, are all valid measures of central tendency depending of the type of data and its distribution. They help to summarize features of the data, so are often called summary statistics. The appropriate statistic for different types of data variables and their distributions is described below.

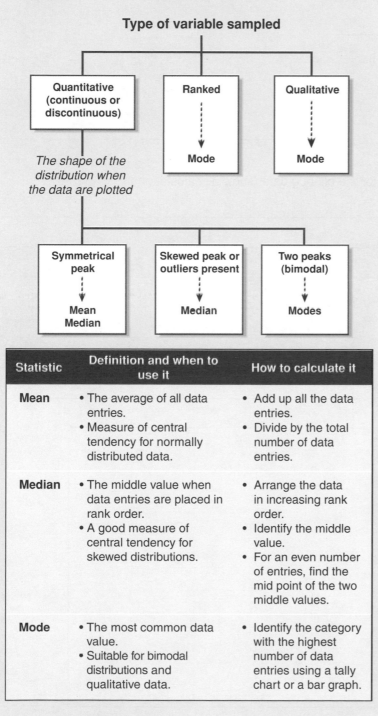

Statistic	Definition and when to use it	How to calculate it
Mean	• The average of all data entries. • Measure of central tendency for normally distributed data.	• Add up all the data entries. • Divide by the total number of data entries.
Median	• The middle value when data entries are placed in rank order. • A good measure of central tendency for skewed distributions.	• Arrange the data in increasing rank order. • Identify the middle value. • For an even number of entries, find the mid point of the two middle values.
Mode	• The most common data value. • Suitable for bimodal distributions and qualitative data.	• Identify the category with the highest number of data entries using a tally chart or a bar graph.

Distribution of data

Variability in continuous data is often displayed as a frequency distribution. There are several types of distribution.

▶ Normal distribution (A): Data has a symmetrical spread about the mean. It has a classical bell shape when plotted.

▶ Skewed data (B): Data is not centered around the middle but has a "tail" to the left or right.

▶ Bimodal data (C): Data which has two peaks.

The shape of the distribution will determine which statistic (mean, median, or mode) should be used to describe the central tendency of the sample data.

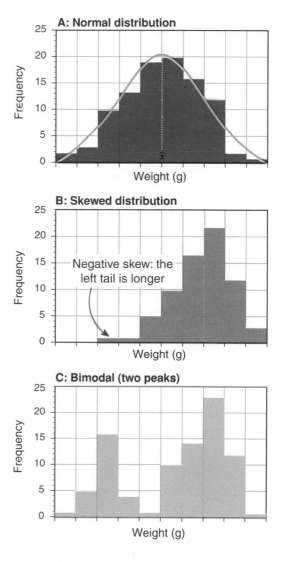

1. The mass of 15 pill-bugs is shown in the table right.

Pill-bug mass (mg)					
10.1	8.2	8.5	8.0	8.8	8.8
7.8	6.7	7.7	8.8	9.8	8.5
8.8	8.9	6.2	8.8	8.4	8.9

(a) Draw up a tally chart in the space provided on the right for the pill-bug masses.

(b) On the graph paper at the bottom of the page, draw a frequency histogram for the pill-bug data.

(c) What type of distribution does the data have?

_____

_____

_____

(d) What would be the best measure of central tendency in the pill-bug data set (mean, median, or mode)?

_____

(e) Explain why you chose your answer in (d).

_____

_____

_____

(f) Calculate the mean, median, and mode for the pill-bug data (show all calculations):

Mean: _____

Median: _____

Mode: _____

(g) What do you notice about your results in (f)? _____

_____

(h) Why do you think this has happened? _____

_____

17 Biological Drawings

Key Idea: Good biological drawings provide an accurate record of the specimen you are studying and enable you to make a record of its important features.

▶ Drawing is a very important skill to have in biology. Drawings record what a specimen looks like and give you an opportunity to record its important features. Often drawing something will help you remember its features at a later date (e.g. in a test).

▶ Biological drawings require you to pay attention to detail. It is very important that you draw what you actually see, and not what you think you should see.

▶ Biological drawings should include as much detail as you need to distinguish different structures and types of tissue, but avoid unnecessary detail which can make your drawing confusing.

▶ Attention should be given to the symmetry and proportions of your specimen. Accurate labeling, a statement of magnification or scale, the view (section type), and type of stain used (if applicable) should all be noted on your drawing.

▶ Some key points for making good biological drawing are described on the example below. The drawing of *Drosophila* (right) is well executed but lacks the information required to make it a good biological drawing.

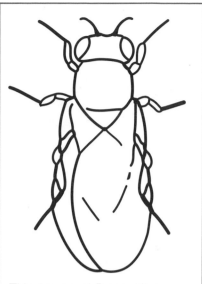

This drawing of *Drosophila* is a fair representation of the animal, but has no labels, title, or scale.

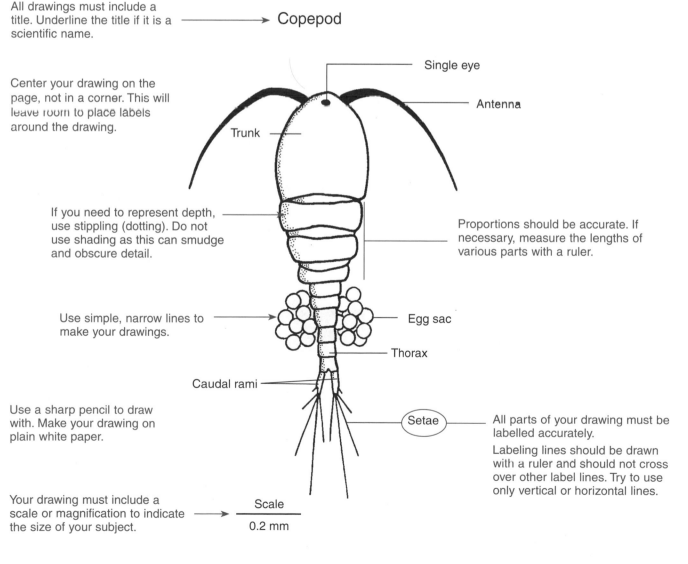

All drawings must include a title. Underline the title if it is a scientific name. ⟶ Copepod

Center your drawing on the page, not in a corner. This will leave room to place labels around the drawing.

If you need to represent depth, use stippling (dotting). Do not use shading as this can smudge and obscure detail.

Use simple, narrow lines to make your drawings.

Use a sharp pencil to draw with. Make your drawing on plain white paper.

Your drawing must include a scale or magnification to indicate the size of your subject.

Single eye
Antenna
Trunk
Proportions should be accurate. If necessary, measure the lengths of various parts with a ruler.
Egg sac
Thorax
Caudal rami
Setae
All parts of your drawing must be labelled accurately.
Labeling lines should be drawn with a ruler and should not cross over other label lines. Try to use only vertical or horizontal lines.
Scale
0.2 mm

CONNECT 18 WEB 17 REFER

Annotated diagrams

An annotated diagram is a diagram that includes a series of explanatory notes. These provide important or useful information about your subject.

Transverse section through collenchyma of *Helianthus* stem. Magnification x 450

Primary wall with secondary thickening.

Cytoplasm
A watery solution containing dissolved substances, enzymes, and the cell organelles.

Nucleus
A large, visible organelle. It contains most of the cell's DNA.

Chloroplast
These are specialized plastids containing the green pigment chlorophyll. Photosynthesis occurs here.

Vacuole containing cell sap.

Plan diagrams

Plan diagrams are drawings made of samples viewed under a microscope at low or medium power. They are used to show the distribution of the different tissue types in a sample without any cellular detail. The tissues are identified, but no detail about the cells within them is included.

The example here shows a plan diagram produced after viewing a light micrograph of a transverse section through a dicot stem.

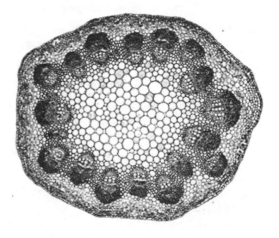

Light micrograph of a transverse section through a dicot stem.

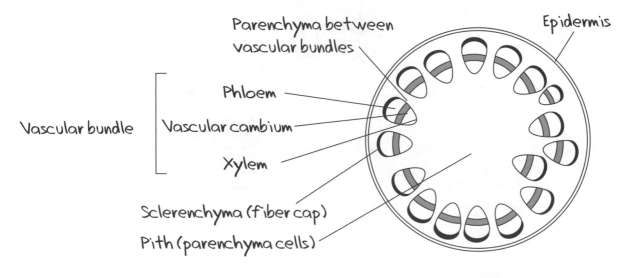

Parenchyma between vascular bundles

Epidermis

Vascular bundle

Phloem

Vascular cambium

Xylem

Sclerenchyma (fiber cap)

Pith (parenchyma cells)

© 2014 **BIOZONE** International
ISBN: 978-1-927173-84-8

18 Practicing Biological Drawings

Key Idea: Attention to detail is vital when making accurate and useful biological drawings.

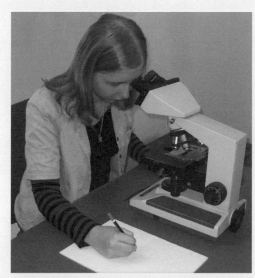

Above: Use relaxed viewing when drawing at the microscope. Use one eye (the left for right handers) to view and the right eye to look at your drawing.

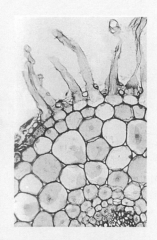

Above: Light micrograph Transverse section (TS) through a *Ranunculus* root.

Right: A biological drawing of the same section.

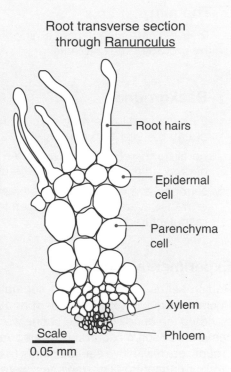

Root transverse section through <u>Ranunculus</u>

Root hairs

Epidermal cell

Parenchyma cell

Xylem

Phloem

Scale
0.05 mm

1. Complete the biological drawing of a cross section through a dicot leaf (below). Use the example above of the *Ranunculus* root as a guide to the detail required in your drawing

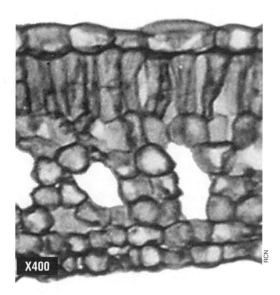

X400

Light micrograph of a cross section through a leaf.

PRAC

19 Test Your Understanding

This activity will test your understanding of science practices. Analyze the data below about the effect of fertilizer on the growth of radishes. Tabulate and graph the data, and draw conclusions about how nitrogen fertilizer effects radish growth.

The aim

To investigate the effect of a nitrogen fertilizer on the growth of radish plants.

Background

Inorganic fertilizers were introduced to crop farming during the late 19th century.

Fertilizer increased crop yields. An estimated 50% of crop yield is attributable to the use of fertilizer.

Nitrogen is a very important element for plant growth. Several types of nitrogen fertilizers are manufactured (e.g. urea).

Radishes

Experimental method

Radish seeds were planted in separate identical pots (5 cm x 5 cm wide x 10 cm deep) and kept together in standard lab conditions. The seeds were planted into a commercial soil mixture and divided randomly into six groups, each with five sample plants (a total of 30 plants in six treatments). The radishes were watered every day at 10 am and 3 pm with 500 mL per treatment per watering. Water soluble fertilizer was added to the 10 am watering on the 1st, 11th and 21st days. The fertilizer concentrations used were: 0.00, 0.06, 0.12, 0.18, 0.24, and 0.30 g L^{-1} and each treatment received a different concentration. The plants were grown for 30 days before being removed from the pots, washed, and the radish root weighed. The results are presented below.

Fertilizer concentration (g L^{-1})	Sample 1	2	3	4	5
0	80.1	83.2	82.0	79.1	84.1
0.06	109.2	110.3	108.2	107.9	110.7
0.12	117.9	118.9	118.3	119.1	117.2
0.18	128.3	127.3	127.7	126.8	DNG*
0.24	23.6	140.3	139.6	137.9	141.1
0.30	122.3	121.1	122.6	121.3	123.1

*DNG = did not germinate

† Based on data from M S Jilani, et al Journal Agricultural Research

1. Identify the independent variable for the experiment and its range: _____

_____

2. Identify the dependent variable for the experiment: _____

3. What is the sample size for each concentration of fertilizer? _____

4. (a) One of the radishes recorded in the table on the previous page did not grow as expected and produced an extreme value. Record the **outlying value** here:

_____

(b) Why should this value not be included in future calculations? _____

_____

_____

5. Use table 1 below to record the raw data from the experiment. You will need to include column and row headings and a title, and complete some simple calculations. Some headings have been entered for you.

Table 1: _____

	Mass of radish root (g)					Total mass	Mean mass

6. The students decided to collect more data by counting the number of leaves on each radish plant at day 30. This data is presented in Table 2.
Use the space below to calculate the **mean**, **median** and **mode** for the leaf data. Add these data to table 2.

Table 2: Number of leaves on radish plant under six different fertilizer concentrations.

Fertilizer concentration (g L⁻¹)	Number of leaves							
	Sample (n)					Mean	Median	Mode
	1	2	3	4	5			
0	9	9	10	8	7			
0.06	15	16	15	16	16			
0.12	16	17	17	17	16			
0.18	18	18	19	18	DNG*			
0.24	6	19	19	18	18			
0.30	18	17	18	19	19			

* DNG: Did not germinate

7. Use the grid below to draw a **line graph** of the experimental results. Plot your calculated mean mass data from Table 1, and remember to include a title and correctly labelled axes.

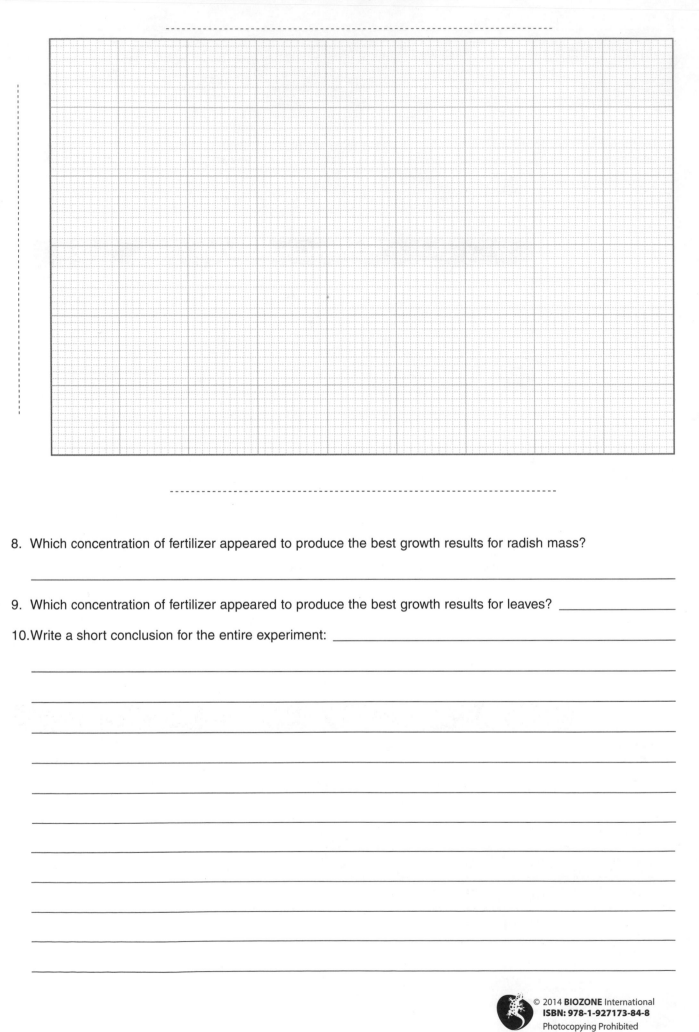

8. Which concentration of fertilizer appeared to produce the best growth results for radish mass?

_____

9. Which concentration of fertilizer appeared to produce the best growth results for leaves? _____

10. Write a short conclusion for the entire experiment: _____

_____

_____

_____

_____

_____

_____

_____

_____

_____

_____

_____

© 2014 **BIOZONE** International
ISBN: 978-1-927173-84-8
Photocopying Prohibited

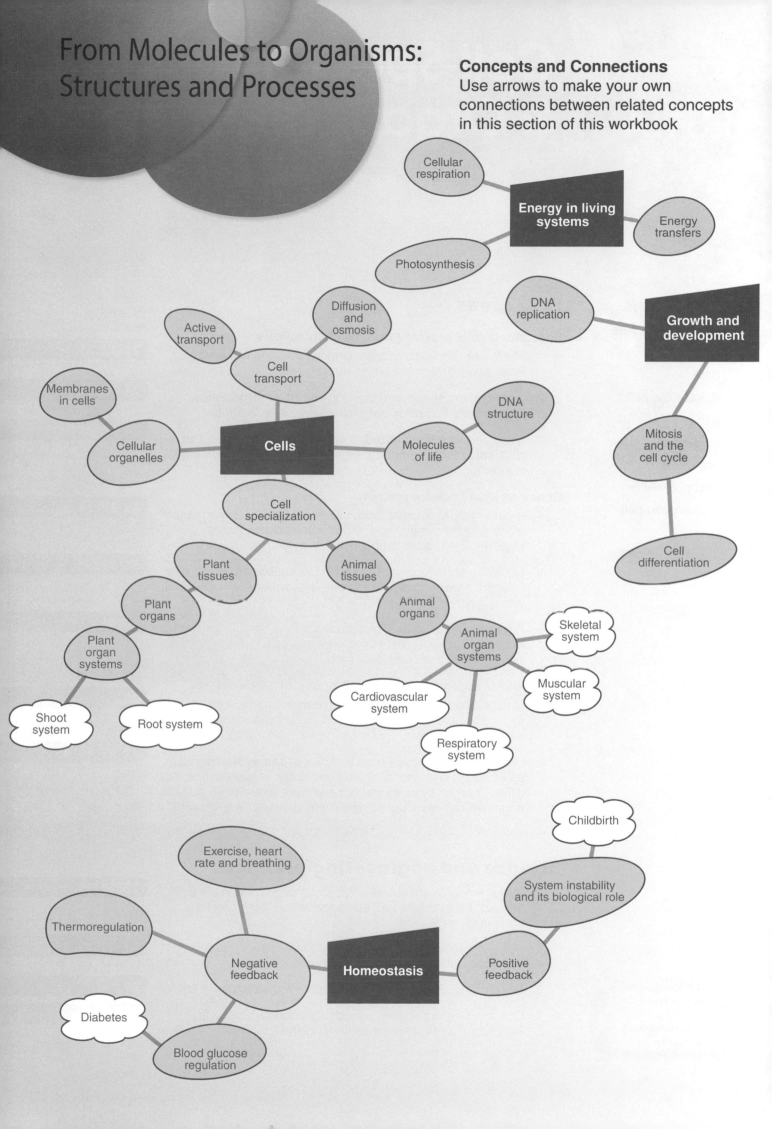

From Molecules to Organisms: Structures and Processes

Concepts and Connections
Use arrows to make your own connections between related concepts in this section of this workbook

- Cellular respiration
- **Energy in living systems**
- Energy transfers
- Photosynthesis
- DNA replication
- **Growth and development**
- Diffusion and osmosis
- Active transport
- Cell transport
- DNA structure
- Membranes in cells
- Cellular organelles
- **Cells**
- Molecules of life
- Mitosis and the cell cycle
- Cell differentiation
- Cell specialization
- Plant tissues
- Animal tissues
- Plant organs
- Animal organs
- Animal organ systems
- Skeletal system
- Plant organ systems
- Muscular system
- Cardiovascular system
- Shoot system
- Root system
- Respiratory system
- Childbirth
- System instability and its biological role
- Exercise, heart rate and breathing
- Thermoregulation
- Negative feedback
- **Homeostasis**
- Positive feedback
- Diabetes
- Blood glucose regulation

Cell Specialization & Organization

Key terms

base-pairing rule

cell

DNA

eukaryotic cell

gene

nucleotide

organelle

organ system

prokaryotic cell

protein

specialized cell

Core ideas

Activity number

Specialized cells provide essential life functions

□ 1. There are two types of **cells**: **prokaryotic** and **eukaryotic**. Plant and animal cells are eukaryotic cells.

`21 22 23`

□ 2. Cells contain specialized components called **organelles**. Each organelle carries out a specialized role in the cell.

`24 - 33`

□ 3. Many cells are **specialized** to carry out specific roles. The size, shape, and number of organelles in a cell depends on the cell's role.

`34 35`

Genes on DNA code for proteins

□ 4. A cell's genetic information is called **DNA**. DNA is made up of many smaller components, called **nucleotides**, joined together. DNA has a double-helix structure.

`36 37 38 39`

□ 5. Nucleotides pair together according to the **base-pairing rule**. Adenine always pairs with thymine and cytosine always pairs with guanine.

`38 39`

□ 6. A **gene** is a region of DNA that codes for a specific **protein**. Proteins carry out most of the work in a cell. The shape of a protein helps it to carry out its job.

`40 41`

Multicellular organisms are organized in a hierarchical way

□ 7. Multicellular organisms exhibit a hierarchical structure of organization. Components at each level of organization are part of the next level.

`20`

□ 8. Multicellular organisms have complex **organ systems**, made up of many components. Each system has a specific role, but different organ systems interact and work together so that the organism can carry out essential life functions. (e.g. growth).

`42 43 44 45`

Science and engineering practices

□ 1. Use a model to show that multicellular organisms have a hierarchical structure and that components from one level contribute to the next.

`20 22 23`

□ 2. Use a model to illustrate that components of a system interact to fulfil essential life functions.

`41 42 43 45`

□ 3. Construct and use a model to illustrate the structure of DNA.

`39`

□ 4. Construct an explanation based on evidence for how the structure of DNA determines the structure of proteins.

`42`

BIOZONE APP
Student Review Series
Cell Specialization & Organization

20 The Hierarchy of Life

Key Idea: The structural organization of multicellular organisms is hierarchical. Components at each level of organization are part of the next level.

All multicellular organisms are organized in a hierarchy of structural levels, where each level builds on the one below it. It is traditional to start with the simplest components (parts) and build from there. Higher levels of organization are more complex than lower levels.

Hierarchical organization enables **specialization** so that individual components perform a specific function or set of related functions. Specialization enables organisms to function more efficiently.

The diagram below explains this hierarchical organization for a human.

The cellular level
Cells are the basic structural and functional units of an organism. Cells are specialized to carry out specific functions, e.g. cardiac (heart) muscle cells (below).

DNA

1

Atoms and molecules

2

The organelle level
Molecules associate together to form the organelles and structural components of cells, e.g. the nucleus (above).

3

The chemical level
All the chemicals essential for maintaining life, e.g. water, ions, fats, carbohydrates, amino acids, proteins, and nucleic acids.

7 **The organism**
The cooperating organ systems make up the organism, e.g. a human.

The tissue level **4**
Groups of cells with related functions form tissues, e.g. cardiac (heart) muscle (above). The cells of tissue often have a similar origin.

6 **The system level**
Groups of organs with a common function form an organ system, e.g. cardiovascular system (right).

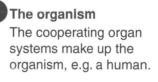

5 **The organ level**
An organ is made up of two or more types of tissues to carry out a particular function. Organs have a definite form and structure, e.g. heart (left).

CONNECT **72** CONNECT **43** CONNECT **35** **REFER**

21 Introduction to Cells

Key Idea: Cells are classified as either prokaryotic cells or eukaryotic and are distinguished on the basis of their size and internal organization and complexity.

▶ The **cell** is the smallest unit of life. Cells are often called the building blocks of life.

▶ Cells are either prokaryotic cells or eukaryotic cells. Within each of these groups, cells may vary greatly in their size, shape, and functional role.

Prokaryotic cells

▶ **Prokaryotic cells** are bacterial cells.

▶ Prokaryotic cells lack a membrane-bound nucleus or any membrane-bound organelles.

▶ They are small (generally 0.5-10 µm) single cells (unicellular).

▶ They are relatively basic cells and have very little cellular organization (their DNA, ribosomes, and enzymes are free floating within the cell cytoplasm).

▶ Single, circular chromosome of naked DNA.

▶ Prokaryotes have a cell wall, but it is different to the cell walls that some eukaryotes have.

Eukaryotic cells

▶ **Eukaryotic cells** have a membrane-bound nucleus, and other membrane-bound organelles.

▶ Plant cells, animals cells, fungal cells, and protists are all eukaryotic cells.

▶ Eukaryotic cells are large (30-150 µm). They may exist as single cells or as part of a multicellular organism.

▶ Multiple linear chromosomes consisting of DNA and associated proteins.

▶ They are more complex than prokaryotic cells. They have more structure and internal organization.

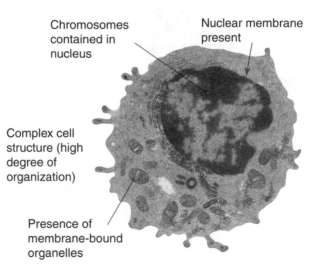

Simple cell structure (limited organization)

Nuclear membrane absent. Single, naked chromosome is free in cytoplasm within a nucleoid region.

Membrane-bound organelles are absent

Nucleoid region (pale)

Peptidoglycan cell wall

A prokaryotic cell: *E.coli*

Chromosomes contained in nucleus

Nuclear membrane present

Complex cell structure (high degree of organization)

Presence of membrane-bound organelles

A eukaryotic cell: a human white blood cell

1. What are the main features of a **prokaryotic cell**? _____

_____

_____

2. (a) What are the main features of a **eukaryotic cell**? _____

_____

_____

(b) Name examples of eukaryotic cells: _____

_____

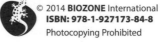 © 2014 **BIOZONE** International
ISBN: 978-1-927173-84-8
Photocopying Prohibited

22 Plant Cells

Key Idea: Plant cells are eukaryotic cells. They have many types of organelles, each of which has a specific role within the cell.

What is an organelle?

The word **organelle** means "small organ". Therefore, organelles are the cell's "organs" and carry out the cell's work.

Organelles represent one level of organization in a multicellular organism. One component (the cell) is made up of many smaller parts (organelles).

Eukaryotic cells contain many different types of organelles. Each type of organelle has a specific role in the cell to help it function.

Plant cells have several types of membrane-bound organelles called plastids. These make and store food and pigments. Some of the organelles found in a plant cell are shown below.

Features of a plant cell

Plant cells are eukaryotic cells. Features that identify plant cells as eukaryotic cells include:

▶ A membrane-bound nucleus.

▶ Membrane-bound organelles (e.g. nucleus, mitochondria, endoplasmic reticulum).

Features that can be used to identify a plant cell include the presence of:

▶ **Cellulose cell wall**

▶ **Chloroplasts and other plastids**

▶ **Large vacuole** (often centrally located)

A generalized plant cell

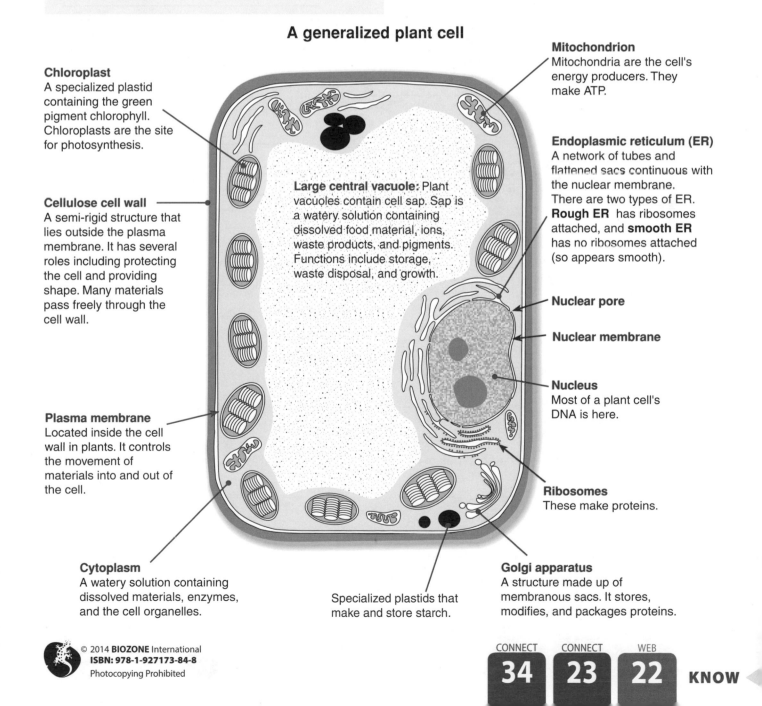

Chloroplast
A specialized plastid containing the green pigment chlorophyll. Chloroplasts are the site for photosynthesis.

Cellulose cell wall
A semi-rigid structure that lies outside the plasma membrane. It has several roles including protecting the cell and providing shape. Many materials pass freely through the cell wall.

Plasma membrane
Located inside the cell wall in plants. It controls the movement of materials into and out of the cell.

Cytoplasm
A watery solution containing dissolved materials, enzymes, and the cell organelles.

Large central vacuole: Plant vacuoles contain cell sap. Sap is a watery solution containing dissolved food material, ions, waste products, and pigments. Functions include storage, waste disposal, and growth.

Specialized plastids that make and store starch.

Mitochondrion
Mitochondria are the cell's energy producers. They make ATP.

Endoplasmic reticulum (ER)
A network of tubes and flattened sacs continuous with the nuclear membrane. There are two types of ER. **Rough ER** has ribosomes attached, and **smooth ER** has no ribosomes attached (so appears smooth).

Nuclear pore

Nuclear membrane

Nucleus
Most of a plant cell's DNA is here.

Ribosomes
These make proteins.

Golgi apparatus
A structure made up of membranous sacs. It stores, modifies, and packages proteins.

© 2014 **BIOZONE** International
ISBN: 978-1-927173-84-8
Photocopying Prohibited

CONNECT **34**   CONNECT **23**   WEB **22**   KNOW

1. Use the diagram of a plant cell on the previous page to become familiar with the features of a plant cell. Use your knowledge to label the ten structures in the transmission electron micrograph (TEM) of the cell below.

Use the following list of terms to help you: *nuclear membrane, cytoplasm, endoplasmic reticulum, mitochondrion, starch granule, nucleus, vacuole, plasma membrane, cell wall, chloroplast.*

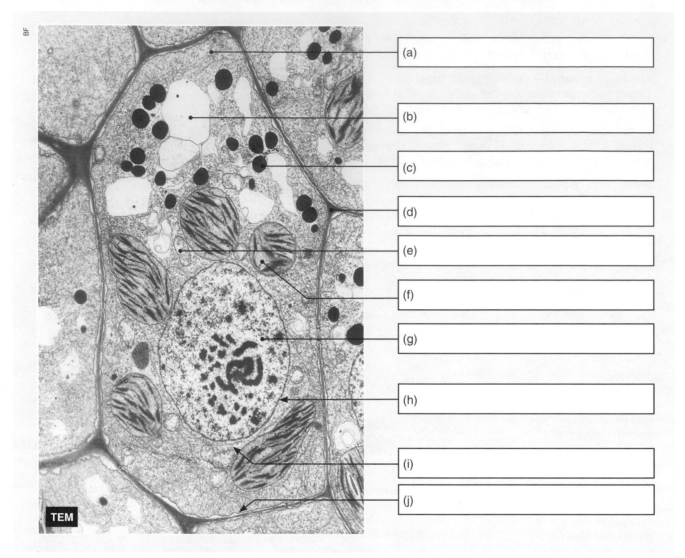

(a)

(b)

(c)

(d)

(e)

(f)

(g)

(h)

(i)

(j)

TEM

2. (a) Which features identify this plant cell as eukaryotic? _____

_____

_____

(b) Use a colored marker to show the cell wall which surrounds this cell.

3. (a) What is an organelle? _____

_____

(b) Why are there so many different types of organelles in eukaryotic cells (e.g. plant and animal cells)?

_____

_____

_____

4. (a) Name the organelle where photosynthesis occurs: _____

(b) How many of these organelles are present in the labeled cell above? _____

 © 2014 **BIOZONE** International
ISBN: 978-1-927173-84-8
Photocopying Prohibited

23 Animal Cells

Key Idea: Animal cells are eukaryotic cells. They lack several of the structures and organelles found in plant cells.

Animal cells are eukaryotic cells. Features that identify them as eukaryotic cells include:

▶ A membrane-bound nucleus.

▶ Membrane-bound organelles.

Did you know?

Animal cells lack the rigid cell wall found in plant cells, so their shape is more irregular and they can sometimes move about or change shape.

Features of an animal cell

Animal cells have many of the same structures and organelles that plant cells have, but several features help to identify them, including:

▶ **No cell wall**

▶ **Often have an irregular shape**

▶ **No chloroplasts or other plastids**

▶ **No large vacuoles (if any)**

▶ **They have centrioles** (not found in the cells of most plants)

A generalized animal cell

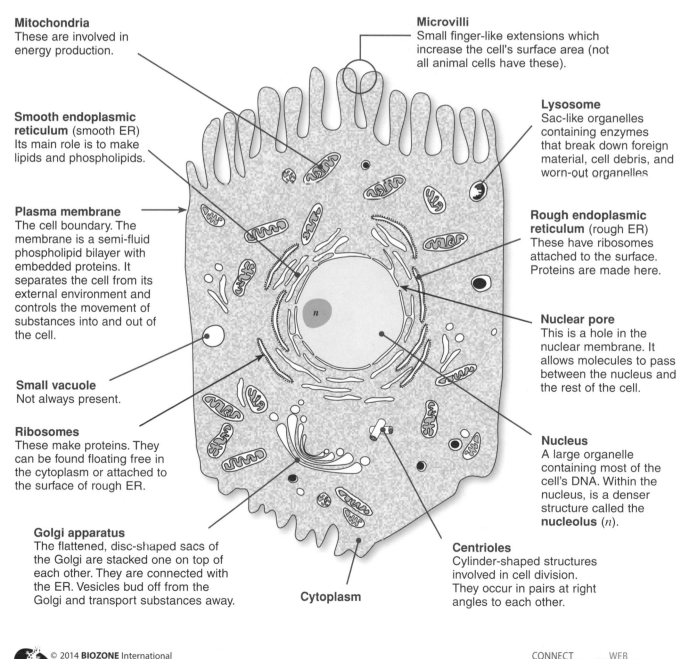

Mitochondria
These are involved in energy production.

Smooth endoplasmic reticulum (smooth ER)
Its main role is to make lipids and phospholipids.

Plasma membrane
The cell boundary. The membrane is a semi-fluid phospholipid bilayer with embedded proteins. It separates the cell from its external environment and controls the movement of substances into and out of the cell.

Small vacuole
Not always present.

Ribosomes
These make proteins. They can be found floating free in the cytoplasm or attached to the surface of rough ER.

Golgi apparatus
The flattened, disc-shaped sacs of the Golgi are stacked one on top of each other. They are connected with the ER. Vesicles bud off from the Golgi and transport substances away.

Microvilli
Small finger-like extensions which increase the cell's surface area (not all animal cells have these).

Lysosome
Sac-like organelles containing enzymes that break down foreign material, cell debris, and worn-out organelles

Rough endoplasmic reticulum (rough ER)
These have ribosomes attached to the surface. Proteins are made here.

Nuclear pore
This is a hole in the nuclear membrane. It allows molecules to pass between the nucleus and the rest of the cell.

Nucleus
A large organelle containing most of the cell's DNA. Within the nucleus, is a denser structure called the **nucleolus** (*n*).

Centrioles
Cylinder-shaped structures involved in cell division. They occur in pairs at right angles to each other.

Cytoplasm

1. Study the diagram of an animal cell on the previous page to become familiar with the features of an animal cell. Use your knowledge to identify and label the structures in the transmission electron micrograph (TEM) of the cell below.

Use the following list of terms to help you: *cytoplasm, plasma membrane, rough endoplasmic reticulum, mitochondrion, nucleus, centriole, Golgi apparatus, lysosome.*

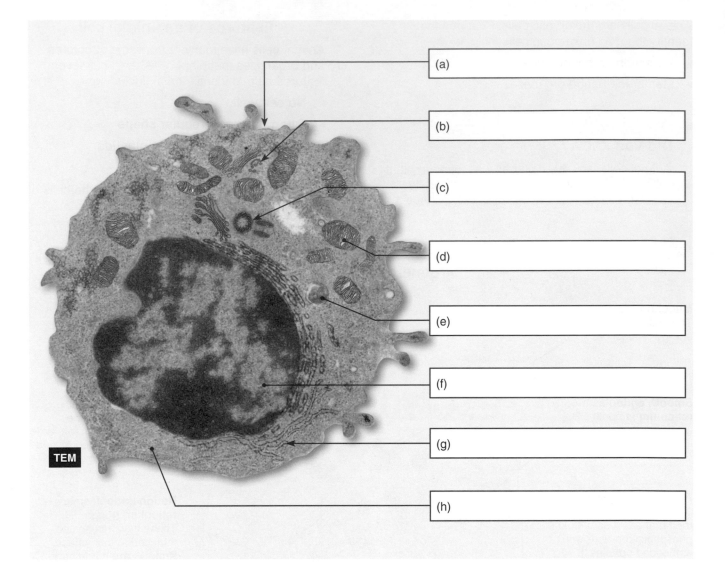

(a)

(b)

(c)

(d)

(e)

(f)

(g)

TEM

(h)

2. Name the features on the cell above that identify it as an animal cell: _____

_____

_____

3. (a) Where is the plasma membrane located on an animal cell? _____

_____

(b) All organelles have a specialized function. What is the function of the plasma membrane?

_____

_____

_____

4. (a) Name the largest organelle visible on the animal cell above: _____

(b) What important material does this organelle contain?_____

24 The Structure of Membranes

Key Idea: Plasma membranes are composed of a lipid bilayer with proteins moving freely within it.

The fluid mosaic model of membrane structure

Glycolipid

Water molecules pass between the phospholipid molecules by osmosis.

Lipid soluble molecules, e.g. gases and steroids, can move through the membrane by diffusion.

Attached carbohydrate

Fatty acid tail is **hydrophobic** (repelled by water)

Phosphate head is **hydrophilic** (attracted to water)

Carrier protein

CO_2

Phospholipids naturally form a **bilayer** with the hydrophobic parts pointing away from the water.

Channel protein

Cholesterol is a packing molecule and is important in cell membrane fluidity.

Glycoprotein

Cell interior

The **fluid-mosaic model** of membrane structure (above) describes a phospholipid bilayer with proteins of different types moving freely within it. The double layer of lipids is quite fluid. It is a dynamic structure and is actively involved in cellular activities.

1. List the important components of the plasma membrane: _____

   _____

2. Identify which kind of molecule on the diagram:

   (a) Can move through the plasma membrane by diffusion: _____

   _____

   (b) Forms a channel through the membrane: _____

   _____

3. List the types of proteins pictured in the diagram:_____

   _____

   _____

4. (a) On the diagram (right) label the hydrophobic and hydrophilic ends of the phospholipid.

   (b) Which end is attracted to water? _____

© 2014 **BIOZONE** International
ISBN: 978-1-927173-84-8
Photocopying Prohibited

WEB
24

 KNOW

25 The Role of Membranes in Cells

Key Idea: Many organelles in a cell are composed of membranes. Membranes control transport and make compartments for reactions.

Isolation of enzymes
Membrane-bound vesicles called lysosomes isolate enzymes to stop them damaging the cell.

Role in lipid synthesis
The smooth endoplasmic reticulum (sER) is a network of interconnected membrane-bound sacs or tubes. Lipids and steroids are made here.

Transport processes
The plasma membrane is involved in the selective transport of substances into and out of the cell.

Role in protein synthesis
The rough endoplasmic reticulum (rER) is similar to the sER in structure but is studded with attached ribosomes. The ribosomes synthesize proteins into the space within the rER.

Containment of DNA
The nuclear envelope is formed by a double-layered membrane. It keeps the DNA within the nucleus.

Energy reactions
The mitochondrion is a membrane-bound organelle where cellular respiration takes place. In plants, photosynthesis occurs in membrane-bound organelles called chloroplasts.

Entry and export of substances
Membrane-bound vesicles move substances from inside the cell to the plasma membrane or vice versa.

Cell communication and recognition
Glycoproteins and glycolipids embedded in the membrane are involved in cell communication and recognition.

Packaging and secretion
The Golgi body is a membrane-bound structure that modifies, sorts, and packages many substances for secretion.

1. List the organelles shown in the diagram above that have membranes: _____

_____

2. (a) Give one example of how membranes are involved in compartmentalization: _____

_____

   (b) For the example you gave, say why compartmentalization is important? _____

_____

_____

3. Give one example of how membranes are involved in the transport of substances: _____

_____

_____

© 2014 **BIOZONE** International
ISBN: 978-1-927173-84-8
Photocopying Prohibited

26 Cell Organelles

Key Idea: Cells contain different types of organelles (small organs) that each carry out a specialized function. Not all cell types contain each organelle.

The table below provides a format to summarize information about the organelles of typical eukaryotic cells. Complete the table by referring to previous pages in this topic. Once you have completed the table, you will have a good understanding of the function and location of the main eukaryotic organelles. The log scale of measurements (top of next page) illustrates the relative sizes of some cellular structures.

Cell Component	Details	Present in	
		Plant cells	Animal cells
(a) Double layer of phospholipids (called the lipid bilayer) — Proteins	**Name:** Plasma (cell surface) membrane **Location:** Surrounding the cell **Function:** Gives the cell shape and protection. It also regulates the movement of substances into and out of the cell.	YES	YES
(b) Outer membrane Inner membrane Matrix — Cristae	**Name:** Mitochondrion **Location:** **Function:**		
(c) Secretory vesicles budding off — Cisternae — Transfer vesicles from the smooth endoplasmic reticulum	**Name:** Golgi apparatus **Location:** In cytoplasm associated with smooth ER **Function:**		
(d) Transport pathway Ribosomes Rough — Smooth — Vesicles budding off — Flattened membrane sacs	**Name:** Smooth and rough endoplasmic reticulum **Location of ER:** Penetrates the whole cytoplasm **Function of smooth ER:** **Function of rough ER:**		

DNA · Plasma membrane · Ribosome · Golgi · Nucleus · Animal cell · Plant cell · Leaf section · Leaf

| 0.1 nm | 1 nm | 10 nm | 100 nm | 1 μm | 10 μm | 100 μm | 1 mm | 10 mm |

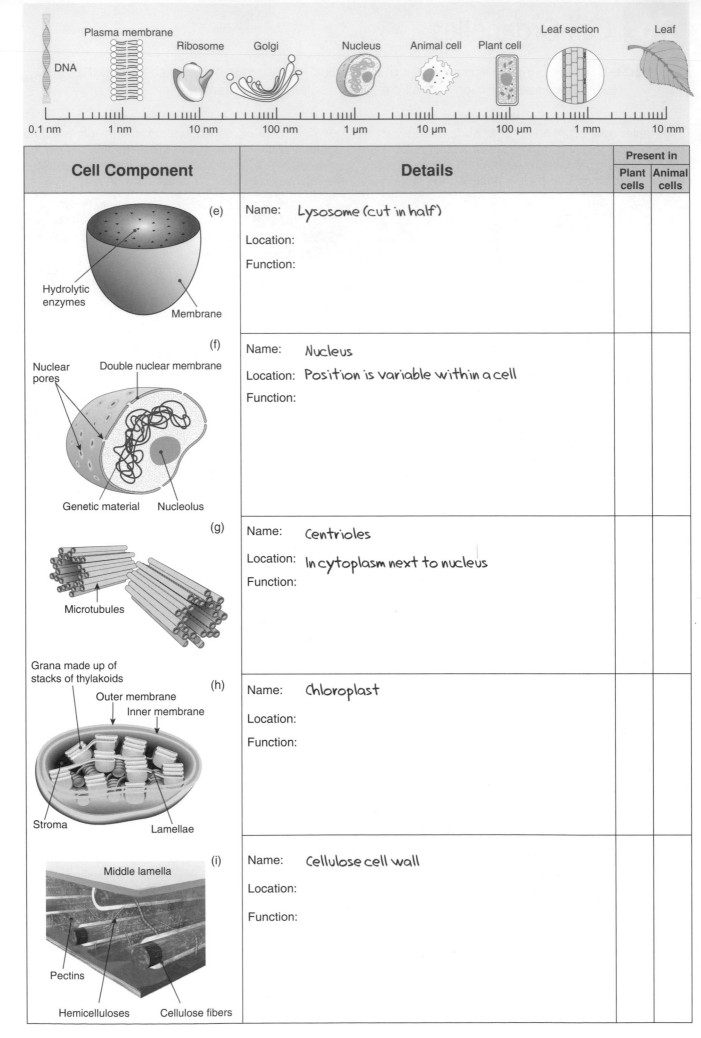

Cell Component	Details	Present in	
		Plant cells	Animal cells
(e) Hydrolytic enzymes / Membrane	**Name:** Lysosome (cut in half) **Location:** **Function:**		
(f) Nuclear pores / Double nuclear membrane / Genetic material / Nucleolus	**Name:** Nucleus **Location:** Position is variable within a cell **Function:**		
(g) Microtubules	**Name:** Centrioles **Location:** In cytoplasm next to nucleus **Function:**		
(h) Grana made up of stacks of thylakoids / Outer membrane / Inner membrane / Stroma / Lamellae	**Name:** Chloroplast **Location:** **Function:**		
(i) Middle lamella / Pectins / Hemicelluloses / Cellulose fibers	**Name:** Cellulose cell wall **Location:** **Function:**		

27 Identifying Organelles

Key Idea: Cellular organelles can be identified in electron micrographs by their specific features.

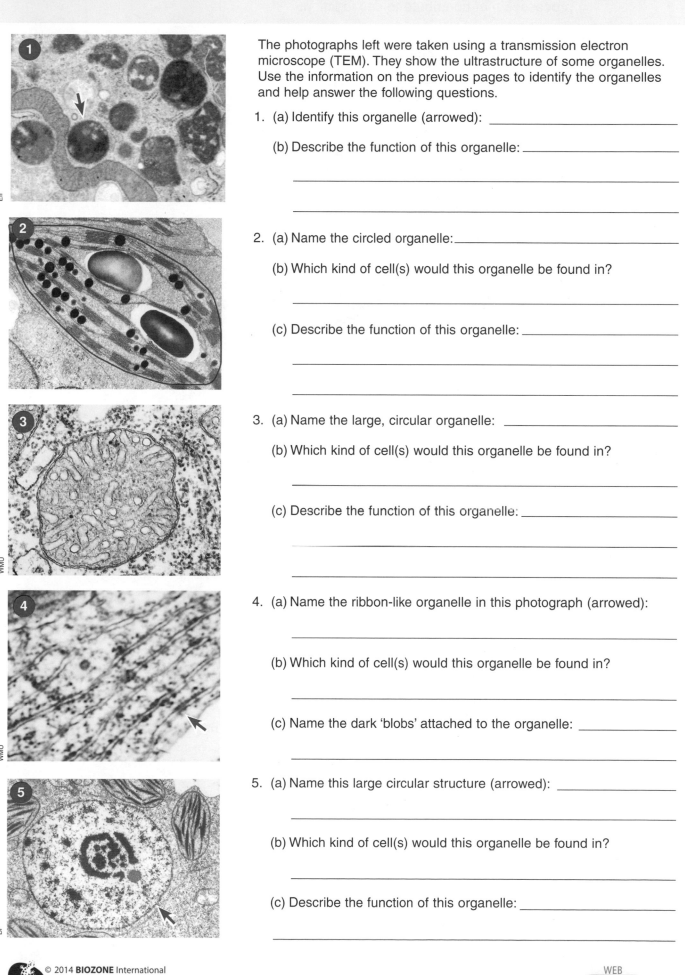

The photographs left were taken using a transmission electron microscope (TEM). They show the ultrastructure of some organelles. Use the information on the previous pages to identify the organelles and help answer the following questions.

1. (a) Identify this organelle (arrowed): _____

   (b) Describe the function of this organelle: _____

   _____

   _____

2. (a) Name the circled organelle: _____

   (b) Which kind of cell(s) would this organelle be found in?

   _____

   (c) Describe the function of this organelle: _____

   _____

   _____

3. (a) Name the large, circular organelle: _____

   (b) Which kind of cell(s) would this organelle be found in?

   _____

   (c) Describe the function of this organelle: _____

   _____

   _____

4. (a) Name the ribbon-like organelle in this photograph (arrowed):

   _____

   (b) Which kind of cell(s) would this organelle be found in?

   _____

   (c) Name the dark 'blobs' attached to the organelle: _____

   _____

5. (a) Name this large circular structure (arrowed): _____

   _____

   (b) Which kind of cell(s) would this organelle be found in?

   _____

   (c) Describe the function of this organelle: _____

   _____

© 2014 **BIOZONE** International
ISBN: 978-1-927173-84-8
Photocopying Prohibited

WEB
27 TEST

28 Processes in Cells

Key Idea: Cells perform essential life processes. Each cell organelle carries out one or more processes that contribute to cell function.

Cell division
Organelle: *centrioles*
Control the movement of chromosomes during cell division.

Autolysis
Organelle: *lysosome*
Destroys unwanted cell organelles and foreign objects.

Animal cell

Transport in and out of the cell
Organelle: *plasma membrane*
Simple diffusion and active transport move substances across the plasma membrane.

Protein synthesis
Organelles: *free ribosomes, rough endoplasmic reticulum*
Genetic information in the nucleus is translated into proteins.

Secretion
Organelle: *Golgi apparatus*
Packages and stores chemicals (e.g. digestive enzymes), and prepares substances (e.g. hormones) for movement out of the cell.

Exocytosis
Organelle: *plasma membrane*
Enables the expulsion of material from the cell).

Endocytosis
Organelle: *plasma membrane*
Can engulf solids or fluid to bring them into the cell.

Cellular respiration
Organelle: *cytoplasm, mitochondria*
Glucose is broken down, supplying the cell with energy to carry out the many other chemical reactions of metabolism.

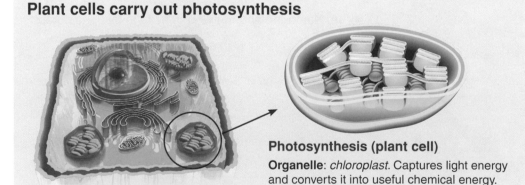

Plant cells carry out photosynthesis

Photosynthesis (plant cell)
Organelle: *chloroplast*. Captures light energy and converts it into useful chemical energy.

1. For each of the processes listed below, identify the organelles or structures associated with that process (there may be more than one associated with a process):

(a) Secretion: _____ (f) Photosynthesis: _____

(b) Respiration: _____ (g) Cell division: _____

(c) Endocytosis: _____ (h) Autolysis: _____

(d) Protein synthesis: _____ (i) Transport in/out of cell: _____

© 2014 **BIOZONE** International
ISBN: 978-1-927173-84-8
Photocopying Prohibited

29 Diffusion in Cells

Key Idea: Diffusion is the movement of molecules from high concentration to a low concentration (i.e. down a concentration gradient).

What is diffusion?

Diffusion is the movement of particles from regions of high concentration to regions of low concentration. Diffusion is a passive process, meaning it needs no input of energy to occur. During diffusion, molecules move randomly about, becoming evenly dispersed.

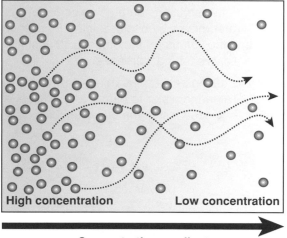

High concentration   **Low concentration**

Concentration gradient

Factors affecting the rate of diffusion

Concentration gradient	The rate of diffusion is higher when there is a greater difference between the concentrations of two regions.
The distance moved	Diffusion over shorter distance occurs at a greater rate than over a larger distance.
The surface area involved	The larger the area across which diffusion occurs, the greater the rate of diffusion.
Barriers to diffusion	Thick barriers have a slower rate of diffusion than thin barriers.
Temperature	Particles at a high temperature diffuse at a greater rate than at a low temperature.

Types of diffusion

Simple diffusion

▶ Molecules move directly through the membrane without any assistance. Example: O_2 diffuses into the blood and CO_2 diffuses out.

Facilitated diffusion

▶ **Carrier-mediated facilitated diffusion**
Carrier proteins allow large lipid-insoluble molecules that cannot cross the membrane by simple diffusion to be transported into the cell. Example: the transport of glucose into red blood cells.

▶ **Channel-mediated facilitated diffusion**
Channels (hydrophilic pores) in the membrane allow inorganic ions pass through the membrane. Example: sodium ions entering nerve cells.

1. What is diffusion? _____

_____

_____

2. What do the three types of diffusion described above all have in common?

_____

_____

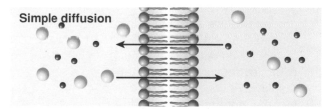

Simple diffusion

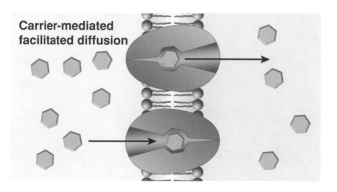

Carrier-mediated facilitated diffusion

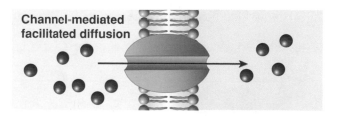

Channel-mediated facilitated diffusion

 © 2014 **BIOZONE** International
ISBN: 978-1-927173-84-8
Photocopying Prohibited

CONNECT WEB

31 **29** **KNOW**

30 Osmosis in Cells

Key Idea: Osmosis is the diffusion of water molecules from a lower solute concentration to a higher solute concentration across a partially permeable membrane.

Osmosis

▸ **Osmosis** is the diffusion of water molecules from regions of lower solute concentration (higher free water concentration) to regions of higher solute concentration (lower free water concentration) across a partially permeable membrane.

▸ A partially permeable membrane lets some, but not all, molecules pass through. The plasma membrane of a cell is an example of a partially permeable membrane.

▸ Osmosis is a passive process (it requires no energy to occur).

Osmotic potential

The presence of solutes (dissolved substances) in a solution increases the tendency of water to move into that solution. This tendency is called the osmotic potential or osmotic pressure. The greater a solution's concentration (i.e. the more total dissolved solutes it contains) the greater the osmotic potential.

Demonstrating osmosis

Osmosis can be demonstrated using the simple experiment described below.

A glucose solution (high solute concentration) is placed into dialysis tubing, and the tubing is placed into a beaker of water (low solute concentration). The difference in concentration of glucose (solute) between the two solutions creates an osmotic gradient. Water moves by osmosis into the glucose solution and the volume of the glucose solution inside the dialysis tubing increases.

The dialysis tubing acts as a partially permeable membrane, allowing water to pass freely, while keeping the glucose inside the dialysis tubing.

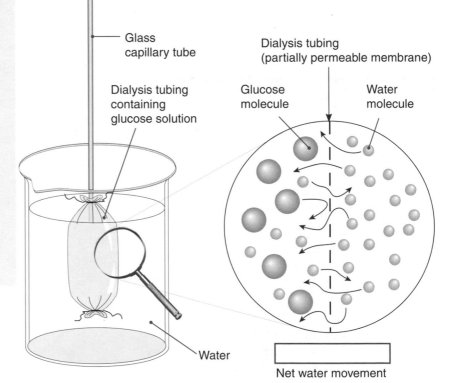

Glass capillary tube

Dialysis tubing (partially permeable membrane)

Dialysis tubing containing glucose solution

Glucose molecule

Water molecule

Water

Net water movement

1. What is osmosis? _____

_____

_____

2. (a) In the blue box on the diagram above, draw an arrow to show the direction of **net water movement**.

(b) Why did water move in this direction? _____

_____

_____

© 2014 **BIOZONE** International
ISBN: 978-1-927173-84-8
Photocopying Prohibited

31 Diffusion and Cell Size

Key Idea: Diffusion is more efficient at delivering materials to the interior of cells when cells have a large surface area relative to their volume.

Single-celled organisms

Single-celled organisms (e.g. *Amoeba*), are small and have a large surface area relative to the cell's volume. The cell's requirements can be met by the diffusion or active transport of materials directly into and out of the cell (below).

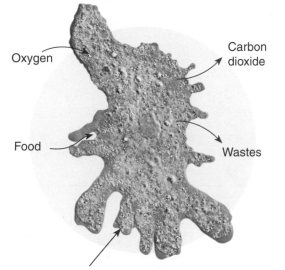

Oxygen

Carbon dioxide

Food

Wastes

The **plasma membrane**, which surrounds every cell, regulates the movement of substances into and out of the cell. For each square micrometer of membrane, only so much of a particular substance can cross per second.

Multicellular organisms

Multicellular organisms (e.g. plants and animals) are often quite large, and large organisms have a small surface area compared to their volume. Diffusion alone is not sufficient to supply their cells with everything they need, so multicellular organisms need specialized systems to transport materials to and from their cells.

In a multicellular organism, such as an elephant, the body's need for respiratory gases cannot be met by diffusion through the skin.

A specialized gas exchange surface (lungs) and circulatory (blood) system are required to supply the body's cells with oxygen and remove carbon dioxide.

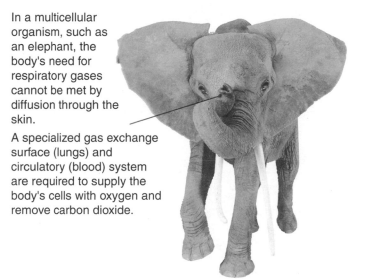

1. Calculate the volume, surface area, and the ratio of surface area to volume for each of the four cubes below (the first has been done for you). Show your calculations as you complete the table below.

2 cm cube

3 cm cube

4 cm cube

5 cm cube

Cube size	Surface area (cm²)	Volume (cm³)	Surface area to volume ratio
2 cm cube	2 x 2 x 6 = 24 cm² (2 cm x 2 cm x 6 sides)	2 x 2 x 2 = 8 cm³ (height x width x depth)	24 to 8 = 3:1
3 cm cube			
4 cm cube			
5 cm cube			

 © 2014 **BIOZONE** International
 ISBN: 978-1-927173-84-8
 Photocopying Prohibited

CONNECT

32 DATA

32 Calculating Diffusion Rates

Key Idea: The surface area to volume ratio decreases as cell volume increases.

1. Use your calculations from the activity *Diffusion and Cell Size* (Q1) to create a graph of the surface area against the volume of each cube, on the grid on the right. Draw a line connecting the points and label axes and units.

2. Which increases the fastest with increasing size: the **volume** or the **surface area**?

   _____

3. Explain what happens to the ratio of surface area to volume with increasing size.

   _____

   _____

   _____

   _____

4. The diffusion of molecules into cells of varying sizes can be modelled using agar cubes infused with phenolphthalein indicator and soaked in sodium hydroxide (NaOH). Phenolphthalein turns pink in the presence of a base (NaOH). As the NaOH diffuses into the agar, the phenolphthalein changes to pink and indicates how far the NaOH has diffused into the agar. Agar blocks are cut into cubes of varying size, so the effect of cell size on diffusion can be studied.

   (a) Use the information below to fill in the table on the right:

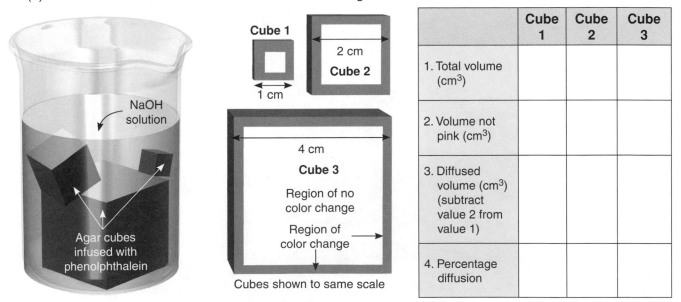

Cube 1

1 cm

2 cm

Cube 2

4 cm

Cube 3

Region of no color change

Region of color change

Cubes shown to same scale

NaOH solution

Agar cubes infused with phenolphthalein

	Cube 1	Cube 2	Cube 3
1. Total volume (cm^3)			
2. Volume not pink (cm^3)			
3. Diffused volume (cm^3) (subtract value 2 from value 1)			
4. Percentage diffusion			

   (b) Diffusion of substances into and out of a cell occurs across the plasma membrane. For a cuboid cell, explain how increasing cell size affects the ability of diffusion to provide the materials required by the cell:

   _____

   _____

   _____

© 2014 **BIOZONE** International
ISBN: 978-1-927173-84-8
Photocopying Prohibited

DATA

33 Active Transport

Key Idea: Active transport uses energy to transport molecules against their concentration gradient across a plasma membrane.

▶ **Active transport** is the movement of molecules (or ions) from regions of low concentration to regions of high concentration across a plasma membrane.

▶ Active transport needs energy to proceed because molecules are being moved against their concentration gradient.

▶ The energy for active transport comes from **ATP** (adenosine triphosphate). Energy is released when ATP is hydrolyzed (water is added) forming ADP (adenosine diphosphate) and inorganic phosphate (Pi).

▶ Transport (carrier) proteins in the plasma membrane are used to actively transport molecules across a membrane (below).

▶ Active transport can be used to move molecules into and out of a cell.

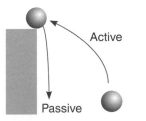

Active

Passive

A ball falling is a passive process (it requires no energy input). Replacing the ball requires active energy input.

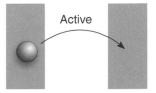

Active

It requires energy to actively move an object across a physical barrier.

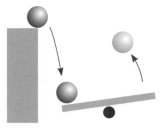

Sometimes the energy of a passively moving object can be used to actively move another. For example, a falling ball can be used to catapult another (left).

Active transport

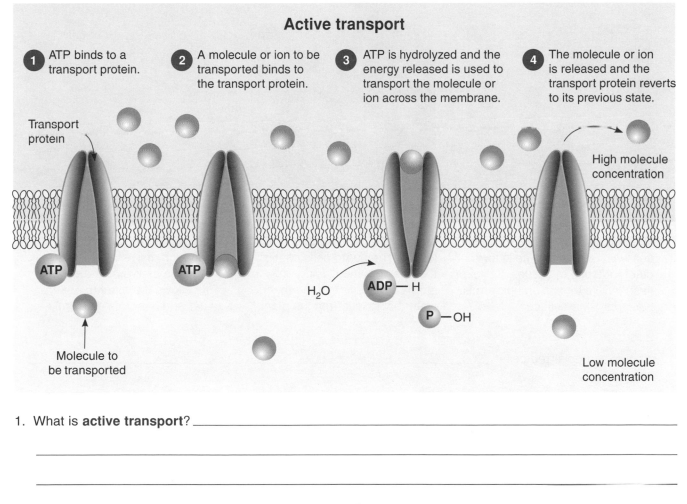

1 ATP binds to a transport protein.

2 A molecule or ion to be transported binds to the transport protein.

3 ATP is hydrolyzed and the energy released is used to transport the molecule or ion across the membrane.

4 The molecule or ion is released and the transport protein reverts to its previous state.

Transport protein

High molecule concentration

ATP

ATP

H_2O

ADP — H

P — OH

Molecule to be transported

Low molecule concentration

1. What is **active transport**? _____

_____

_____

2. Where does the energy for active transport come from? _____

_____

_____

 © 2014 **BIOZONE** International
ISBN: 978-1-927173-84-8
Photocopying Prohibited

34 Specialization in Plant Cells

Key Idea: The specialized cells in a plant have specific features associated with their particular roles in the plant.

Cell specialization

▶ A **specialized cell** is a cell with the specific features needed to perform a particular function in the organism.

▶ Cell specialization occurs during development when specific genes are switched on or off.

▶ Multicellular organisms have many types of specialized cells. These work together to carry out the essential functions of life.

▶ The size and shape of a cell allows it to perform its function. The number and type of organelles in a cell is also related to the cell's role in the organism.

Cells in the leaves of plants are often green because they contain the pigment chlorophyll which is needed for photosynthesis.

Specialized cells in vascular tissue are needed to transport water and sugar around the plant.

Some cells are strengthened to provide support for the plant, allowing it to keep its form and structure.

Plants have root-hair cells so they can get water and nutrients (mineral ions) from the soil.

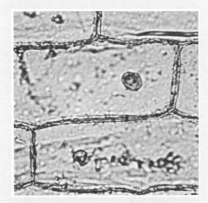

Many plant cells have a regular shape because of their semi-rigid cell wall. Specialized cells form different types of tissues. Simple tissues, like this onion epidermis, have only one cell type.

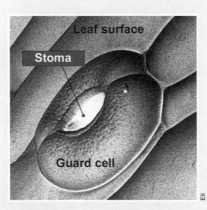

Specialized guard cells surround the stomata (pores) on plant leaves. The guard cells control the opening and closing of stomata and prevent too much water being lost from the plant.

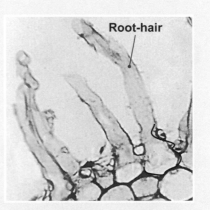

A plant root-hair is a tube-like outgrowth of a plant root cell. Their long, thin shape greatly increases their surface area. This allows the plant to absorb water and minerals efficiently.

1. What is a specialized cell? _____

_____

2. (a) Name the specialized cell that helps to prevent water loss in plants: _____

(b) How does this cell prevent water from being lost in plants? _____

_____

3. How do specialized root hairs help plants to absorb more water and minerals from the soil?

_____

_____

 © 2014 **BIOZONE** International
ISBN: 978-1-927173-84-8
Photocopying Prohibited

35 Specialization in Animal Cells

Key Idea: There are many different types of animal cells, each with a specific role in the body. Animal cells are often highly modified for their specific role.

Specialization in animal cells

▶ There are over 200 different types of cells in the human body.

▶ Animal cells lack a cell wall, so they can take on many different shapes. Therefore, there many more types of animal cells than there are plant cells.

▶ Specialized cells often have modifications or exaggerations to a normal cell feature to help them do their job. For example, nerve cells have long, thin extensions to carry nerve impulses over long distances in the body.

▶ Specialization improves efficiency because each cell type is highly specialized to perform a particular task.

Fat cell

Thin, flat epithelial cells line the walls of blood vessels (arrow). Large fat cells store lipid.

Some nerve cells are over 1 m long.

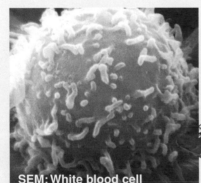

Louisa Howard, Katherine Connolly Dartmouth College

SEM: White blood cell

Some animal cells can move or change shape. A sperm cell must be able to swim so that it can fertilize an egg. A white blood cell changes its shape to engulf and destroy foreign materials (e.g. bacteria).

TEM: Cellular projections of intestinal cell

RBC

Cells that line the intestine have extended cell membranes. This increases their surface area so that more food (nutrients) can be absorbed. Red blood cells (RBCs) have no nucleus so they have more room inside to carry oxygen around the body.

SEM: Egg and sperm

The egg (ovum) is the largest human cell. It is about 0.1 mm in diameter and can be seen with the naked eye. The smallest human cells are sperm cells and red blood cells.

1. What is the advantage of cell specialization in a multicellular organism? _____

_____

_____

2. For each of the following specialized animal cells, name a feature that helps it carry out its function:

(a) White blood cell: _____

(b) Sperm cell: _____

(c) Nerve cell: _____

(d) Red blood cell: _____

© 2014 **BIOZONE** International
ISBN: 978-1-927173-84-8
Photocopying Prohibited

CONNECT CONNECT

44 **43** **KNOW**

36 What is DNA?

Key Idea: A cell's genetic information is called DNA. In eukaryotic cells, DNA is located in the cell nucleus.

About DNA

▶ **DNA** stands for **d**eoxyribo**n**ucleic **a**cid.

▶ DNA is called the blueprint for life because it contains all of the information an organism needs to develop, function, and reproduce.

▶ DNA stores and transmits genetic information.

▶ DNA is found in every cell of all living organisms.

▶ DNA has a double-helix structure (left). If the DNA in a single human cell was unwound, it would be more than two meters long! The long DNA molecules are tightly packed in an organized way so that they can fit into the nucleus.

DNA is found in every cell of all living organisms: animals, plants, fungi, protists, and bacteria.

DNA contains the instructions an organism needs to develop, survive, and reproduce. Small differences in DNA cause differences in appearance.

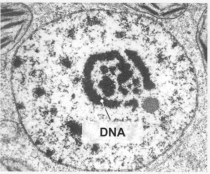

DNA

In eukaryotes, most of the cell's DNA is located in the nucleus (above). A very small amount is located in mitochondria and in the chloroplasts of plants.

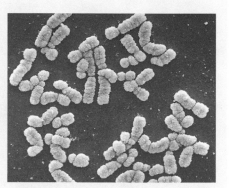

The DNA in eukaryotes is packaged into chromosomes (above). Each chromosome is made up of a DNA molecule and associated proteins. The proteins help to package the DNA into the nucleus.

1. (a) What does DNA stand for? _____

   (b) Where is most of the DNA found in eukaryotes? _____

   (c) What does DNA do? _____

   _____

   _____

2. Why does DNA have to be tightly packaged up? _____

   _____

   _____

CONNECT 38 CONNECT 145 CONNECT 153 CONNECT 157

KNOW

© 2014 **BIOZONE** International
ISBN: 978-1-927173-84-8
Photocopying Prohibited

37 Nucleotides

Key Idea: Nucleotides are the building blocks of DNA and RNA. A nucleotide has three components; a base, a sugar, and a phosphate group.

The structure of a nucleotide

Nucleotides are the building blocks of nucleic acids (DNA and RNA). Nucleotides have three parts to their structure (see diagrams below):

► A nitrogen containing base
► A five carbon sugar
► A phosphate group

Symbolic form of a nucleotide
(showing positions of the 5 C atoms on the sugar)

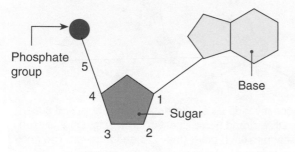

Chemical structure of a nucleotide

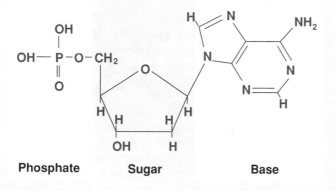

Phosphate   Sugar   Base

Nucleotide bases

Five different kinds of nitrogen bases are found in nucleotides. These are:

Adenine (A)
Guanine (G)
Cytosine (C)
Thymine (T)
Uracil (U)

DNA contains adenine, guanine, cytosine, and thymine.

RNA also contains adenine, guanine, and cytosine, but uracil (U) is present instead of thymine.

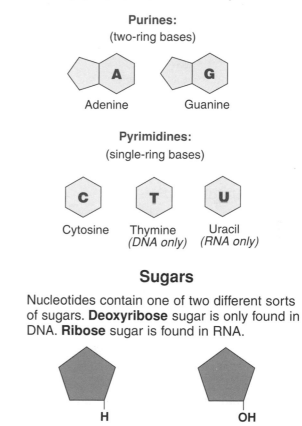

Purines:
(two-ring bases)

Adenine   Guanine

Pyrimidines:
(single-ring bases)

Cytosine   Thymine   Uracil
            (DNA only)   (RNA only)

Sugars

Nucleotides contain one of two different sorts of sugars. **Deoxyribose** sugar is only found in DNA. **Ribose** sugar is found in RNA.

Deoxyribose sugar   Ribose sugar
(found in DNA)   (found in RNA)

1. What are the three components of a nucleotide? _____

_____

2. List the nucleotide bases present:

(a) In DNA: _____

(b) In RNA: _____

3. Name the sugar present: (a) In DNA: _____   (b) In RNA: _____

 © 2014 **BIOZONE** International
ISBN: 978-1-927173-84-8
Photocopying Prohibited

CONNECT
38 KNOW

38 DNA and RNA

Key Idea: DNA and RNA are nucleic acids made up of long chains of nucleotides which store and transmit genetic information. DNA is double-stranded. RNA is single-stranded.

The structure of DNA

Nucleotides join together to form **nucleic acids**. **Deoxyribonucleic acid** (DNA) is a nucleic acid.

DNA consists of a two strands of nucleotides linked together to form a **double helix**. A double helix is like a ladder twisted into a corkscrew shape. The rungs of the ladder are the two nitrogen bases joined by hydrogen bonds. The double helix is 'unwound' to show its structure in the diagram (right).

Who discovered the DNA double helix?

Two scientists, James Watson and Francis Crick, are credited with discovering the structure of DNA. However, they used X-ray pictures of DNA from another scientist, Rosalind Franklin, to confirm their hypothesis.

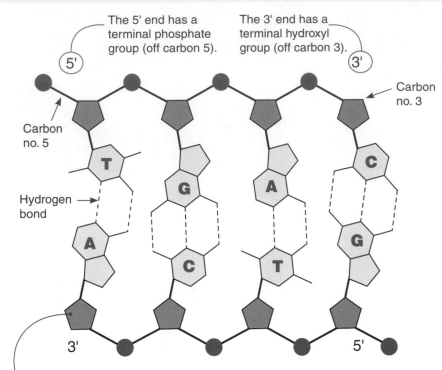

The 5' end has a terminal phosphate group (off carbon 5).

The 3' end has a terminal hydroxyl group (off carbon 3).

Carbon no. 5

Carbon no. 3

Hydrogen bond

The DNA backbone is made up of alternating phosphate and sugar molecules. Each DNA strand has a direction. The single strands run in the opposite direction to each other (they are anti-parallel). The ends of a DNA strand are labeled 5' (five prime) and 3' (three prime).

The structure of RNA

Ribonucleic acid (RNA) is a type of nucleic acid. Like DNA, the nucleotides are linked together through a condensation reaction. RNA is a single stranded, and has many functions including protein synthesis, and cell regulation. There are 3 types of RNA:

▶ Messenger RNA (mRNA)

▶ Transfer RNA (tRNA)

▶ Ribosomal RNA (rRNA)

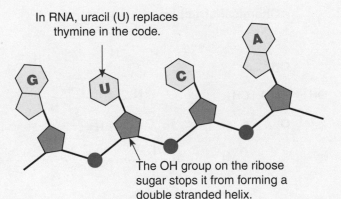

In RNA, uracil (U) replaces thymine in the code.

The OH group on the ribose sugar stops it from forming a double stranded helix.

1. The diagram on the right shows a double-stranded DNA molecule. Label the following:

   (a) Sugar group   (d) Purine bases
   (b) Phosphate group (e) Pyrimidine bases
   (c) Hydrogen bonds

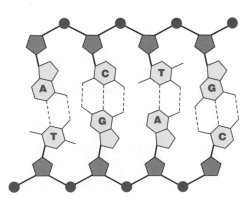

2. If you wanted to use a radioactive or fluorescent tag to label only the RNA in a cell and not the DNA, what compound(s) would you label?

   _____

3. If you wanted to use a radioactive or fluorescent tag to label only the DNA in a cell and not the RNA, what compound(s) would you label?

   _____

© 2014 **BIOZONE** International
ISBN: 978-1-927173-84-8
Photocopying Prohibited

39 Modeling the Structure of DNA

Key Idea: Nucleotides pair together in a specific way called the base pairing rule. In DNA, adenine always pairs with thymine, and cytosine always pairs with guanine.

The exercise on the following pages is designed to help you understand the structure of DNA, and learn the base pairing rule for DNA.

The way the nucleotide bases pair up between strands is very specific. The chemistry and shape of each base means they can only bond with one other DNA nucleotide. Use the information in the table on the right to if you need help remembering the base pairing rule while you are constructing your DNA molecules.

DNA base pairing rule			
Adenine	always pairs with	**Thymine**	A ⟷ T
Thymine	always pairs with	**Adenine**	T ⟷ A
Cytosine	always pairs with	**Guanine**	C ⟷ G
Guanine	always pairs with	**Cytosine**	G ⟷ C

1. Cut out each of the nucleotides on page 55 by cutting along the columns and rows (see arrows indicating two such cutting points). Although drawn as geometric shapes, these symbols represent chemical structures.

2. Place one of each of the four kinds of nucleotide on their correct spaces below:

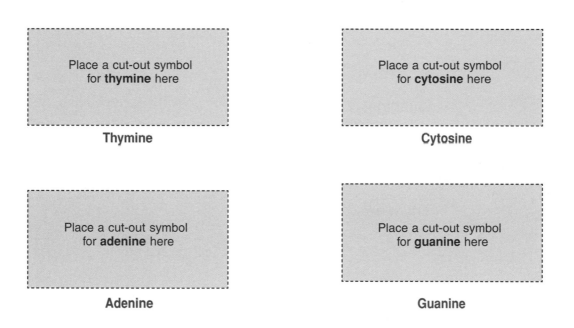

Place a cut-out symbol for **thymine** here

Thymine

Place a cut-out symbol for **cytosine** here

Cytosine

Place a cut-out symbol for **adenine** here

Adenine

Place a cut-out symbol for **guanine** here

Guanine

3. Identify and label each of the following features on the adenine nucleotide immediately above: **phosphate, sugar, base, hydrogen bonds**.

4. Create one strand of the DNA molecule by placing the 9 correct 'cut out' nucleotides in the labelled spaces on the following page (DNA molecule). Make sure these are the right way up (with the **P** on the left) and are aligned with the left hand edge of each box. Begin with thymine and end with guanine.

5. Create the complementary strand of DNA by using the base pairing rule above. Note that the nucleotides have to be arranged upside down.

6. Once you have checked that the arrangement is correct, glue, paste, or tape these nucleotides in place.

© 2014 **BIOZONE** International
ISBN: 978-1-927173-84-8
Photocopying Prohibited

PRAC

DNA Molecule

Put the named nucleotides on the left hand side to create the template strand

Thymine

Thymine

Put the matching **complementary** nucleotides opposite the template strand

Cytosine

Adenine

Adenine

Guanine

Thymine

Thymine

Cytosine

Guanine

Nucleotides

Tear out this page along the perforation and separate each of the 24 nucleotides by cutting along the columns and rows (see arrows indicating the cutting points).

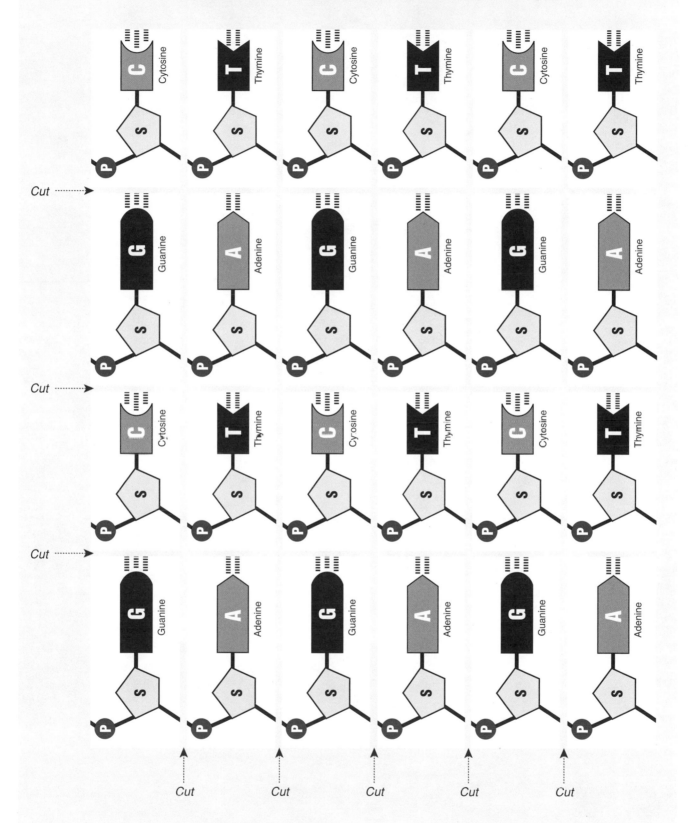

Cut ·······▶

Cut ·······▶

Cut ·······▶

Cut   *Cut*   *Cut*   *Cut*   *Cut*

This page is left blank deliberately

40 Genes Code for Proteins

Key Idea: Genes are sections of DNA that code for proteins. Genes are expressed when they are transcribed into mRNA and then translated into a protein.

▶ A **gene** is a section of DNA that codes for a protein. **Gene expression** is the process of rewriting a gene into a protein. It involves two stages: **transcription** of the DNA and **translation** of the mRNA into protein.

▶ A gene is bounded by a start (promoter) region, upstream of the gene, and a terminator region, downstream of the gene. These regions control transcription by telling RNA polymerase where to start and stop.

▶ RNA polymerase binds to the promoter region to begin transcription of the gene.

▶ The one gene-one protein model is helpful for visualizing the processes involved in gene expression, although it is overly simplistic for eukaryotes. The information flow for gene to protein is shown below.

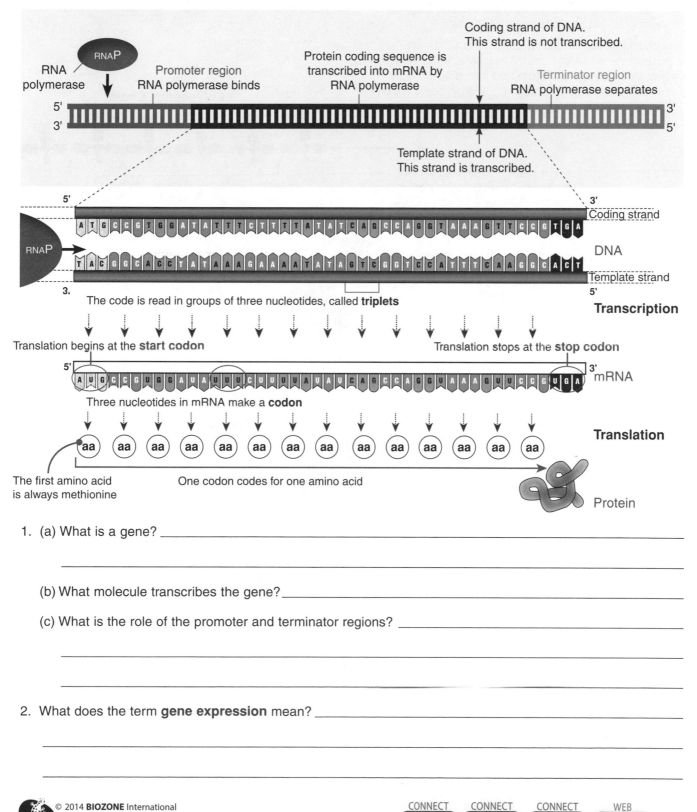

1. (a) What is a gene? _____

_____

(b) What molecule transcribes the gene?_____

(c) What is the role of the promoter and terminator regions? _____

_____

_____

2. What does the term **gene expression** mean? _____

_____

_____

© 2014 **BIOZONE** International
ISBN: 978-1-927173-84-8
Photocopying Prohibited

CONNECT CONNECT CONNECT WEB

146 **42** **41** **40** **KNOW**

41 Proteins

Key Idea: The structure of a protein allows it to carry out its specific role in an organism. Chemical interactions between amino acids determine a protein's shape.

Proteins are made up of amino acids

Proteins are large molecules made up of many smaller units called **amino acids** joined together. The amino acids are joined together by peptide bonds.

All amino acids have a common structure (right) consisting of an amine group, a carboxyl group, a hydrogen atom, and an 'R' group. Each type of amino acid has a different 'R' group (side chain). Each "R" group has a different chemical property.

The chemical properties of the amino acids are important because the chemical interactions between amino acids cause a protein fold into a specific three dimensional shape. The protein's shape helps it carry out its specialized role.

The general structure of an amino acid

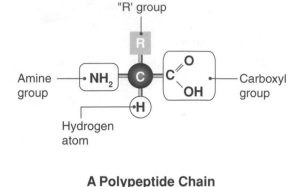

The order of amino acids in a protein is directed by the order of nucleotides in DNA and mRNA.

A Polypeptide Chain

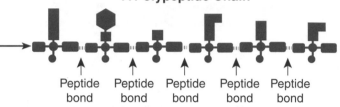

Peptide bond | Peptide bond | Peptide bond | Peptide bond | Peptide bond

Proteins carry out specialized roles

Proteins have important roles in all of the essential life processes an organisms carries out. As a result, there are many different types of proteins within an organism, each with its own specialized role.

The shape of a protein determines its role. Proteins generally fall into two groups, globular and fibrous (right). The shape of a protein is so important to its function, that if the structure of a protein is destroyed (**denatured**) it can no longer carry out its function.

Insulin: a globular protein

Collagen: a fibrous protein

Globular proteins

▶ Globular proteins are round and water soluble. Their functions include:
- Catalytic (e.g. enzymes)
- Regulation (e.g. hormones)
- Transport (e.g. hemoglobin)
- Protective (e.g. antibodies)

Fibrous proteins

▶ Fibrous proteins are long and strong Their functions include:
- Support and structure (e.g. connective tissue)
- Contractile (e.g. myosin, actin)

1. (a) Name the four components of an amino acid: _____

_____

(b) What makes each type of amino acid unique? _____

2. Why are proteins important in organisms? _____

_____

3. (a) Why is the shape of a protein important? _____

_____

(b) What happens to a protein if it loses its shape? _____

© 2014 **BIOZONE** International
ISBN: 978-1-927173-84-8
Photocopying Prohibited

42 Cracking the Genetic Code

Key Idea: Scientists used mathematics and scientific experiments to unlock the genetic code. A a series of three nucleotides, called a triplet, codes for a single amino acid.

The genetic code

Once it was discovered that DNA carries the genetic code needed to produce proteins, the race was on to "crack the code" and find out how it worked.

The first step was to find out how many nucleotide bases code for an amino acid. Scientists knew that there were four nucleotide bases in mRNA, and that there are 20 amino acids. Simple mathematics (right) showed that a one or two base code did not produce enough amino acids, but a triplet code produced more amino acids than existed. The triplet code was accepted once scientists confirmed that some amino acids have multiple codes.

Number of bases in the code	Working	Number of amino acids produced
Single (4^1)	4	4 amino acids
Double (4^2)	4 x 4	16 amino acids
Triple (4^3)	4 x 4 x 4	64 amino acids

A triplet (three nucleotide bases) codes for a single amino acid. The triplet code on mRNA is called a codon.

How was the genetic code cracked?

Once the triplet code was discovered, the next step was to find out which amino acid each codon produced. Two scientists, Marshall Nirenberg and Heinrich Matthaei, developed an experiment to crack the code. Their experiment is shown on the right.

Over the next few years, similar experiments were carried out using different combinations of nucleotides until all of the codes were known.

In a test tube, Nirenberg and Matthaei added all of the components needed to make proteins (except mRNA).

They then made an mRNA strand containing only one repeated nucleotide. The strand below shows cytosine (C).

C C C C C C C C C C C C C C

Once the components were added together an amino acid was produced. The codon CCC produced the amino acid proline (Pro).

Pro Pro Pro Pro

---

1. (a) How many types of nucleotide bases are there in mRNA? _____

   (b) How many types of amino acids are there? _____

   (c) Why did scientists reject a one or two base code when trying to work out the genetic code?

   _____

   _____

2. A triplet code could potentially produce 64 amino acids. Why are only 20 amino acids produced?

   _____

   _____

© 2014 **BIOZONE** International
ISBN: 978-1-927173-84-8
Photocopying Prohibited

CONNECT

1 KNOW

43 Organ Systems Work Together

Key Idea: The different organ systems of an organism have specific roles and interact to bring about the efficient functioning of the body.

An organ system is a group of organs that work together to perform a certain group of tasks. Although each system has a specific job (e.g. digestion, reproduction, internal transport, or gas exchange) the organ systems must interact to maintain the functioning of the organism.

There are 11 organ (body) systems in humans. Plants have fewer organ systems.

The example on this page shows how the muscular and skeletal organ systems in humans work together to achieve movement in the arm.

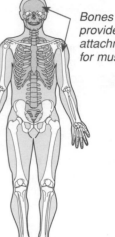

Skeletal muscles provide support and pull on bones to create movement.

Bones provide attachments for muscles.

Muscular system

Skeletal system

Bones act with muscles to form levers that enable movement

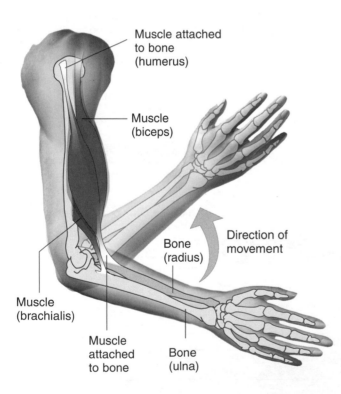

Muscle attached to bone (humerus)

Muscle (biceps)

Direction of movement

Bone (radius)

Muscle (brachialis)

Muscle attached to bone

Bone (ulna)

While you are sitting at your desk, lay your forearm on the desk, palm facing upwards. Now put your other hand on your upper arm and raise your resting arm towards you, bending at the elbow. You would have felt the muscles in your upper arm move. This required two organ systems, the skeletal and muscular systems, to work together.

What you felt was the muscle contracting (shortening), causing the bone attached to it to move at its joint, creating movement. A pair of muscles, each with opposing actions, can work in opposition to create movement of a body part.

1. What is an organ system? _____

_____

2. (a) What is the role of the skeletal system in movement? _____

_____

(b) What is the role of the muscular system in movement? _____

_____

CONNECT CONNECT CONNECT

KNOW 44 45 72

© 2014 **BIOZONE** International
ISBN: 978-1-927173-84-8
Photocopying Prohibited

44 Interacting Systems in Animals

Key Idea: The circulatory and respiratory systems interact together to provide the body's tissues with oxygen and remove carbon dioxide.

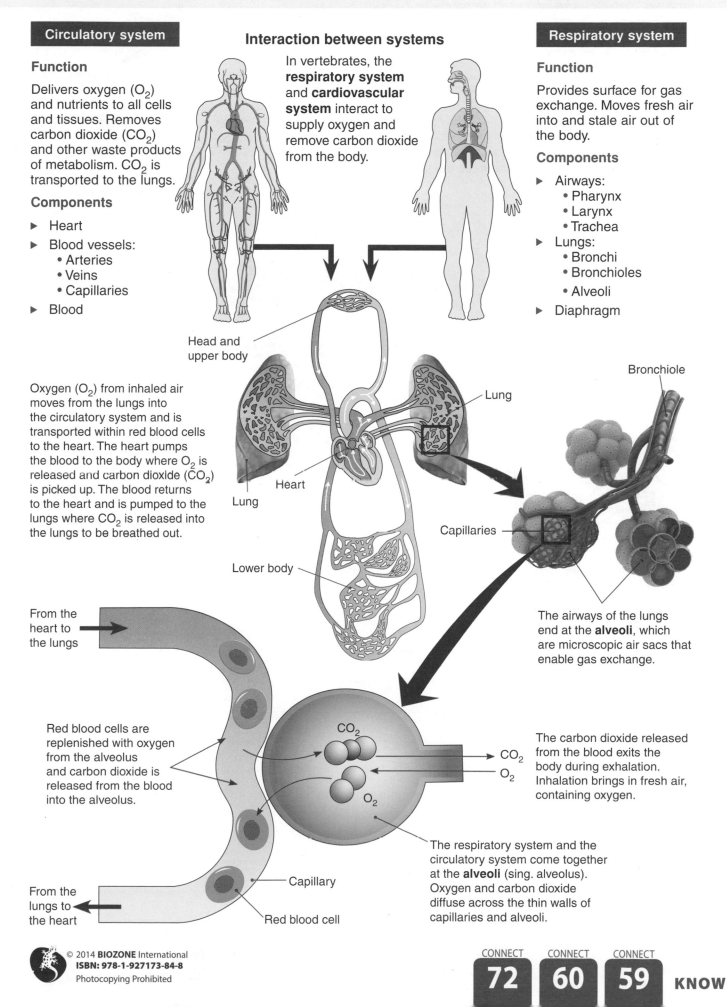

Circulatory system

Function

Delivers oxygen (O_2) and nutrients to all cells and tissues. Removes carbon dioxide (CO_2) and other waste products of metabolism. CO_2 is transported to the lungs.

Components

▶ Heart
▶ Blood vessels:
  • Arteries
  • Veins
  • Capillaries
▶ Blood

Interaction between systems

In vertebrates, the **respiratory system** and **cardiovascular system** interact to supply oxygen and remove carbon dioxide from the body.

Respiratory system

Function

Provides surface for gas exchange. Moves fresh air into and stale air out of the body.

Components

▶ Airways:
  • Pharynx
  • Larynx
  • Trachea
▶ Lungs:
  • Bronchi
  • Bronchioles
  • Alveoli
▶ Diaphragm

Head and upper body

Oxygen (O_2) from inhaled air moves from the lungs into the circulatory system and is transported within red blood cells to the heart. The heart pumps the blood to the body where O_2 is released and carbon dioxide (CO_2) is picked up. The blood returns to the heart and is pumped to the lungs where CO_2 is released into the lungs to be breathed out.

Bronchiole

Lung

Heart

Lung

Capillaries

Lower body

The airways of the lungs end at the **alveoli**, which are microscopic air sacs that enable gas exchange.

From the heart to the lungs

Red blood cells are replenished with oxygen from the alveolus and carbon dioxide is released from the blood into the alveolus.

CO_2

CO_2
O_2

O_2

The carbon dioxide released from the blood exits the body during exhalation. Inhalation brings in fresh air, containing oxygen.

Capillary

From the lungs to the heart

Red blood cell

The respiratory system and the circulatory system come together at the **alveoli** (sing. alveolus). Oxygen and carbon dioxide diffuse across the thin walls of capillaries and alveoli.

CONNECT
72

CONNECT
60

CONNECT
59

KNOW

Responses to exercise

During exercise, your body needs more oxygen to meet the extra demands placed on the muscles, heart, and lungs. At the same time, more carbon dioxide must be expelled. To meet these increased demands, blood flow must increase. This is achieved by increasing the rate of heart beat. As the heart beats faster, blood is circulated around the body more quickly, and exchanges between the blood and tissues increase.

The arteries and veins must be able to resist the extra pressure of higher blood flow and must expand (dilate) to accommodate the higher blood volume. If they didn't, they could rupture (break). During exercise, the muscular, cardiovascular, and nervous systems interact to maintain the body's systems in spite of increased demands (right).

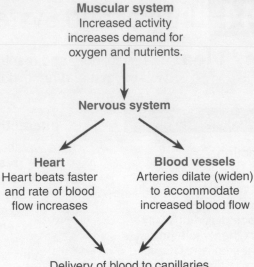

Muscular system
Increased activity increases demand for oxygen and nutrients.

Nervous system

Heart
Heart beats faster and rate of blood flow increases

Blood vessels
Arteries dilate (widen) to accommodate increased blood flow

Delivery of blood to capillaries of working muscle increases

A thick layer of elastic tissue and smooth muscle. When the smooth muscle relaxes, the artery expands to allow more blood to flow.

Capillaries dilate during exercise to increase the rate of exchanges of gases, nutrients, and wastes between the blood and the tissues.

Endothelium is in contact with the blood

Elastic outer layer prevents the artery over-expanding.

Muscular activity helps return blood to the heart

Valves stop back-flow of blood

Artery
The strong stretchy structure of arteries enables them to respond to increases in blood flow and pressure as more blood is pumped from the heart.

Vein
Veins return blood to the heart. They are less muscular than arteries, but valves and the activity of skeletal muscles, especially during exercise, help venous return.

1. In your own words, describe how the circulatory system and respiratory system work together to provide the body with oxygen and remove carbon dioxide:

_____

_____

_____

_____

_____

2. (a) What happens to blood flow during exercise? _____

   (b) How do body systems interact to accommodate the extra blood flow needed when a person exercises?

_____

_____

_____

_____

© 2014 **BIOZONE** International
ISBN: 978-1-927173-84-8

45 Plant Organ Systems

Key Idea: The plant body is divided into the shoot system (stems, leaves, and other above-ground parts) and the below-ground root system.

Plants have fewer organ systems than animals because they are simpler and have lower energy demands. The two primary plant organ systems are the shoot system and the root system.

Shoot system

The above-ground parts of the plant: including organs such as leaves, buds, stems, and the flowers and fruit (or cones) if present. All parts of the shoot system produce hormones.

Leaves
▸ Manufacture food via photosynthesis.
▸ Exchange gases with the environment.
▸ Store food and water.

Stems
▸ Transport water and nutrients between roots and leaves.
▸ Support and hold up the leaves, flowers and fruit.
▸ Produce new tissue for photosynthesis and support.
▸ Store food and water.

Structures for sexual reproduction
▸ Reproductive structures are concerned with passing on genes to the next generation.
▸ Flowers or cones are the reproductive structures of seed plants (angiosperms and gymnosperms).
▸ Fruits provide flowering plants with a way to disperse the seeds.

The shoot and root systems of plants are connected by transport tissues (xylem and phloem) that are continuous throughout the plant.

Root system

The below-ground parts of the plant, including the roots and root hairs.

Roots
▸ Anchor the plant in the soil
▸ Absorb and transport minerals and water
▸ Store food
▸ Produce hormones
▸ Produce new tissue for anchorage and absorption.

1. Describe how each of the following systems provides for the essential functions of life for the plant:

(a) Root system: _____

_____

(b) Shoot system: _____

_____

2. In the following list of plant functions, circle in blue the functions that are shared by the root and shoot system, circle in red those unique to the shoot system, and circle in black those unique to the root system:

Photosynthesis, transport, absorption, anchorage, storage, sexual reproduction, hormone production, growth

CONNECT
46 **KNOW**

46 Interacting Systems in Plants

Key Idea: The shoot and root systems of plants interact to balance water uptake and loss, so that the plant can maintain the essential functions of life.

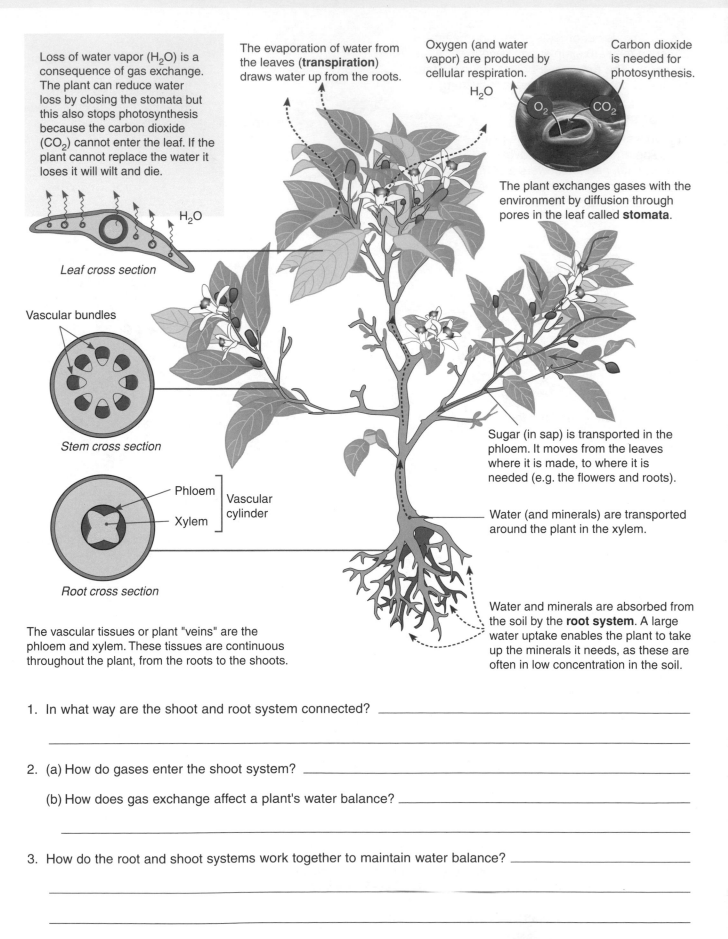

Loss of water vapor (H_2O) is a consequence of gas exchange. The plant can reduce water loss by closing the stomata but this also stops photosynthesis because the carbon dioxide (CO_2) cannot enter the leaf. If the plant cannot replace the water it loses it will wilt and die.

H_2O

Leaf cross section

The evaporation of water from the leaves (**transpiration**) draws water up from the roots.

Oxygen (and water vapor) are produced by cellular respiration.

H_2O

O_2 CO_2

Carbon dioxide is needed for photosynthesis.

The plant exchanges gases with the environment by diffusion through pores in the leaf called **stomata**.

Vascular bundles

Stem cross section

Phloem ⎤
          ⎬ Vascular cylinder
Xylem ⎦

Root cross section

The vascular tissues or plant "veins" are the phloem and xylem. These tissues are continuous throughout the plant, from the roots to the shoots.

Sugar (in sap) is transported in the phloem. It moves from the leaves where it is made, to where it is needed (e.g. the flowers and roots).

Water (and minerals) are transported around the plant in the xylem.

Water and minerals are absorbed from the soil by the **root system**. A large water uptake enables the plant to take up the minerals it needs, as these are often in low concentration in the soil.

1. In what way are the shoot and root system connected? _____

   _____

2. (a) How do gases enter the shoot system? _____

   (b) How does gas exchange affect a plant's water balance? _____

   _____

3. How do the root and shoot systems work together to maintain water balance? _____

   _____

   _____

 © 2014 **BIOZONE** International
ISBN: 978-1-927173-84-8
Photocopying Prohibited

47 Chapter Review

Summarize what you know about this topic so far under the headings provided. You can draw diagrams or mind maps, or write short notes to organize your thoughts. Use the images and hints and guidelines included to help you:

The structure and function of organelles

HINT: Include characteristic features of plant and animal cells.

Differences between eukaryotic and prokaryotic cells

HINT: These differences should distinguish the cell types.

The importance of specialized cells

HINT: How does the size, shape, and organelles of a cell help it carry out its role in an organism?

Interactions between organ systems

HINT: How do the interactions help to maintain life's functions? Use plant and animal examples.

© 2014 **BIOZONE** International
ISBN: 978-1-927173-84-8
Photocopying Prohibited

REVISE

The structure and function of the plasma membrane

HINT: How does the membrane's structure contribute to its function?
You may wish to draw a diagram to show all the components of a membrane.

DNA

The structure of DNA and RNA

HINT: Compare the features of each molecule, and explain the role of the base pairing rule.

Cellular transport

HINT: Include both active and passive transport. Include reference to the concentration gradient.

Genes code for proteins

HINT: Include information about the structure of proteins and how the genetic code was cracked.

48 KEY TERMS: Did You Get It?

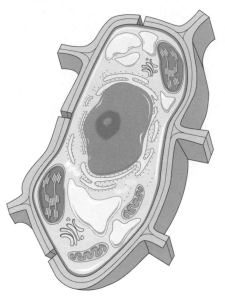

1. (a) Is the cell on the right an animal or plant cell?

   _____

   (b) List the features of the cell that support your answer:

   _____

   _____

   _____

   _____

   _____

   _____

   (c) What word is used to describe cells that carry out a very specific job?

   _____

2. Organelles are the cell's "organs": they carry out the cell's work. Draw a line to match the organelle in the left hand column with its correct description in the right hand column.

Cell wall	A lipid bilayered membrane surrounding a cell. It controls the movement of substances into and out of the cell.
Chloroplast	A structure present in plant cells but not animal cells. It is found outside the plasma membrane and gives rigidity to the cell.
Nucleus	Membrane-bound area within a eukaryotic cell where most of a cell's DNA is found.
Mitochondrion	These structures are involved in making proteins in a cell.
Plasma membrane	An organelle found in plants which contains chlorophyll and is the site of photosynthesis.
Ribosome	This organelle is involved in producing the cell's energy.

3. Fill in the missing words in the paragraph below. Use the word list below to help you.
   Guanine, nucleotides, protein, shape, double-helix, base, DNA, phosphate, denatured, cytosine, base pairing

   All living cells contain genetic material called _____, which stores and transmits the information an

   organism needs to develop, function, and reproduce. DNA is very large. It is made up of building blocks

   called _____ joined together. A nucleotide has three parts, a_____ , a sugar, and a

   _____ group. There are four different types of nucleotides in DNA, adenine,_____ ,

   _____ , and thymine. They pair together in a very specific way called the _____ _____ rule.

   This pairing gives DNA its characteristic _____ shape. Segments of DNA, called genes,

   code for a specific _____. Proteins are very important because they control every aspect of an

   organism's structure and function. The _____ of a protein determines its functional role. If the

   protein loses its shape, becomes _____ , it can no longer perform its role.

© 2014 **BIOZONE** International
ISBN: 978-1-927173-84-8
Photocopying Prohibited

VOCAB

LS1.A

Feedback Mechanisms

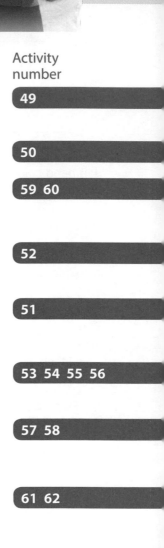

Key terms

diabetes mellitus
homeostasis
negative feedback
positive feedback
thermoregulation
transpiration

Core ideas

Feedback mechanisms maintain an organism's internal environment within certain limits

Activity number

☐ 1. **Homeostasis** mechanisms help the body maintain a constant internal environment, even when external conditions are changing.

`49`

☐ 2. The body's systems must maintain homeostasis so essential life process can be carried out to maintain life.

`50`

☐ 3. During exercise, the circulatory and respiratory systems are mainly responsible for maintaining homeostasis.

`59 60`

Feedback mechanisms can be positive or negative

☐ 4. **Positive feedback** mechanisms amplify a response, usually to achieve a certain outcome. Labor is an example of a physiological process involving positive feedback.

`52`

☐ 5. **Negative feedback** mechanisms have a stabilizing effect, are self correcting, and encourage a return to the steady state. Negative feedback regulates most natural systems, e.g. in physiology and ecology.

`51`

☐ 6. **Thermoregulation** (regulation of body temperature) is controlled by negative feedback. Failure to maintain a constant body temperature can be life threatening.

`53 54 55 56`

☐ 7. Blood sugar levels are tightly regulated by negative feedback mechanisms. **Diabetes mellitus** is a life-threatening disease which can occur when normal regulatory controls no longer work.

`57 58`

☐ 8. Plants lose water by a process called **transpiration**. They must take up enough water to balance the water loss so they can continue to carry out essential life processes.

`61 62`

Science and engineering practices

☐ 1. Develop and use a model to show that the hierarchical organization of interacting systems provides specific functions in a multicellular organism.

`50`

☐ 2. Use a model based on evidence to show how negative feedback mechanisms maintain homeostasis.

`51 55 57`

☐ 3. Carry out an investigation to show how the body maintains homeostasis, for example, during exercise.

`60`

☐ 4. Carry out an investigation to show how plants maintain water balance in changing environmental conditions.

`62`

49 Homeostasis

Key Idea: Homeostasis is the ability to maintain a constant internal environment despite changes in the external environment.

What is homeostasis?

Homeostasis literally means "constant state". Organisms maintain homeostasis, i.e. a relatively constant internal environment, even when the external environmental is changing. This takes energy.

For example, when you exercise (right), your body must keep your body temperature constant at about 37.0 °C despite the increased heat generated by activity. Similarly, you must regulate blood sugar levels and blood pH, water and electrolyte balance, and blood pressure. Your body's organ systems carry out these tasks.

To maintain homeostasis, the body must detect changes in the environment (through receptors), process this sensory information, and respond to it appropriately. The response provides new feedback to the receptor. These three components are illustrated below.

How homeostasis is maintained

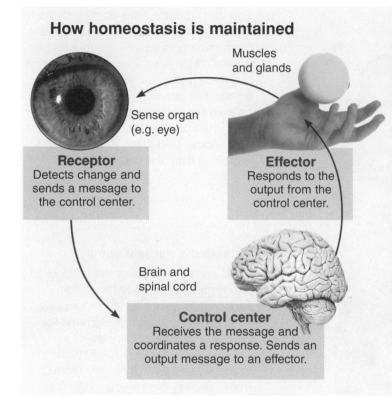

Sense organ (e.g. eye)

Muscles and glands

Receptor
Detects change and sends a message to the control center.

Effector
Responds to the output from the control center.

Brain and spinal cord

Control center
Receives the message and coordinates a response. Sends an output message to an effector.

The analogy of a thermostat on a heater is a good way to understand how homeostasis is maintained. A heater has sensors (a receptor) to monitor room temperature. It also has a control center to receive and process the data from the sensors. Depending on the data it receives, the control center activates the effector (heating unit), switching it on or off. When the room is too cold, the heater switches on. When it is too hot, the heater switches off. This maintains a constant temperature.

1. What is **homeostasis**? _____

_____

2. What is the role of the following components in maintaining homeostasis:

(a) **Receptor**: _____

_____

(b) **Control center**: _____

_____

(c) **Effector**: _____

_____

KNOW

50 Keeping in Balance

Key Idea: Essential life processes, such as growth, require the body's systems to be kept in balance. Many organ systems work together to maintain homeostasis.

Why is homeostasis important?

▶ An organism must constantly regulate its internal environment in order to carry out essential life processes, such as growing and responding to the environment. Changes outside of normal levels for too long can stop the body systems working properly, and can result in illness or death.

▶ Homeostasis relies on monitoring all the information received from the internal and external environment and coordinating appropriate responses. This often involves many different organ systems working together.

▶ Most of the time an organism's body systems are responding to changes at the subconscious level, but sometimes homeostasis is achieved by changing a behavior (e.g. finding shade if the temperature is too high).

Some examples of how the body keeps in balance

Regulating respiratory gases

All of the body's cells need oxygen to carry out cellular respiration to produce usable energy. Oxygen must be delivered to all cells, and carbon dioxide (a waste product of cellular respiration) must be removed. Oxygen demand and carbon dioxide production change with activity level and changes in the environment. For example, the body needs more oxygen during exercise than when a person is resting quietly, so respiration rates increase to meet the increased demand.

Maintaining fluid balance

Without enough water, the body becomes dehydrated and essential life processes are disrupted. Dehydration quickly results in death if untreated. The amount of water lost in urine, feces, breathing, and sweat, must be balanced by consuming enough water (as food and drink) to cover the losses.

Coping with disease-causing organisms

We are under constant attack from disease-causing organisms (pathogens) which can cause damage to the body systems. The body's immune system produces substances and cells to prevent the entry of pathogens and limit the damage they cause if they do enter. The cardiovascular system circulates these components through the body.

Coordinating responses

The body is constantly bombarded by stimuli from the environment. The brain must prioritize its responses and decide which stimuli are important and require a response, and which ones do not. For example, if a person steps on a sharp object and experiences pain, their body coordinates the responses needed to lift the foot and remove it from the source of pain.

Maintaining nutrient supply

Food and drink provide the energy and nutrients the body needs to carry out its essential life processes. Factors that change the demand for nutrients include activity level (e.g. the body burns more energy while it is active) and environmental factors (more energy is required to maintain homeostasis in cold environments).

Repairing injuries

Damage to the body's tissues triggers responses to repair it and return it to a normal state as quickly as possible. For example, blood clotting stops too much blood from leaving the body from an open wound. If too much blood is lost, a person will go into fatal shock.

CONNECT 43   CONNECT 44   CONNECT 46

KNOW

© 2014 **BIOZONE** International
ISBN: 978-1-927173-84-8
Photocopying Prohibited

The body's organ systems work together to maintain homeostasis

Organ systems work together to maintain the environment necessary for the functioning of the body's cells. A constant internal environment allows an organism to be somewhat independent of its external environment, so that it can move about even as its environment changes. The simplified example below illustrates how organ systems interact to exchange material with each other to maintain a constant internal environment.

Once food has been digested in the digestive system, it is absorbed (taken up) into the blood of the circulatory system.

The **digestive system** is responsible for the digestion (break down) and absorption of food. Ultimately, it provides the nutrients required by all body systems.

Unabsorbed digestive matter expelled from the digestive system as feces.

Food provides the energy and nutrients needed to provide energy to all of the body's systems.

Urine is produced by the kidneys of the urinary system. It contains the waste products of metabolism, particularly nitrogen-containing wastes and excess ions.

Heart

The **circulatory system** distributes gases, nutrients, and other substances (e.g. hormones) around the body in the blood.

The **urinary system** has several roles including disposal of nitrogen-containing and other waste products from the body, regulating ion balance, and controlling the volume and pressure of the blood.

Once nutrients have been absorbed, they are carried in the blood and delivered to cells all around the body (solid arrows). Wastes (dashed arrow) move from the cells back into the blood and are transported and removed.

1. Why is it important that the body systems are kept in balance? _____

_____

_____

_____

2. Why is it important that the brain prioritizes the importance of the incoming stimuli? _____

_____

_____

3. Using an example, briefly explain why homeostasis often involves more than one body system:

_____

_____

_____

_____

51 Negative Feedback Mechanisms

Key Idea: Negative feedback mechanisms detect changes in the internal environment away from the normal and act to return the internal environment back to a steady state.

Negative feedback is a control system which maintains the body's internal environment at a steady state. Negative feedback has a stabilizing effect and acts to discourage variations from a set point. It works by returning internal conditions back to a steady state when variations are detected (right).

Most body systems achieve homeostasis through negative feedback. Body temperature, blood glucose levels, and blood pressure are all controlled by negative feedback mechanisms.

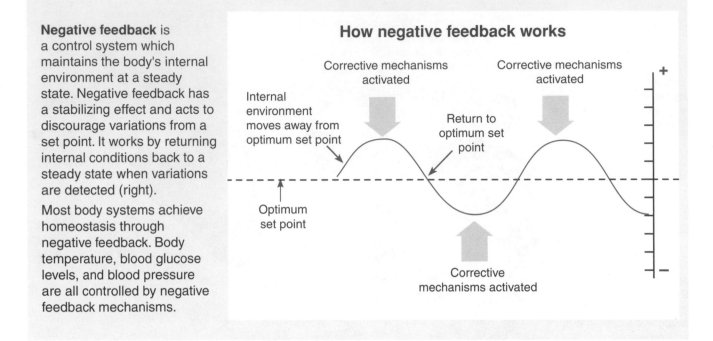

How negative feedback works

Corrective mechanisms activated

Corrective mechanisms activated

Internal environment moves away from optimum set point

Return to optimum set point

Optimum set point

Corrective mechanisms activated

Stomach emptying: an example of negative feedback

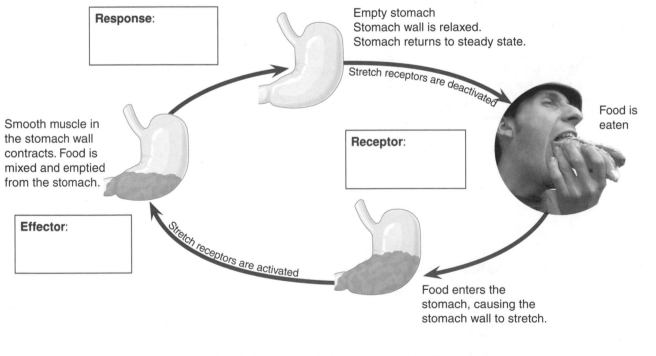

Response:

Empty stomach
Stomach wall is relaxed.
Stomach returns to steady state.

Stretch receptors are deactivated

Food is eaten

Smooth muscle in the stomach wall contracts. Food is mixed and emptied from the stomach.

Receptor:

Effector:

Stretch receptors are activated

Food enters the stomach, causing the stomach wall to stretch.

1. How does negative feedback maintain homeostasis? _____

_____

_____

2. (a) On the diagram of stomach emptying, name the receptor, effector, and response in the spaces provided.

   (b) What is the steady state for this example? _____

© 2014 **BIOZONE** International
ISBN: 978-1-927173-84-8
Photocopying Prohibited

52 Positive Feedback Mechanisms

Key Idea: Positive feedback mechanisms amplify a physiological response in order to achieve a particular outcome.

Positive feedback mechanisms amplify (increase) or speed up a physiological response, usually to achieve a particular outcome. Examples of positive feedback include fruit ripening, fever, blood clotting, childbirth (labor) and lactation (production of milk). A positive feedback mechanism stops when the end result is achieved (e.g. the baby is born, a pathogen is destroyed by a fever, or ripe fruit falls off a tree).

Positive feedback is less common than negative feedback because it creates an escalation in response, which is unstable. This response can be dangerous (or even cause death) if it is prolonged.

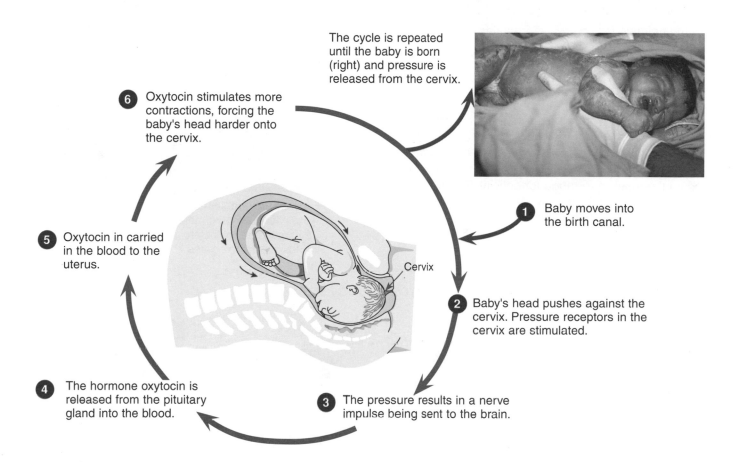

The cycle is repeated until the baby is born (right) and pressure is released from the cervix.

6 Oxytocin stimulates more contractions, forcing the baby's head harder onto the cervix.

5 Oxytocin in carried in the blood to the uterus.

4 The hormone oxytocin is released from the pituitary gland into the blood.

Cervix

1 Baby moves into the birth canal.

2 Baby's head pushes against the cervix. Pressure receptors in the cervix are stimulated.

3 The pressure results in a nerve impulse being sent to the brain.

1. (a) Why is positive feedback much less common than negative feedback in body systems?

_____

_____

(b) Why can positive feedback be dangerous if it continues on for too long? _____

_____

_____

(c) How is a positive feedback loop normally stopped? _____

_____

2. (a) Name the regulatory factor in childbirth: _____

(b) What event brings an end to the positive feedback loop in childbirth? _____

_____

53 Sources of Body Heat

Key Idea: An optimal body temperature is required for essential life processes. Ectotherms obtain heat from the environment. Endotherms generate heat from metabolism.

Where do animals get their body heat from?

Animals are classified into two groups based on the source of their body heat.

▶ **Ectotherms** depend on the environment for their heat energy (e.g. heat from the sun).

▶ **Endotherms** generate most of their body heat from internal metabolic processes.

Many animals fall somewhere between the two extremes.

Why is body heat important?

An optimal body temperature is needed for essential life processes. Many ectotherms cannot function optimally until their body temperature has reached a certain level. For example, they cannot move quickly in the early morning and at night when their body temperature is low.

The enzymes involved in metabolic pathways all have an optimal temperature range for activity. Below the optimum temperature, metabolic reactions proceed very slowly. Above the optimum temperature, the enzymes may become damaged and the reaction does not proceed.

Reptiles are ectotherms. This lizard is basking (laying in the sun) on warm rocks to increase its body temperature. Until has warmed up, it may not be able to capture prey or avoid predators.

Birds are endotherms. Even in the cold Antarctic temperatures, the metabolic activity of these Emperor penguins provides enough warmth to sustain life processes.

Source of body heat

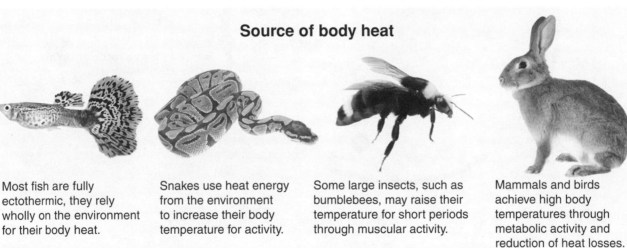

Most fish are fully ectothermic, they rely wholly on the environment for their body heat.

Snakes use heat energy from the environment to increase their body temperature for activity.

Some large insects, such as bumblebees, may raise their temperature for short periods through muscular activity.

Mammals and birds achieve high body temperatures through metabolic activity and reduction of heat losses.

Increasingly endothermic ⟶

1. Distinguish between **ectotherms** and **endotherms** in terms of their sources of body heat:

_____

_____

2. (a) Why are the movements of many ectotherms slow in the morning?_____

_____

_____

   (b) Why could this be a disadvantage? _____

_____

© 2014 **BIOZONE** International
ISBN: 978-1-927173-84-8
Photocopying Prohibited

54 Thermoregulation

Key Idea: Thermoregulation is the regulation of body temperature independently of changes in the environmental temperature.

Thermoregulation is a term that describes the regulation of body temperature in the face of changes in the temperature of the external environment. When we look at temperature regulation in animals, we can consider two extremes of body temperature tolerance:

▶ **Homeotherms** maintain a constant body temperature. They are usually strict thermoregulators.

▶ **Poikilotherms** allow their body temperature to vary with the temperature of the environment. They usually thermoregulate to avoid overheating.

In reality, many animals fall somewhere on a continuum between these two extremes. We have seen in the previous activity that animals are classed as ectotherms or endotherms depending on their sources of heat energy. Most, but not all, endotherms are also homeothermic and most, but not all, ectotherms allow their body temperatures to vary somewhat. Thermoregulation relies on physical, physiological, and behavioral mechanisms.

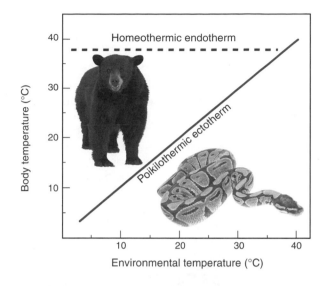

Mechanisms of thermoregulation

Homeothermic endotherm (mammal)

Wool, hair, or fur traps air next to the skin providing an insulating layer to reduce heat loss and slow heat gain.

Heat can be generated by shivering.

In cold weather, many mammals cluster together to retain body heat.

Panting and sweating cool through evaporation. Mammals usually sweat or pant but not both.

Poikilothermic ectotherm (reptile)

Basking in the sun is common in lizards and snakes. The sun warms the body up and they seek shade to cool down.

Increasing blood flow to the surface can help lose heat quickly.

Some lizards reduce points of contact with hot ground (e.g. standing on two legs instead of four) reducing heat uptake via conduction.

1. What is **thermoregulation**? _____

_____

2. The graph (top of page) shows temperature regulation in a homeothermic endotherm and a poikilothermic ectotherm. Describe how each responds to changes in environmental temperature:

_____

_____

_____

3. Thermoregulation can be aided by both physical features and behavior. Give an example of each:

(a) Behavior: _____

(b) Physical features: _____

KNOW

55 Thermoregulation in Humans

Key Idea: The hypothalamus regulates body temperature in humans. It coordinates nervous and hormonal responses to keep the body temperature within its normal range.

The hypothalamus regulates temperature

In humans, the temperature regulation center of the body is a region of the brain called the **hypothalamus**. It has a 'set-point' temperature of 36.7°C (98.6°F).

The hypothalamus acts like a thermostat. Changes in the core body temperature or in the skin temperature are registered by the hypothalamus, which then coordinates the appropriate nervous and hormonal responses to counteract the changes and restore normal body temperature. When normal temperature is restored, the corrective mechanisms are switched off. This is an example of a negative feedback regulation.

Infection can reset the set-point of the hypothalamus to a higher temperature. The body temperature then increases above the normal range, resulting in a **fever**. Fever is an important defense against infection.

Counteracting heat loss

The hypothalamus monitors blood temperature and receives input from thermoreceptors in the skin. The heat promoting center in the hypothalamus detects a fall in skin or core temperature below 35.8°C and coordinates responses that generate and conserve heat.

Increased metabolic rate produces heat.

Body hairs become raised and increase the insulating air layer around the body.

In extreme cold, two hormones (epinephrine and thyroxine) increase the energy-releasing activity of the liver.

The flow of blood to the skin decreases, keeping warm blood near the core (where the vital organs are).

Shivering (fast contraction and relaxation of muscles) produces internal heat.

Counteracting heat gain

The heat losing center in the hypothalamus monitors any rise in skin or core temperature above 37.5°C and coordinates responses that increase heat loss.

Sweating occurs. This cools the body by evaporation.

Decreased metabolic rate. This reduces the amount of heat generated by the body.

Body hairs become flattened against the skin. This reduces the insulating air layer around the body and helps heat loss.

The flow of blood to the skin increases. Warm blood from the body core is transported to the skin and the heat is lost from the skin surface.

Factors causing heat loss

► Wind
► Cold external temperature
► Not wearing enough clothing
► Being wet or in cold water
► Dehydration or being in "shock"

Factors causing heat gain

► Warm external temperature
► High humidity
► Excessive fat deposits
► Wearing too much clothing

© 2014 **BIOZONE** International
ISBN: 978-1-927173-84-8
Photocopying Prohibited

Thermoregulation in newborns

Newborn babies cannot fully thermoregulate until six months of age. They can become too cold or too hot very quickly.

Newborns minimize heat loss by reducing the blood supply to the periphery (skin, hands, and feet). This helps to maintain the core body temperature. Increased brown fat activity and general metabolic activity generates heat. Newborns are often dressed in a hat to reduce heat loss from the head, and tightly wrapped to trap heat next to their bodies.

Newborns lower their temperature by increasing peripheral blood flow. This allows heat to be lost, cooling the core temperature. Newborns can also reduce their body temperature by sweating, although their sweat glands are not fully functional until four weeks after birth.

A baby's body surface is three times greater than an adult's. There is greater surface area for heat to be lost from.

Newborns cannot shiver to produce heat.

Heat losses from the head are high because the head is very large compared to the rest of the body.

Newborns have thin skin, and blood vessels that run close to the skin, these features allow heat to be lost easily.

Newborns have very little white fat beneath their skin to insulate them against heat loss.

1. Where is the temperature regulation center in humans located? _____

2. (a) Why does infection result in an elevated core body temperature? _____

   _____

   (b) What is the purpose of this? _____

   _____

3. Describe the role of the following in maintaining a constant body temperature in humans:

   (a) The skin: _____

   _____

   (b) The hypothalamus: _____

   _____

   (c) Sweating: _____

   _____

   (d) Shivering: _____

   _____

4. Describe the features of a newborn that can cause it to lose heat quickly: _____

   _____

   _____

   _____

5. How can newborns control body temperature by altering blood flow to the skin? _____

   _____

   _____

© 2014 BIOZONE International
ISBN: 978-1-927173-84-8
Photocopying Prohibited

56 Hypothermia and Hyperthermia

Key Idea: Failure of normal thermoregulatory mechanisms can result in hypothermia (low body temperature) or hyperthermia (high body temperature).

Hypothermia

Hypothermia means low body temperature. It occurs when the body cannot generate enough heat and the core body temperature drops below 35°C. In hypothermia, the body loses heat faster than it can produce it.

At temperatures below 35°C, metabolic reactions and body functions are impaired. People with hypothermia experience a loss of coordination, difficulty in moving, and mental fatigue.

Hypothermia is caused by exposure to low temperatures, and results from the body's inability to replace the heat being lost to the environment.

Hypothermia is treated by rewarming the body. However, rewarming the body too quickly can actually cause the body to attempt to remove the sudden excess of heat and cause more heat loss.

Hyperthermia

Hyperthermia means high body temperature. It occurs when the body cannot dissipate excess heat and the core body temperature exceeds 38.5°C. In hyperthermia, heat is produced more quickly than it can be lost.

Prolonged hyperthermia is potentially fatal and is treated as a medical emergency.

Causes include dehydration, prolonged exposure to excessive heat or humidity, and strenuous exercise. Fever is not hyperthermia because the hypothalamus set-point of a person with hyperthermia has not been reset.

Hyperthermia is treated by cooling the body. Mild cases (above) are treated by drinking water, removing excess clothing, and resting in a cool place. Medical treatment is needed for hyperthermia above 40°C.

1. (a) What is **hypothermia**?_____

   _____

   (b) How is it treated?_____

   _____

2. (a) What is **hyperthermia**? _____

   _____

   (b) How is it treated?_____

   _____

3. What is the common link in both hypothermia and hyperthermia?

   _____

   _____

4. On the thermometer (right), mark the normal thermoregulatory set-point, and the temperatures at which hypothermia and hyperthermia occur.

40°
39°
38°
37°
36°
35°
34°
°C

© 2014 **BIOZONE** International
ISBN: 978-1-927173-84-8
Photocopying Prohibited

57 Controlling Blood Sugar Levels

Key Idea: Blood glucose levels are regulated by negative feedback. Two hormones, insulin and glucagon, control blood glucose levels.

The importance of blood glucose

Glucose is the body's main energy source. It is chemically broken down during cellular respiration to generate ATP, which is used to power the chemical reactions of the cell. Glucose is the main sugar circulating in blood, so it is often called blood sugar. Blood glucose levels are tightly controlled because cells must receive an adequate and regular supply of fuel. Prolonged high or low blood glucose causes serious physiological problems and even death. Normal activities, such as eating and exercise, alter blood glucose levels, but the body's control mechanisms regulate levels so that fluctuations are minimized.

Controlling blood glucose levels

Blood glucose (BG) is controlled by two hormones produced by special islet cells in the pancreas. The hormones work antagonistically and levels are tightly controlled by **negative feedback**.

▶ **Insulin** lowers blood glucose by promoting glucose uptake by cells and glycogen storage in the liver.

▶ **Glucagon** increases blood glucose by promoting release of glucose from the breakdown of stored glycogen.

▶ When normal blood glucose levels are restored, negative feedback stops hormone secretion.

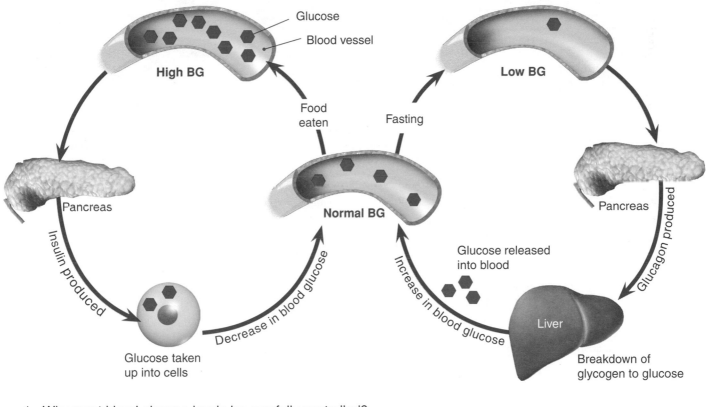

1. Why must blood glucose levels be carefully controlled? _____

2. For the following two scenarios, describe how normal blood glucose level is restored:

   (a) Low blood glucose: _____

   (b) High blood glucose: _____

3. What mechanism regulates the secretion of the hormones controlling BG? _____

 CONNECT 58 WEB 57 **KNOW**

58 Diabetes

Key Idea: Diabetes mellitus is a condition where blood glucose levels are too high. It may be caused by a lack of insulin (type 1) or by resistance to insulin's effects (type 2).

Diabetes mellitus (often just called diabetes) is a condition in which blood glucose is too high because the body's cells cannot take up glucose in the normal way. It is usually detected by glucose appearing in the urine (glucose is normally reabsorbed and does not enter the urine). The two types of diabetes, type 1 and type 2, have different causes and treatment, but both are life threatening conditions if untreated.

Type 1 diabetes

No insulin is produced because the insulin-producing cells of the pancreas are damaged.

Type 2 diabetes

Insulin is produced. However, either not enough insulin is made, or the body's cells do not react to it.

The **beta cells** of the pancreatic islets (outlined above) produce insulin.

The body does not produce insulin. Also called insulin dependent diabetes.

Age at onset: Early in life (often in childhood).

Symptoms: Frequent urination, thirst, weight loss, fatigue, frequent infections, hunger.

Cause: The beta cells are destroyed by the body's own immune system. A genetic predisposition and environmental factors (e.g. a viral infection) may be the trigger.

Treatment: Blood glucose is monitored regularly. Insulin injections combined with dietary management keep blood sugar levels stable.

The body does not respond to insulin. Also called insulin resistant diabetes.

Age at onset: > 40 years old, but is becoming common in younger adults and obese children.

Symptoms: As for type 1 but usually milder.

Cause: Lifestyle factors (obesity, a sedentary lifestyle, high blood pressure, high blood lipids, and poor diet) increase susceptibility to type 2 diabetes.

Treatment: Increased physical activity, losing weight, and improving diet may control type 2 diabetes in many cases. Drugs and insulin may be required if lifestyle changes are insufficient on their own.

1. Why does diabetes mellitus result in high blood sugar levels? _____

_____

2. Discuss the differences between type 1 and type 2 diabetes, including causes and treatments:

_____

_____

_____

_____

_____

© 2014 **BIOZONE** International
ISBN: 978-1-927173-84-8
Photocopying Prohibited

59 Homeostasis during Exercise

Key Idea: The circulatory and respiratory systems are primarily responsible for maintaining homeostasis during exercise.

During exercise, greater metabolic demands are placed on the body, and it must work harder to maintain homeostasis.

Maintaining homeostasis during exercise is principally the job of the circulatory and respiratory systems, although the skin, kidneys, and liver are also important.

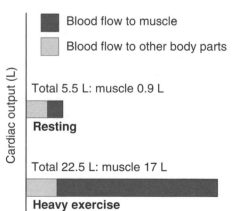

Working muscles need more ATP than muscles at rest.

Increased body temperature
During exercise, the extra heat produced by muscle contraction must be dispersed to prevent overheating. Thermoregulatory mechanisms, such as sweating and increased blood flow to the skin, release excess heat into the surrounding environment and help cool the body.

Increased heart rate
An increased heart rate circulates blood around the body more quickly. This increases the rate at which exchanges can be made between the blood and the working tissues. Oxygen and glucose are delivered and metabolic wastes (e.g. carbon dioxide) are removed.

Increased breathing rate
Exercise increases the body's demand for energy (ATP). Oxygen is required for cellular respiration and ATP production. Increasing the rate of breathing delivers more oxygen to working tissues and enables them to make the ATP they need to keep working. An increased breathing rate also increases the rate at which carbon dioxide is expelled from the body.

Increased glucose production
During exercise, working muscles quickly use up freely available blood glucose. Glucose is mobilized from glycogen stores in the liver and supplies the body with fuel to maintain ATP production.

1. The graph (right) compares the change in cardiac output (a measure of total blood flow in liters) during rest and during exercise. The color of the bars indicates the proportion of blood flow in skeletal muscle compared to other body parts.

   ■ Blood flow to muscle
   □ Blood flow to other body parts

   Total 5.5 L: muscle 0.9 L
   Resting

   Total 22.5 L: muscle 17 L
   Heavy exercise

   Cardiac output (L)

   (a) What percentage of the blood goes to the muscles at rest?

   _____

   _____

   (b) What percentage of the blood goes to the muscles during exercise?

   _____

   _____

2. (a) What happens to the total blood flow during heavy exercise compared to at rest? _____

   _____

   (b) Why does this change occur? _____

   _____

   _____

© 2014 **BIOZONE** International
ISBN: 978-1-927173-84-8
Photocopying Prohibited

CONNECT

44 **KNOW**

60 Exercise and Heart Rate

Key Idea: Heart rate and breathing rate both increase during exercise to meet the body's increased metabolic demands.

In this practical, you will work in groups of three to see how exercise affects breathing and heart rates. Choose one person to carry out the exercise and one person each to record heart rate and breathing rate.

Heart rate (beats per minute) is obtained by measuring the pulse (right) for 15 seconds and multiplying by four.

Breathing rate (breaths per minute) is measured by counting the number of breaths taken in 15 seconds and multiplying it by four.

CAUTION: The person exercising should have no known pre-existing heart or respiratory conditions.

Measuring the carotid pulse

Gently press your index and middle fingers, not your thumb, against the carotid artery in the neck (just under the jaw) or the radial artery (on the wrist just under the thumb) until you feel a pulse.

Measuring the radial pulse

Procedure

Resting measurements

Have the person carrying out the exercise sit down on a chair for 5 minutes. They should try not to move. After 5 minutes of sitting, measure their heart and breathing rates. Record the resting data on the table (right).

Exercising measurements

Choose an exercise to perform. Some examples include step ups onto a chair, skipping rope, jumping jacks, and running in place.
Begin the exercise, and take measurements after 1, 2, 3, and 4 minutes of exercise. The person exercising should stop just long enough for the measurements to be taken. Record the results in the table.

Post exercise measurements

After the exercise period has finished, have the exerciser sit down in a chair. Take their measurements 1 and 5 minutes after finishing the exercise. Record the results on the table.

	Heart rate (beats minute^{-1})	Breathing rate (breaths minute^{-1})
Resting		
1 minute		
2 minutes		
3 minutes		
4 minutes		
1 minute after		
5 minutes after		

1. (a) Graph your results on separate piece of paper. You will need to use one axis for heart rate and another for breathing rate. When you have finished answering the questions below, attach it to this page.

   (b) Analyze your graph and describe what happened to heart rate and breathing rate **during exercise**:

   _____

   _____

   _____

2. (a) Describe what happened to heart rate and breathing rate **after exercise**: _____

   _____

   _____

   (b) Why did this change occur?_____

   _____

CONNECT 44 CONNECT 72

© 2014 **BIOZONE** International
ISBN: 978-1-927173-84-8
Photocopying Prohibited

61 Homeostasis in Plants

Key Idea: In plants, evaporative water loss from stomata drives a transpiration stream that ensures plants have a constant supply of water to support essential life processes.

Maintaining water balance

Like animals, plants need water for life processes. Water gives cells turgor, transports dissolved substances, and is a medium in which metabolic reactions can take place. Maintaining water balance is an important homeostatic function in plants.

Vascular plants obtain water from the soil. Water enters the plant via the roots, and is transported throughout the plant by a specialized tissue called xylem. Water is lost from the plant by evaporation. This evaporative water loss is called **transpiration**.

Transpiration has several important functions:

▶ Provides a constant supply of water needed for essential life processes (such as photosynthesis).

▶ Cools the plant by evaporative water loss.

▶ Helps the plant take up minerals from the soil.

However, if too much water is lost by transpiration, a plant will become dehydrated and may die.

The role of stomata

Water loss occurs mainly through **stomata** (pores in the leaf). The rate of water loss can be regulated by specialized guard cells either side of the stoma, which open or close the pore.

▶ Stomata open: transpiration rate increases.

▶ Stomata closed: transpiration rates decrease.

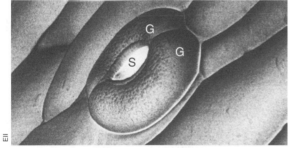

Guard cells (G) control the size of the stoma (S).

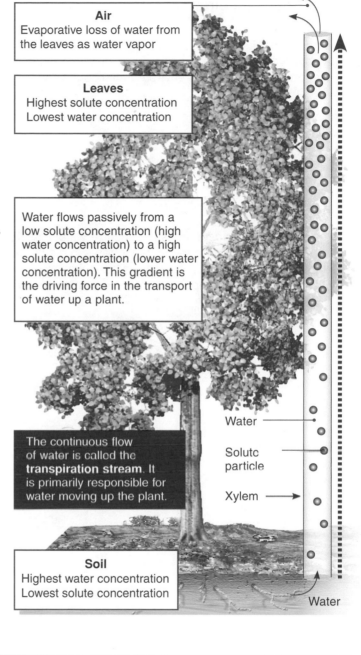

Air
Evaporative loss of water from the leaves as water vapor

Leaves
Highest solute concentration
Lowest water concentration

Water flows passively from a low solute concentration (high water concentration) to a high solute concentration (lower water concentration). This gradient is the driving force in the transport of water up a plant.

The continuous flow of water is called the **transpiration stream**. It is primarily responsible for water moving up the plant.

Water
Solute particle
Xylem

Soil
Highest water concentration
Lowest solute concentration

Water

1. (a) What is transpiration? _____

(b) How does transpiration provide water for essential life processes in plants? _____

_____

_____

2. How do plants regulate the amount of water lost from the leaves? _____

_____

3. (a) What would happen if too much water was lost by transpiration? _____

_____

(b) When might this happen? _____

CONNECT 62 CONNECT 46 WEB 61 **KNOW**

62 Measuring Transpiration in Plants

Key Idea: Physical factors in the environment such as humidity, temperature, light level, and air movement affect transpiration. Transpiration can be measured with a potometer.

Measuring transpiration

▶ Transpiration rate (water loss per unit of time) can be measured using a simple instrument called a potometer. A basic potometer can easily be moved around so that transpiration rate can be measured under different environmental conditions.

▶ Potometers are commonly used to investigate the effect of the following environmental conditions on transpiration rate:

- Humidity
- Water supply
- Temperature
- Light level
- Air movement

▶ Many plants have adaptations to minimize water loss. Potometers can be used to compare transpiration rates in plants with different adaptations. For example comparing transpiration rates in plants with narrow leaves compared to rates in plants with broad leaves.

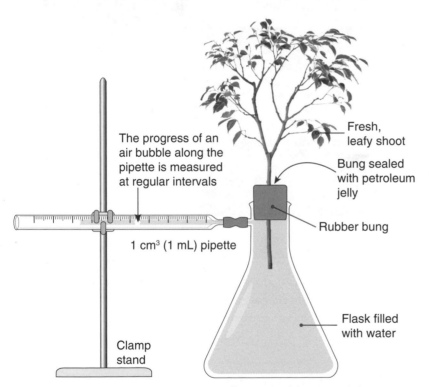

The progress of an air bubble along the pipette is measured at regular intervals

1 cm³ (1 mL) pipette

Clamp stand

Fresh, leafy shoot

Bung sealed with petroleum jelly

Rubber bung

Flask filled with water

How a potometer works

A bubble potometer, like the one shown on the left, measures the rate of water loss indirectly.

Transpiration rate is measured by measuring how much the bubble moves over a period of time. The movement of the bubble is assumed to be caused by the plant taking up water to replace the water lost by transpiration.

In transpiration experiments, it is important that the system is sealed (watertight). You can then be confident that the water losses you are recording are the result of transpiration, not leakage from the system.

Transpiration experiment

▶ This activity describes the results of a plant transpiration experiment investigating the effect of specific environmental conditions on transpiration rate.

▶ The environmental conditions investigated were ambient (standard room conditions), wind (fan), humidity (mist), and bright light (lamp).

▶ A potometer was used to measure transpiration.

▶ Once set up, the apparatus was equilibrated for 10 minutes, and the position of the air bubble in the pipette was recorded. This is the time 0 reading.

▶ The plant was then exposed to one of the environmental conditions described above. Students recorded the bubble position every two minutes over a 20 minute period. Results are given in Table 1.

© 2014 **BIOZONE** International
ISBN: 978-1-927173-84-8
Photocopying Prohibited

1. The distance the bubble travelled for each environmental condition is given in Table 1 below.
   Convert the distance the bubble travelled (in mm) into water loss (mL). For every mm the bubble moved, 0.1 mL of water was lost by transpiration (e.g. 1 mm = 0.1 mL water loss).
   Determine the water loss for each environmental condition (the first has been completed for you).

Table 1. Water loss under different environmental conditions

Treatment \ Time (min)	0	2	4	6	8	10	12	14	16	18	20
Ambient (mm)	0	0	0	1	1	2	3	3	4	5	5
Ambient (mL)	O	O	O	O.l	O.l	0.2	0.3	0.3	0.4	0.5	0.5
Fan (mm)	0	4	7	9	14	17	25	28	31	33	34
Fan (mL)											
High humidity (mm)	0	0	0	0	1	1	2	2	2	3	3
High humidity (mL)											
Bright light (mm)	0	1	3	5	6	7	8	9	10	11	13
Bright light (mL)											

2. Using an appropriate graph, plot the water loss (in mL) for each environmental condition on the grid below.

3. (a) What is the control for this experiment? _____

   (b) Name the environmental conditions that increased water loss: _____

   _____

   (c) How do the environmental conditions in (b) cause water loss? _____

   _____

   _____

   _____

4. Why is it important that the potometer has no leaks in it? _____

   _____

© 2014 **BIOZONE** International
ISBN: 978-1-927173-84-8
Photocopying Prohibited

63 Chapter Review

Summarize what you know about this topic under the headings given. You can draw diagrams or mind maps, or write short notes to organize your thoughts. Use the image, hints, and guidelines included to help you.

Homeostatic regulation

HINT: Include reference to feedback mechanisms and the importance of homeostasis.

Homeostasis in plants

HINT: Include reference to transpiration and the interaction between plant organ systems.

Homeostasis in humans

HINT: Examples could include control of blood glucose or thermoregulation. What happens when homeostatic mechanisms fail?

© 2014 **BIOZONE** International
ISBN: 978-1-927173-84-8
Photocopying Prohibited

64 KEY TERMS: Vocabulary and Comprehension

1. Test your vocabulary by matching each term to its definition, as identified by its preceding letter code.

diabetes mellitus _____

homeostasis _____

negative feedback _____

positive feedback _____

thermoregulation _____

transpiration _____

A A mechanism in which the output of a system acts to oppose changes to the input of the system. The net effect is to stabilize the system and dampen fluctuations.

B The loss of water vapor by plants, mainly from leaves via the stomata.

C A destabilizing mechanism in which the output of the system causes an escalation in the initial response.

D Regulation of the internal environment to maintain a stable, constant condition.

E A condition in which the blood glucose level is elevated above normal levels, either because the body doesn't produce enough insulin, or because the cells do not respond to the insulin that is produced.

F The regulation of body temperature.

2. Test your knowledge about feedback mechanisms by studying the two graphs below, and answering the questions about them. In your answers, use biological terms appropriately to show your understanding.

A

B

Type of feedback mechanism: _____

Mode of action: _____

Biological examples of this mechanism:

Type of feedback mechanism: _____

Mode of action: _____

Biological examples of this mechanism:

© 2014 **BIOZONE** International
ISBN: 978-1-927173-84-8
Photocopying Prohibited

VOCAB

LS1.B | Growth & Development

Key terms

cell cycle

cell differentiation

cell division

DNA replication

interphase

mitosis

nucleotides

semi-conservative

stem cell

tissue

zygote

Core ideas

Organisms grow and develop through mitosis

Activity number

☐ 1. Multicellular organisms develop from a single cell (a fertilized egg) called a **zygote**.

`65`

☐ 2. The **cell cycle** describes the events in a cell leading to its division into two daughter cells. The cell cycle consists of two main phases, **interphase** and the mitotic or M phase (**mitosis** and cytokinesis).

`68`

☐ 3. During mitosis (**cell division**) a cell divides to produce two genetically identical cells.

`68`

☐ 4. Mitosis has three functions: growth of an organism, replacement of damaged cells, and asexual reproduction (in some organisms).

`69`

☐ 5. **DNA replication** must take place before a cell can divide. DNA replication produces two identical copies of DNA. A copy goes to each new cell produced during mitosis.

`66  67`

☐ 6. DNA replication is **semi-conservative**. Each replicated DNA molecule consists of one 'old' (parent) strand of DNA, and one 'new' (daughter) strand of DNA.

`66  67`

Cells become differentiated to carry out specialized roles

☐ 7. A multicellular organism is made up of many different types of specialized cells. Specialized cells have specific roles in the organism. They arise through **cellular differentiation**, a process involving the activation of specific genes within a cell.

`70  71`

☐ 8. **Stem cells** are unspecialized cells that can give rise to many different cell types.

`71`

☐ 9. Differentiation and specialization of cells produces **tissues** and organs, which work together to meet the needs of the organism.

`72`

Science and engineering practices

☐ 1. Use a model to illustrate the role of mitosis and cellular differentiation in producing and maintaining a multicellular organism.

`65  70`

☐ 2. Use a model to show how the components of blood (a liquid tissue) are produced through cellular differentiation from stem cells.

`71`

65 Growth and Development of Organisms

Key Idea: Multicellular organisms develop from a single cell. The cell divides by mitosis many times, and produces genetically identical copies of the original cell.

Organisms grow and develop through mitosis

▶ Multicellular organisms begin as a single cell (a fertilized egg) and develop into complex organisms made up of many cells. This is achieved by the process of **mitosis** or cell division.

▶ In multicellular organisms, mitosis is responsible for growth and for the replacement of old and damaged cells. In some eukaryotic organisms, mitosis is also responsible for reproduction (e.g. yeast cells).

▶ Two important processes must occur in order for new cells to be produced. The first is the duplication of the genetic material (DNA). The second is the division (splitting) of the parent cell into two identical daughter cells. The two daughter cells have the same genetic material as the parent cell.

▶ In multicellular animals, mitosis only occurs in body cells (somatic cells). Sperm and egg cells (gametes) are produced by a different process (meiosis).

The role of mitosis in human development

Every cell in an adult's body has the exact same genetic material as the fertilized egg it has developed from.

An egg and sperm join together in fertilization and produce a zygote.

The zygote begins to divide by mitosis. The size of the embryo begins to increase as more cells are formed. This early embryo contains about 100 cells.

Once born, mitosis continues to be important in the growth of an organism until they are fully grown.

In adults, cell division is involved in the replacement of old cells rather than growth. This continues through the adult's lifetime.

Definitions

Zygote: A fertilized egg cell.

Embryo: The ball of cells that forms four to five days after fertilization.

1. Briefly explain how multicellular organisms can develop from a single cell: _____

_____

_____

2. What two things must occur for a new cell to be produced? _____

_____

3. Explain the role of mitosis in:

(a) A developing embryo: _____

(b) An adult: _____

CONNECT

68 KNOW

66 DNA Replication

Key Idea: Before a cell can divide, the DNA must be copied. DNA replication produces two identical copies of DNA. A copy goes to each new daughter cell.

▶ Before a cell can divide, its DNA must be copied (replicated). **DNA replication** ensures that the two daughter cells receive identical genetic information.

▶ In eukaryotes, DNA is organized into structures called chromosomes in the nucleus.

▶ After the DNA has replicated, each chromosome is made up of two chromatids, which are joined at the centromere. A chromatid is simply one half of a replicated chromosome, when joined at the centromere. The two chromatids will become separated during cell division to form two separate chromosomes.

▶ DNA replication takes place in the time between cell divisions. The process is **semi-conservative**, meaning that each chromatid contains half original (parent) DNA and half new (daughter) DNA.

1. What is the purpose of DNA replication?

_____

_____

_____

_____

_____

2. What would happen if DNA was not replicated prior to cell division?

_____

_____

_____

_____

3. Explain what semi-conservative replication means:

_____

_____

_____

_____

_____

_____

DNA replication duplicates chromosomes

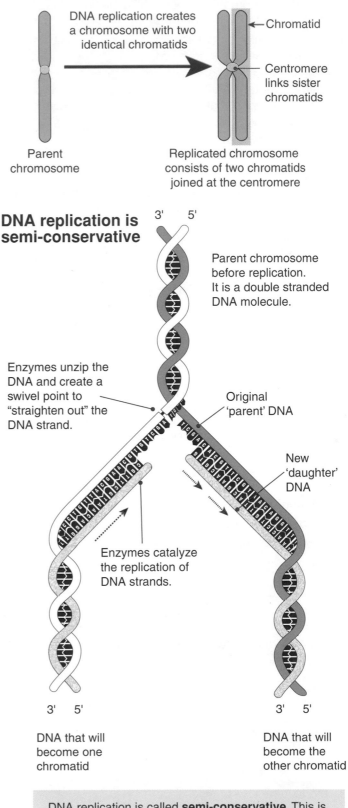

DNA replication creates a chromosome with two identical chromatids

Chromatid

Centromere links sister chromatids

Parent chromosome

Replicated chromosome consists of two chromatids joined at the centromere

DNA replication is semi-conservative

3'   5'

Parent chromosome before replication. It is a double stranded DNA molecule.

Enzymes unzip the DNA and create a swivel point to "straighten out" the DNA strand.

Original 'parent' DNA

New 'daughter' DNA

Enzymes catalyze the replication of DNA strands.

3'   5'

3'   5'

DNA that will become one chromatid

DNA that will become the other chromatid

DNA replication is called **semi-conservative**. This is because each resulting DNA molecule is made up of one parent strand and one daughter strand of DNA.

 © 2014 **BIOZONE** International
ISBN: 978-1-927173-84-8
Photocopying Prohibited

67 Details of DNA Replication

Key Idea: DNA replication is achieved by enzymes attaching new nucleotides to the growing DNA strand at the replication fork.

▶ During DNA replication, new nucleotides (the units that make up the DNA molecule) are added at a region called the **replication fork**. The replication fork moves along the chromosome as replication progresses.

▶ Nucleotides are added in by complementary base-pairing: Nucleotide A is always paired with nucleotide T. Nucleotide C is always paired with nucleotide G.

▶ The DNA strands can only be replicated in one direction, so one strand has to be copied in short segments, which are joined together later.

▶ This whole process occurs simultaneously for each chromosome of a cell and the entire process is tightly controlled by enzymes.

1. How are the new strands of DNA lengthened?

_____

_____

_____

_____

2. What rule ensures that the two new DNA strands are identical to the original strand?

_____

_____

3. Why does one strand of DNA need to copied in segments?

_____

_____

_____

_____

4. Describe three activities carried out by enzymes during DNA replication:

_____

_____

_____

_____

Stages in DNA replication

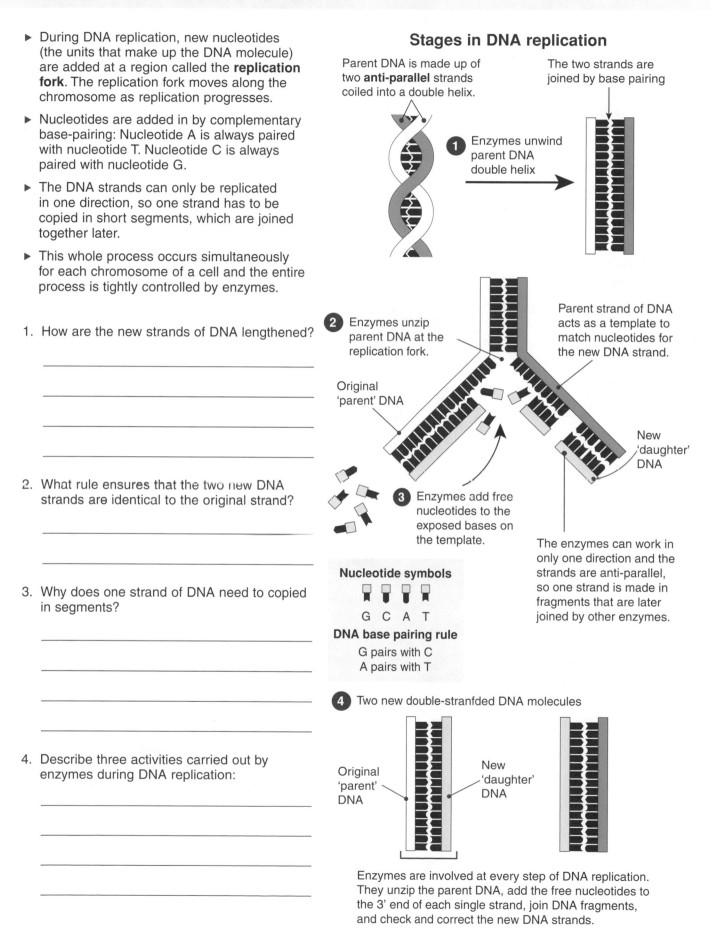

Parent DNA is made up of two **anti-parallel** strands coiled into a double helix.

The two strands are joined by base pairing

1 Enzymes unwind parent DNA double helix

2 Enzymes unzip parent DNA at the replication fork.

Original 'parent' DNA

Parent strand of DNA acts as a template to match nucleotides for the new DNA strand.

New 'daughter' DNA

3 Enzymes add free nucleotides to the exposed bases on the template.

The enzymes can work in only one direction and the strands are anti-parallel, so one strand is made in fragments that are later joined by other enzymes.

Nucleotide symbols

G C A T

DNA base pairing rule

G pairs with C
A pairs with T

4 Two new double-stranfded DNA molecules

Original 'parent' DNA

New 'daughter' DNA

Enzymes are involved at every step of DNA replication. They unzip the parent DNA, add the free nucleotides to the 3' end of each single strand, join DNA fragments, and check and correct the new DNA strands.

CONNECT

39 KNOW

68 Mitosis and the Cell Cycle

Key Idea: Mitosis is an important part of the cell cycle in which the replicated chromosomes are separated and the cell divides, producing two new cells.

Mitosis is a stage in the cell cycle

▸ The life cycle of a cell is called the **cell cycle**. It is divided into a number of stages in which specific activities occur. Mitosis is one stage of the cell cycle.

▸ The activities carried out during the cell cycle include growth of the cell, replication of the DNA, condensation of the chromosomes, mitosis, and cytokinesis.

▸ M-phase (**mitosis** and cytokinesis) is the part of the cell cycle in which the parent cell divides in two to produce two genetically identical daughter cells (right).

▸ Mitosis is one of the shortest stages of the cell cycle. When a cell is not undergoing mitosis, it is said to be in **interphase**. Interphase accounts for 90% of the cell cycle.

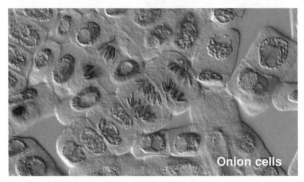
Onion cells

At any one time, only a small proportion of the cells in an organism will be undergoing mitosis. The majority of the cells will be in interphase.

An overview of mitosis

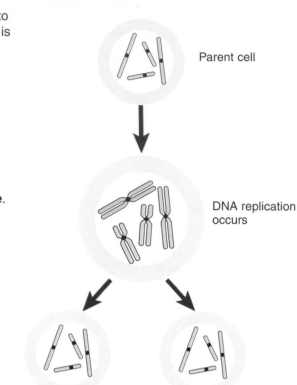

Parent cell

DNA replication occurs

The cell divides forming two identical daughter cells. The chromosome number remains the same as the parent cell.

The cell cycle

Interphase

Cells spend most of their time in interphase. Interphase is divided into three stages (right):

▸ The first gap phase.

▸ The S-phase.

▸ The second gap phase.

During interphase the cell grows, carries out its normal activities, and replicates its DNA in preparation for cell division. Interphase is not a stage in mitosis.

Mitosis and cytokinesis (M-phase)

Mitosis and cytokinesis occur during M-phase. During mitosis, the cell nucleus (containing the replicated DNA) divides in two equal parts. Cytokinesis occurs at the end of M-phase. During cytokinesis the cell cytoplasm divides, and two new daughter are produced.

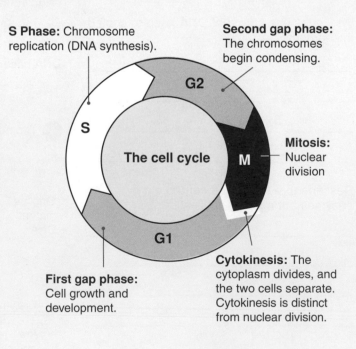

S Phase: Chromosome replication (DNA synthesis).

Second gap phase: The chromosomes begin condensing.

Mitosis: Nuclear division

Cytokinesis: The cytoplasm divides, and the two cells separate. Cytokinesis is distinct from nuclear division.

First gap phase: Cell growth and development.

G2

S

The cell cycle

M

G1

© 2014 **BIOZONE** International
ISBN: 978-1-927173-84-8
Photocopying Prohibited

The cell cycle and stages of mitosis

► Mitosis is continuous, but is divided into stages for easier reference (1-6 below). Enzymes are critical at key stages. The example below illustrates the cell cycle in an animal cell.

► In animal cells, centrioles form the spindle. During **cytokinesis** (division of the cytoplasm) a constriction forms that divides the cell in two. Cytokinesis is part of M-phase, but it is distinct from mitosis.

► Plant cells lack centrioles, and the spindle is organized by structures associated with the plasma membrane. In plant cells, cytokinesis involves construction of a cell plate in the middle of the cell. This will eventually form a new cell wall.

The animal cell cycle and stages of mitosis

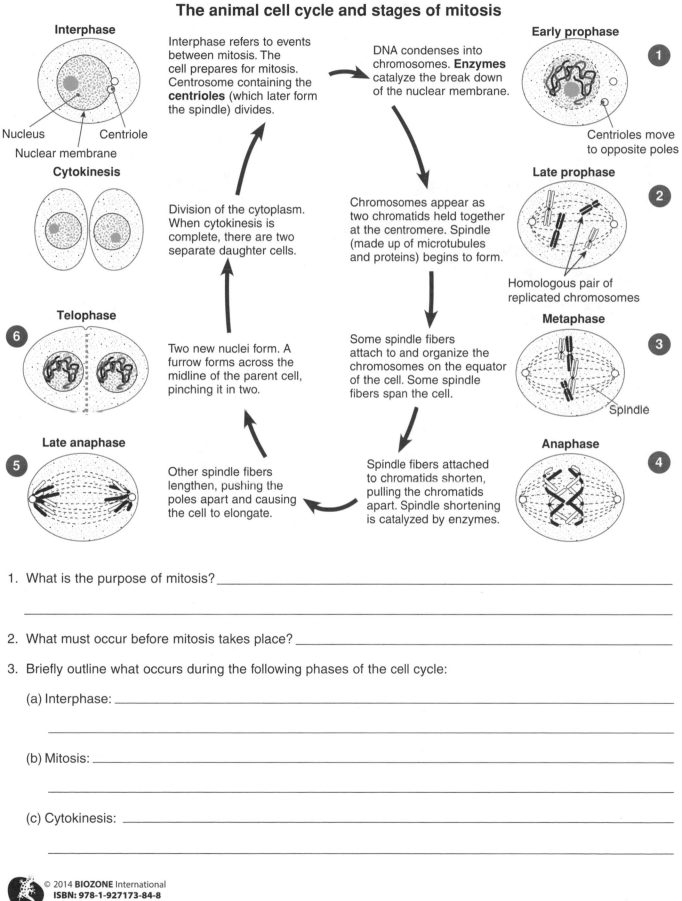

Interphase

Nucleus
Centriole
Nuclear membrane

Interphase refers to events between mitosis. The cell prepares for mitosis. Centrosome containing the **centrioles** (which later form the spindle) divides.

DNA condenses into chromosomes. **Enzymes** catalyze the break down of the nuclear membrane.

Early prophase 1

Centrioles move to opposite poles

Cytokinesis

Division of the cytoplasm. When cytokinesis is complete, there are two separate daughter cells.

Chromosomes appear as two chromatids held together at the centromere. Spindle (made up of microtubules and proteins) begins to form.

Late prophase 2

Homologous pair of replicated chromosomes

Telophase 6

Two new nuclei form. A furrow forms across the midline of the parent cell, pinching it in two.

Some spindle fibers attach to and organize the chromosomes on the equator of the cell. Some spindle fibers span the cell.

Metaphase 3

Spindle

Late anaphase 5

Other spindle fibers lengthen, pushing the poles apart and causing the cell to elongate.

Spindle fibers attached to chromatids shorten, pulling the chromatids apart. Spindle shortening is catalyzed by enzymes.

Anaphase 4

1. What is the purpose of mitosis? _____

_____

2. What must occur before mitosis takes place? _____

3. Briefly outline what occurs during the following phases of the cell cycle:

(a) Interphase: _____

_____

(b) Mitosis: _____

_____

(c) Cytokinesis: _____

_____

© 2014 **BIOZONE** International
ISBN: 978-1-927173-84-8
Photocopying Prohibited

69 The Functions of Mitosis

Key Idea: Mitosis has three primary functions: growth of the organism, replacement of damaged or old cells, and asexual reproduction (in some organisms).

Mitotic cell division has three purposes.

▶ **Growth**: Multicellular organisms grow from a single fertilized cell into a mature organism. Depending on the organism, the mature form may consist of several thousand cells up to several trillion cells.

▶ **Repair**: Damaged and old cells are replaced with new cells.

▶ **Asexual reproduction**: Some unicellular eukaryotes (such as yeasts) and some multicellular organisms (e.g. *Hydra*) reproduce asexually by mitotic division.

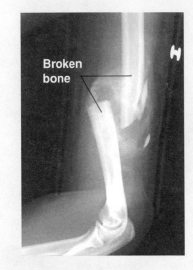
Broken bone

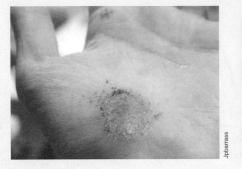

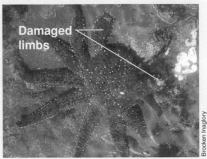
Damaged limbs

Brocken Inaglory

Jpbarrass

Repair

Mitosis is vital in the repair and replacement of damaged cells. When you break a bone, or graze your skin, new cells are generated to repair the damage. Some organisms, like this sea star (above right) are able to generate new limbs if they are broken off.

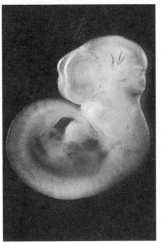

Matthias Zepper

Growth

Multicellular organisms develop from a single cell. Organisms, such as this 12 day old mouse embryo (left), grow by increasing their cell number. Cell growth is highly regulated and once the mouse reaches its adult size (above), physical growth stops.

Asexual reproduction

Some simple eukaryotic organisms reproduce asexually by mitosis. Yeasts (such as baker's yeast, used in baking) can reproduce by budding. The parent cell buds to form a daughter cell (right). The daughter cell continues to grow, and eventually separates from the parent cell.

Parent cell

Daughter cell

1. Use examples to explain the role of mitosis in:

   (a) Growth of an organism: _____

   _____

   (b) Replacement of damaged cells: _____

   _____

   (c) Asexual reproduction: _____

   _____

 © 2014 **BIOZONE** International
ISBN: 978-1-927173-84-8
Photocopying Prohibited

KNOW

70 Differentiation of Cells

Key Idea: Many different cell types arise during development of the embryo. Activation of specific genes determines what type of cell will develop.

Recall that when a cell divides by mitosis, it produces genetically identical cells. However, a multicellular organism is made up of many different types of cells, each specialized to carry out a particular role. How can it be that all of an organism's cells have the same genetic material, but the cells have a wide variety of shapes and functions? The answer is through **cellular differentiation**. Although each cell has the same genetic material (genes), different genes are turned on (activated) or off in different patterns during development in particular types of cells. The differences in gene activation controls what type of cell forms. Once the developmental pathway of a cell is determined, it cannot alter its path and change into another cell type.

▶ All of the cells in your body have been derived from a single fertilized cell (zygote).

▶ From the human zygote, 230 different cell types arise.

▶ Some types of cells have a shorter cell cycle than others.

▶ Often the length of the cell cycle reflects adaptation to the cell's environment. For example, skin cells and epithelial cells lining the intestine are constantly worn away so must be replaced by regenerating cells.

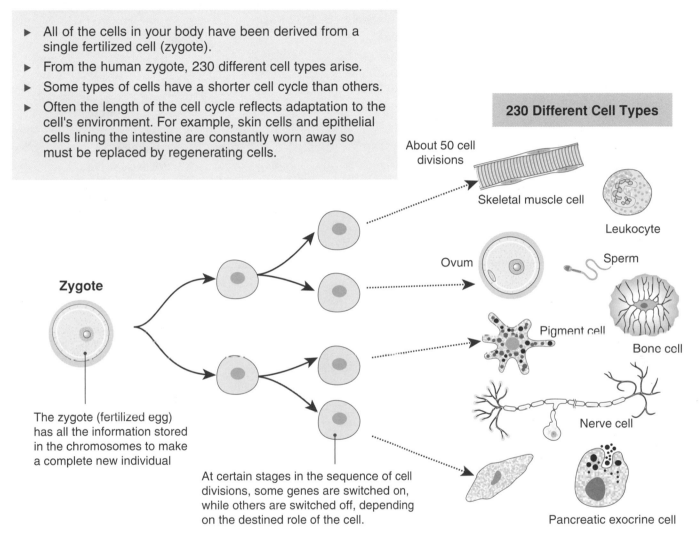

230 Different Cell Types

About 50 cell divisions — Skeletal muscle cell, Leukocyte, Ovum, Sperm, Pigment cell, Bone cell, Nerve cell, Pancreatic exocrine cell

Zygote

The zygote (fertilized egg) has all the information stored in the chromosomes to make a complete new individual

At certain stages in the sequence of cell divisions, some genes are switched on, while others are switched off, depending on the destined role of the cell.

1. Name the cell from which all other cells are derived: _____

2. Explain how so many different types of cells can be formed, even though all cells have the same DNA:

_____

_____

_____

3. Why do some types of cells have shorter cell cycles than others? _____

_____

_____

71 Blood Cell Differentiation

Key Idea: Stem cells are undifferentiated cells, which can develop into many different cell types. Related cell types come together to form tissues, such as blood.

A zygote can differentiate into many different types of cells because early on it divides into **stem cells**. Stem cells are unspecialized and can give rise to many types of cells. The different categories of stem cells, have different properties:

Totipotent stem cells can differentiate into all the cells in an organism. In humans, only the zygote and its first few divisions are totipotent.

Pluripotent stem cells can become any cells of the body, except extra-embryonic cells, such as the placenta. Embryonic stem cells are pluripotent.

Multipotent stem cells give rise a limited number of cell types, usually related to their tissue of origin. This is illustrated for blood cells, which collectively make up the cellular components of blood - a liquid tissue.

Properties of stem cells

Self renewal: The ability to divide many times while maintaining an unspecialized state.

Potency: The ability to differentiate into specialized cells.

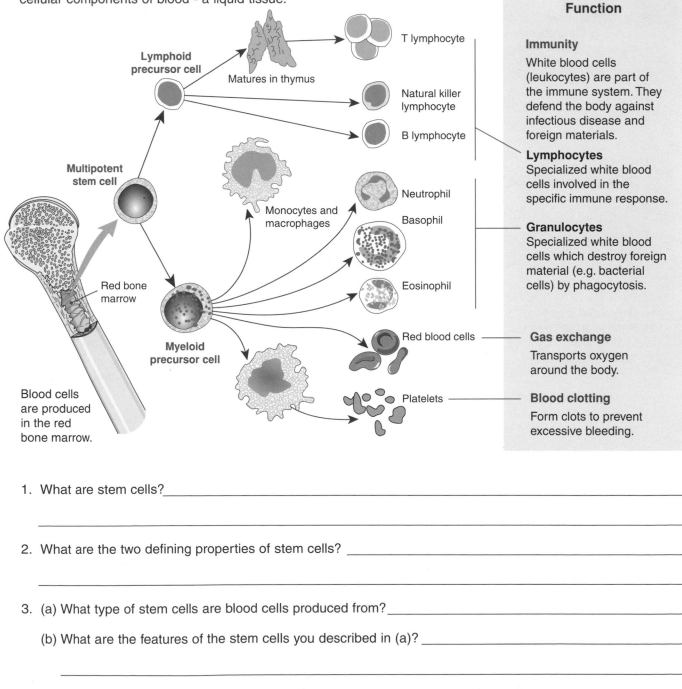

Function

Immunity
White blood cells (leukocytes) are part of the immune system. They defend the body against infectious disease and foreign materials.

Lymphocytes
Specialized white blood cells involved in the specific immune response.

Granulocytes
Specialized white blood cells which destroy foreign material (e.g. bacterial cells) by phagocytosis.

Gas exchange
Transports oxygen around the body.

Blood clotting
Form clots to prevent excessive bleeding.

1. What are stem cells? _____

   _____

2. What are the two defining properties of stem cells? _____

   _____

3. (a) What type of stem cells are blood cells produced from? _____

   (b) What are the features of the stem cells you described in (a)? _____

   _____

© 2014 **BIOZONE** International
ISBN: 978-1-927173-84-8
Photocopying Prohibited

72 Tissues Work Together

Key Idea: Tissues are specialized to perform particular tasks. Different tissue types work together to meet the body's needs efficiently.

A tissue is a collection of related cell types that work together to carry out a specific function. Different tissues come together to form organs. The cells, tissues, and organs of the body interact to meet the needs of the entire organism. This activity explains the role of the four tissue types (below) in humans.

Muscle tissue	Epithelial tissue	Nervous tissue	Connective tissue
▶ Contractile tissue ▶ Produces movement of the body or its parts ▶ Includes smooth, skeletal, and cardiac muscle	▶ Lining tissue ▶ Covers the body and lines internal surfaces ▶ Can be modified to perform specific roles	▶ Receives and responds to stimuli ▶ Makes up the structures of the nervous system ▶ Regulates function of other tissues	▶ Supports, protects, and binds other tissues ▶ Contains cells in an extracellular matrix ▶ Can be hard or fluid

Cilia **Mucus-producing cell**

The upper respiratory tract is lined with ciliated epithelium to move irritants before they reach the lungs. The lungs and cardiovascular system work together to respond to changes in oxygen demand.

Processes **Neuron**

Nervous tissue is made up of nerve cells (neurons) and supporting cells. The long processes of neurons control the activity of muscles and glands.

E CT SM

The digestive tract is lined with epithelial tissue (E) and held in place by connective tissue (CT). It is moved by smooth muscle (SM) in response to messages from neurons.

CT E SM

Epithelial tissue (E) lines organs such as the bladder. Connective tissue (CT) supports the organ. This epithelium is layered so that it can stretch. The bladder's activity is controlled by smooth muscle (SM) and neurons.

Bone is a type of connective tissue. It provides shape to the body and works with muscle to produce movement. Ligaments are also connective tissue structures. They hold bones together.

Skeletal muscle tissue contracts to pull on the rigid bones of the skeleton to bring about movement of the body. Tendons are connective tissue structures that attach muscles to bones.

CONNECT **60** CONNECT **44** WEB **72** KNOW

Tissues work together and make up organs, which perform specific functions

The body's tissues work together in order for the body to function. Tissues also group together to form organs. For example, epithelial tissues are found associated with other tissues in every organ of the body.

Other examples include:

Nerves, muscles, and movement

▶ Nerves stimulate muscles to move.

▶ Connective tissue binds other tissues together and holds them in place (e.g. skeletal muscle tissue is held together by connective tissue sheaths to form discrete muscle, neurons are bundled together by connective tissue to form nerves).

▶ Bones are held together by connective tissue ligaments at joints, allowing the skeleton to move. Skeletal muscles are attached to bone by connective tissue tendons. Muscle contraction causes the tendon to pull on the bone, moving it.

Heart, lungs, blood vessels, and blood

▶ Cardiac muscle pumps blood (a specialized connective tissue) around the body within blood vessels.

▶ In the lungs, blood vessels surround the epithelium of the tiny air sacs to enable the exchange of gases between the blood and the air in the lungs.

▶ Neurons regulate the activity of heart and lungs to respond to changes in oxygen demand, as when a person is exercising.

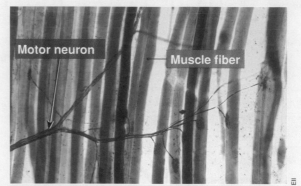

This image shows a neuron branching to supply muscle fibers. Impulses from the neuron will cause the muscle to contract. Neurons are bundled together by connective tissue wrappings to form nerves.

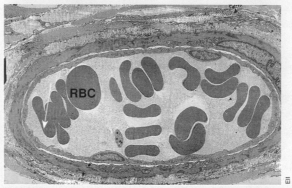

This image shows red blood cells within a vein. Veins return blood to the heart. When oxygen demand increases, heart and breathing rates increase and blood is delivered more quickly to working tissues.

1. Describe the main function of each of the following types of tissues:

(a) Epithelial tissue: _____

_____

(b) Connective tissue: _____

_____

(c) Nervous tissue: _____

_____

(d) Muscle tissue: _____

_____

2. Describe how different tissues interact to bring about movement of a body part: _____

_____

_____

_____

_____

_____

 © 2014 **BIOZONE** International
ISBN: 978-1-927173-84-8
Photocopying Prohibited

73 Chapter Review

Summarize what you know about this topic under the headings provided. You can draw diagrams or mind maps, or write short notes to organize your thoughts. Use the images (right) and the hints and guidelines included to help you:

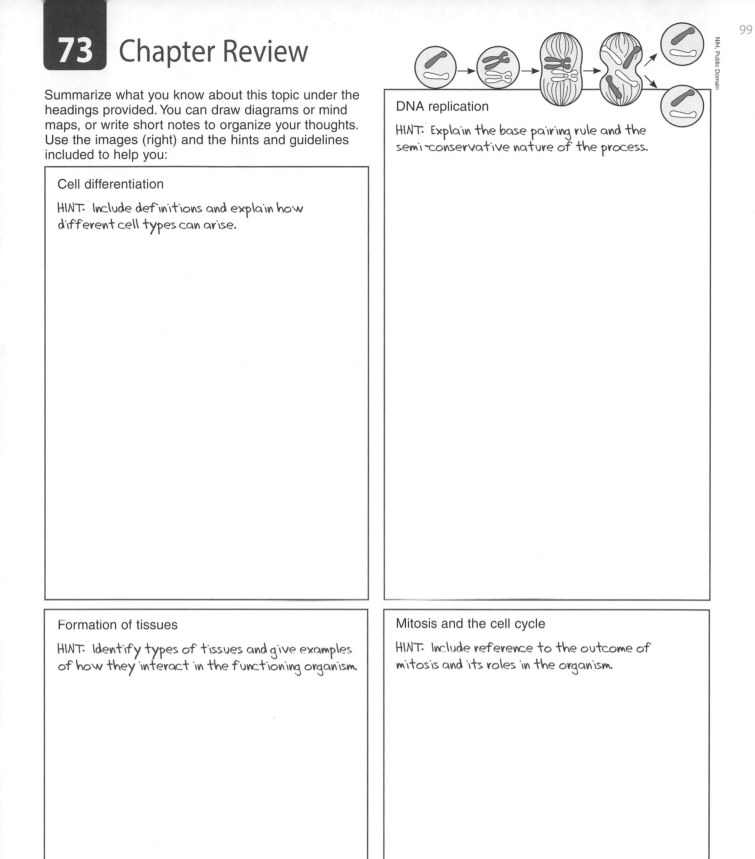

Cell differentiation

HINT: Include definitions and explain how different cell types can arise.

DNA replication

HINT: Explain the base pairing rule and the semi-conservative nature of the process.

Formation of tissues

HINT: Identify types of tissues and give examples of how they interact in the functioning organism.

Mitosis and the cell cycle

HINT: Include reference to the outcome of mitosis and its roles in the organism.

© 2014 **BIOZONE** International
ISBN: 978-1-927173-84-8
Photocopying Prohibited

REVISE

74 KEY TERMS: Vocabulary and Comprehension

1. Test your vocabulary by matching each term to its definition, as identified by its preceding letter code.

cell cycle

cell differentiation

cell division

DNA replication

interphase

mitosis

nucleotides

semi-conservative

stem cell

tissue

zygote

A Process by which a parent cell divides into two or more daughter cells.

B The stage in the cell cycle between divisions.

C The phase of a cell cycle resulting in nuclear division.

D The process by which a less specialized cell becomes a more specialized cell type.

E An undifferentiated cell, with the properties of self renewal and potency.

F The changes that take place in a cell in the period between its formation as a product of cell division and its own subsequent division.

G DNA replication is said to be this because each new DNA molecule is made up of one parent strand and one daughter strand of DNA.

H The process occurring prior to cell division to produce a copy of all the DNA in the nucleus.

I A fertilized egg cell.

J Name given to a group of related cells organized together to perform a specific function.

K The units that make up nucleic acids (DNA and RNA).

2. DNA replication occurs during the S (synthesis) phase of the cell cycle.

The light micrograph (right) shows a section of cells in an onion root tip. These cells have a cell cycle of approximately 24 hours. The cells can be seen to be in various stages of the cell cycle. By counting the number of cells in the various stages it is possible to calculate how long the cell spends in each stage of the cycle.

Count and record the number of cells in the image which are undergoing mitosis and those that are in interphase. Estimate the amount of time a cell spends in each phase.

Onion root tip cells

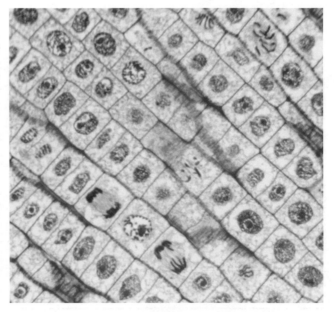

Stage	No. of cells	% of total cells	Estimated time in stage
Interphase			
Mitosis			
Total		100	

VOCAB

LS1.C / PS3.D — Energy in Living Systems

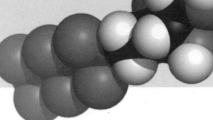
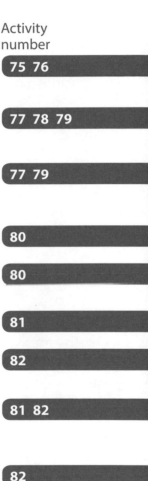

Key terms

ATP

cellular respiration

chloroplast

glucose

heat energy

mitochondria

photosynthesis

Core ideas

Activity number

Photosynthesis converts light energy into stored chemical energy

☐ 1. **ATP** is an energy transfer molecule. It provides the energy to drive cellular reactions. Energy is released during the hydrolysis of ATP.
75 76

☐ 2. **Photosynthesis** is the process that captures light energy and converts it into stored chemical energy. In plants, photosynthesis occurs in organelles called **chloroplasts**.
77 78 79

☐ 3. In photosynthesis, carbon dioxide and water are converted into glucose and oxygen.
77 79

Glucose can be used to make other macromolecules

☐ 4. **Glucose** consists of carbon, oxygen, and hydrogen atoms.
80

☐ 5. Glucose is used as a precursor to make many other biologically important molecules (e.g. DNA and proteins).
80

Matter and energy flow between different living systems

☐ 6. The glucose produced during photosynthesis is used in cellular respiration.
81

☐ 7. In **cellular respiration**, glucose and oxygen are used to produce ATP, which provides the energy needed to perform cellular work, such as muscle contraction.
82

☐ 8. Cellular respiration takes place in the cell cytoplasm and in the **mitochondrion**.
81 82

Heat is released during chemical reactions

☐ 9. The **heat energy** released from chemical reactions is lost to the surrounding environment and can be used to maintain body temperature.
82

Science and engineering practices

☐ 1. Use a model to show how the ATP molecule provides energy to carry out life's functions.
76

☐ 2. Use a model to show how photosynthesis transforms light energy into stored chemical energy.
77

☐ 3. Construct an explanation based on evidence for the fate of glucose in living systems.
80

☐ 4. Use a model to illustrate how energy is transferred in living systems.
81

BIOZONE APP
Student Review Series
Energy in Living Systems

75 Energy in Cells

Key Idea: Cells need energy to perform the functions essential to life. This energy is provided by cellular respiration and stored in the molecule ATP.

Energy for metabolism

▶ All organisms require energy to be able to perform the metabolic processes required for them to function and reproduce.

▶ This energy is obtained by **cellular respiration**, a set of metabolic reactions which ultimately convert biochemical energy from 'food' into the energy-carrying molecule **adenosine triphosphate** (**ATP**).

▶ ATP is considered to be a universal energy carrier, transporting chemical energy within the cell for use in metabolic processes such as biosynthesis, cell division, cell signaling, thermoregulation, cell movement, and active transport of substances across membranes.

The mitochondrion

Cellular respiration, which produces ATP, occurs in the mitochondria. A mitochondrion is bounded by a double membrane. The inner and outer membranes are separated by an inter-membrane space, compartmentalizing the regions of the mitochondrion in which the different reactions of cellular respiration occur.

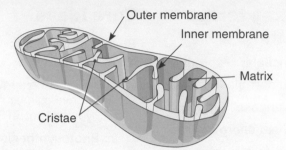

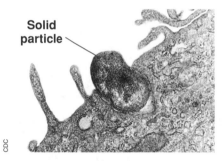

Energy is needed to actively transport molecules and substances across the cellular membrane such as engulfing solid particles, phagocytosis (above).

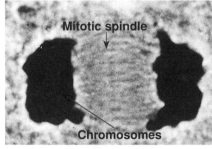

Cell division (mitosis) (above), requires energy to proceed. ATP provides energy for the mitotic spindle formation and chromosome separation.

The maintenance of body temperature requires energy. Both heating and cooling the body require energy by shivering and secretion of sweat.

1. What process produces usable energy in cells? _____

_____

2. How is energy carried around the cell? _____

3. (a) Describe the general role of mitochondria in the cell: _____

_____

_____

(b) What is the purpose of the folded inner membrane in mitochondria? _____

_____

_____

_____

4. (a) What energy-using process helps warm the body? _____

(b) What energy-using process helps cool the body? _____

 © 2014 **BIOZONE** International
ISBN: 978-1-927173-84-8
Photocopying Prohibited

76 ATP

Adenosine Triphosphate (ATP)

▶ The ATP molecule is a nucleotide derivative. It consists of three components; a purine base (**adenine**), a pentose sugar (**ribose**), and three **phosphate groups**. The three dimensional structure of ATP is shown right

▶ ATP acts as a store of energy within the cell. The bonds between the phosphate groups contain electrons in a high energy state, which store a large amount of energy that is released during a chemical reaction. The removal of one phosphate group from ATP results in the formation of adenosine diphosphate (ADP).

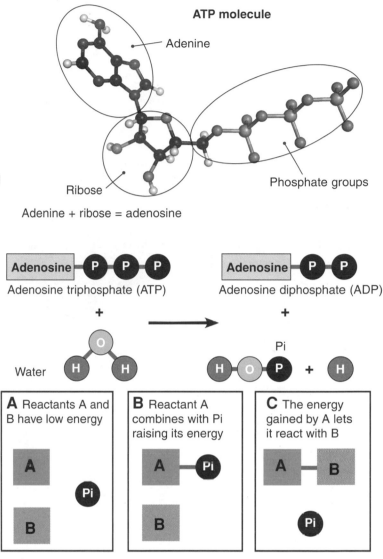

ATP molecule

Adenine

Ribose

Phosphate groups

Adenine + ribose = adenosine

How does ATP provide energy?

▶ The bonds between the phosphate groups of ATP are unstable and very little energy is needed to break them. The energy in the ATP molecule is transferred to a target molecule (e.g. a protein) by a hydrolysis reaction. Water is split during the reaction and added to the terminal phosphate on ATP, forming ADP and an inorganic phosphate molecule (Pi).

▶ When the Pi molecule combines with a target molecule, energy is released. Most of the energy (about 60%) is lost as heat (this helps keep you warm). The rest of the energy is transferred to the target molecule, allowing it to do work, e.g. joining with another molecule (right).

Note! The phosphate bonds in ATP are often referred to as being high energy bonds. This can be misleading. The bonds contain *electrons in a high energy state* (making the bonds themselves relatively weak). A small amount of energy is required to break the bonds, but when the intermediaries recombine and form new chemical bonds a large amount of energy is released. The final product is less reactive than the original reactants.

---

1. What are the three components of ATP? _____

2. (a) What is the biological role of ATP? _____

   _____

   (b) Where is the energy stored in ATP? _____

   _____

   (c) What products are formed during hydrolysis of ATP? _____

   _____

3. Why does the conversion of ATP to ADP help keep us warm? _____

   _____

 © 2014 **BIOZONE** International
ISBN: 978-1-927173-84-8
Photocopying Prohibited

CONNECT **82**   CONNECT **81**   WEB **76**   **KNOW**

77 Introduction to Photosynthesis

Key Idea: Photosynthesis is the process of converting sunlight, carbon dioxide, and water into glucose and oxygen.

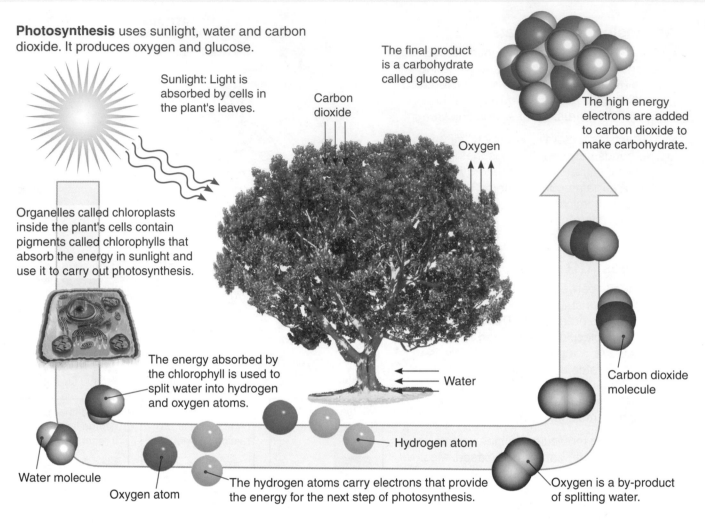

Photosynthesis uses sunlight, water and carbon dioxide. It produces oxygen and glucose.

Sunlight: Light is absorbed by cells in the plant's leaves.

Carbon dioxide

The final product is a carbohydrate called glucose

The high energy electrons are added to carbon dioxide to make carbohydrate.

Oxygen

Organelles called chloroplasts inside the plant's cells contain pigments called chlorophylls that absorb the energy in sunlight and use it to carry out photosynthesis.

The energy absorbed by the chlorophyll is used to split water into hydrogen and oxygen atoms.

Water

Carbon dioxide molecule

Hydrogen atom

Water molecule

Oxygen atom

The hydrogen atoms carry electrons that provide the energy for the next step of photosynthesis.

Oxygen is a by-product of splitting water.

1. Complete the schematic diagram of photosynthesis below:

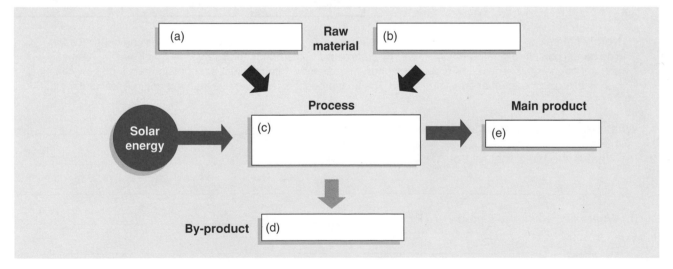

(a) _____

Raw material

(b) _____

Solar energy → Process (c) _____ → Main product (e) _____

By-product (d) _____

2. Use the diagram above to write a word equation for photosynthesis: _____

_____

3. Where does photosynthesis take place in a plant? _____

4. Where does the oxygen released during photosynthesis come from? _____

 © 2014 **BIOZONE** International
ISBN: 978-1-927173-84-8
Photocopying Prohibited

78 Chloroplasts

Key Idea: Photosynthesis occurs in organelles called chloroplasts. Chloroplasts contain the pigment chlorophyll, which captures light energy.

▶ Photosynthesis takes place in disk-shaped organelles called **chloroplasts** (4-6 μm in diameter). The inner structure of chloroplasts is characterized by a system of membrane-bound compartments called **thylakoids** arranged into stacks called **grana** linked together by **stroma lamellae**. The light dependent reactions of photosynthesis occur in the thylakoids.

▶ Pigments on these membranes called **chlorophylls** capture light energy by absorbing light of specific wavelengths. Chlorophylls reflect green light, giving leaves their green color.

▶ Chloroplasts are usually aligned with their broad surface parallel to the cell wall to maximize the surface area for light absorption.

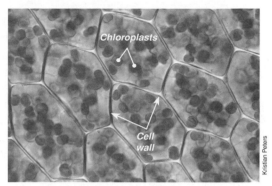

A mesophyll leaf cell contains 50-100 chloroplasts.

Chloroplast structure

Chloroplast is enclosed by an inner and outer membrane.

Thylakoid membranes provide a large surface area for light absorption. They are organized so as not to shade each other.

Liquid **stroma** contains the enzymes for the light independent phase.

Starch granule

Lipid droplet

Grana (*sing.* granum)

Stroma lamellae connect grana. They account for 20% of the thylakoid membranes.

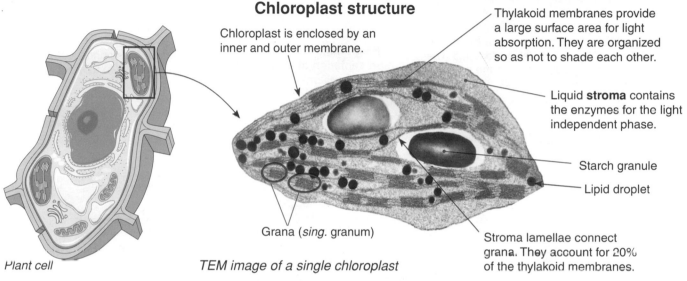

Plant cell

TEM image of a single chloroplast

1. Based on the information above, label the transmission electron micrograph (TEM) of a chloroplast below:

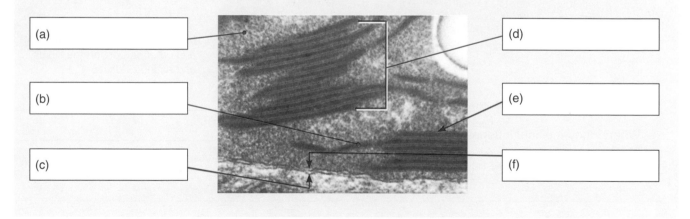

(a)

(b)

(c)

(d)

(e)

(f)

2. What does chlorophyll do? _____

_____

3. What features of chloroplasts help maximize the amount of light that can be absorbed? _____

_____

_____

© 2014 **BIOZONE** International
ISBN: 978-1-927173-84-8
Photocopying Prohibited

CONNECT
CONNECT
27
22
KNOW

79 Stages in Photosynthesis

Key Idea: Photosynthesis consists of two phases, the light dependent phase and the light independent phase.

▶ Photosynthesis has two phases, the light dependent phase and the light independent phase.

▶ In the **light dependent reactions**, light energy is converted to chemical energy (ATP and NADPH). This phase occurs in the in the thylakoid membranes of the chloroplasts.

▶ In the **light independent reactions**, the chemical energy is used to synthesize carbohydrate. This phase occurs in the stroma of chloroplasts.

Photosynthesis can be summarized in the equation:

$$6CO_2 + 12H_2O \xrightarrow[\text{Chlorophyll}]{\text{Light}} C_6H_{12}O_6 + 6O_2 + 6H_2O$$

The apparently extra 6 H_2Os are to show that O_2 comes from the H_2O, not the CO_2.

Light dependent phase (LDP):
In the first phase of photosynthesis, chlorophyll captures light energy, which is used to split water, producing O_2 gas (waste) and H^+ ions that are transferred to the molecule NADPH. ATP is also produced. The light dependent phase occurs in the thylakoid membranes of the grana.

Light independent phase (LIP):
The second phase of photosynthesis occurs in the stroma and uses the NADPH and the ATP to drive a series of enzyme-controlled reactions (the **Calvin cycle**) that fix carbon dioxide to produce triose phosphate. This phase does not need light to proceed.

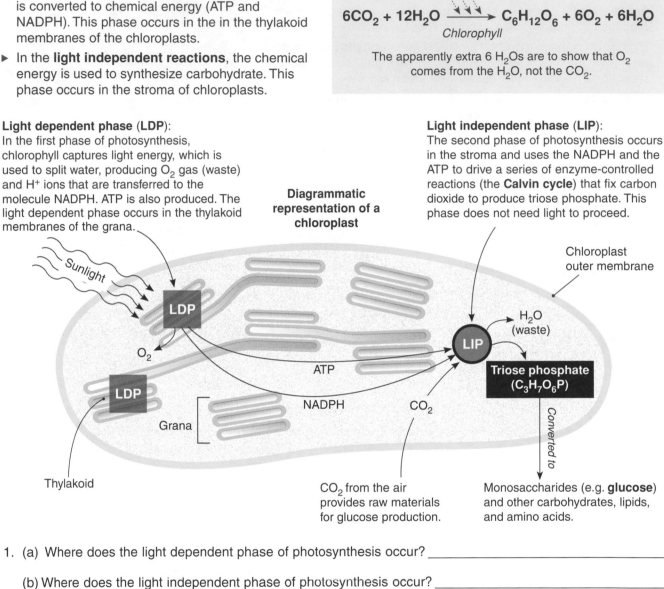

Diagrammatic representation of a chloroplast

CO₂ from the air provides raw materials for glucose production.

Monosaccharides (e.g. **glucose**) and other carbohydrates, lipids, and amino acids.

1. (a) Where does the light dependent phase of photosynthesis occur? _____

   (b) Where does the light independent phase of photosynthesis occur? _____

2. How are the light dependent and light independent phases linked? _____

   _____

   _____

3. In two experiments, radioactively-labeled oxygen (shown in blue) was used to follow oxygen through the photosynthetic process. The results of the experiment are shown below:

   $6CO_2 + 12H_2O + \text{sunlight energy} \rightarrow C_6H_{12}O_6 + 6O_2 + 6H_2O$

   $6CO_2 + 12H_2O + \text{sunlight energy} \rightarrow C_6H_{12}O_6 + 6O_2 + 6H_2O$

   What would you conclude from the results of this experiment? _____

   _____

© 2014 **BIOZONE** International
ISBN: 978-1-927173-84-8
Photocopying Prohibited

80 The Fate of Glucose

Key Idea: Glucose is an important precursor molecule used to produce a wide range of other molecules.

Glucose is a multipurpose molecule

▶ Glucose is a versatile biological molecule. It contains the elements carbon, oxygen, and hydrogen, which are used to build many other molecules produced by plants, animals, and other living organisms.

▶ Plants make their glucose directly through the process of photosynthesis and use it to build all the molecules they require. Animals obtain their glucose (as carbohydrates) by consuming plants or other animals. Other molecules (e.g. amino acids and fatty acids) are also obtained by animals this way.

▶ Glucose has three main fates: immediate use to produce ATP molecules (available energy for work), storage for later ATP production, or for use in building other molecules.

The fate of glucose

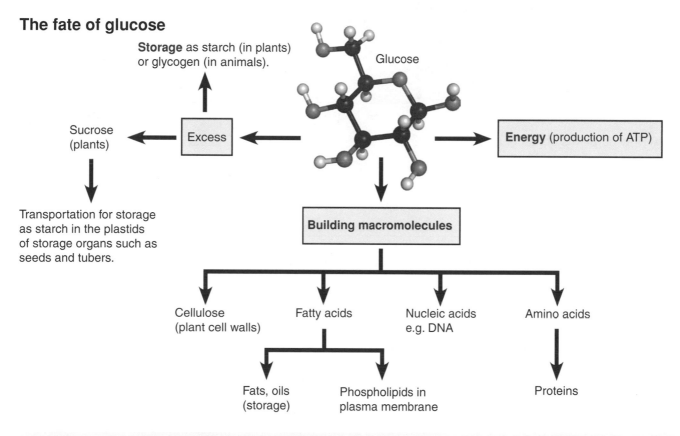

Storage as starch (in plants) or glycogen (in animals).

Glucose

Sucrose (plants)

Excess

Energy (production of ATP)

Transportation for storage as starch in the plastids of storage organs such as seeds and tubers.

Building macromolecules

Cellulose (plant cell walls)

Fatty acids

Nucleic acids e.g. DNA

Amino acids

Fats, oils (storage)

Phospholipids in plasma membrane

Proteins

How do we know how glucose is used?

▶ Labeling the carbon atoms in a glucose molecule with isotopes shows how glucose is incorporated into other molecules.

▶ An isotope is an element (e.g. carbon) whose atoms have a particular number of neutrons in their nucleus. The different number of neutrons allows the isotopes to be identified by their density (e.g. a carbon atom with 13 neutrons is denser than a carbon atom with 12 neutrons).

▶ Some isotopes are radioactive. These radioactive isotopes can be traced using X-ray film or devices that detect the disintegration of the isotopes, such as Geiger counters.

The carbon atom

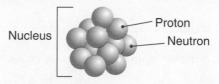

Nucleus

Proton

Neutron

The nucleus of an atom is made up of neutrons and protons. For any element, the number of protons remains the same, but the number of neutrons can vary. Electrons (not shown) are found outside the nucleus.

Naturally occurring C isotopes

^{12}C	^{13}C	^{14}C
6 protons	6 protons	6 protons
6 neutrons	7 neutrons	8 neutrons
Stable. 99.9% of all C isotopes.	Stable	Radioactive

CONNECT

41

CONNECT

1

KNOW

Isotope experiments with animals

Experiments using ^{13}C isotopes to identify the fate of glucose in guinea pigs showed that 25% of the glucose intake was respired. The rest of the glucose was incorporated into proteins, fats, and glycogen.

Corals are small sea anemone-like organisms that live in a symbiotic relationship with algae. The algae transfer sugars to the coral in return for the safe environment provided by the coral. Experiments with ^{13}C showed that the major molecule being transferred to the coral was glucose.

Isotope experiments with plants

^{13}C isotopes were used to trace the movement of glucose in plant leaves. It was found that some glucose is converted to fructose (a sugar molecule similar to glucose). Fructose molecules can be joined together to be stored as fructan in plant vacuoles. Fructose is also added to glucose to produce sucrose. Sucrose is transported out of the leaf.

Four molecule model of glucose use in a plant.

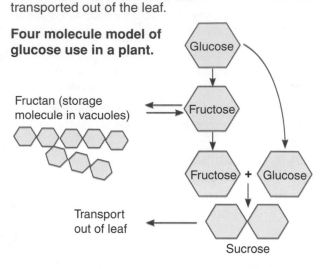

1. (a) How do plants obtain glucose? _____

   (b) How do animals obtain glucose? _____

2. What are the three main fates of glucose? _____
   _____

3. Identify a use for glucose in a plant that does not occur in animals: _____
   _____
   _____

4. (a) How can isotopes of carbon be separated? _____
   _____
   _____

   (b) How can this help trace how glucose is used in an organism? _____
   _____
   _____

5. How is glucose used to make other molecules needed by an organism? _____
   _____
   _____
   _____

6. Describe the fate of glucose in the glucose, fructose, sucrose system shown above: _____
   _____
   _____
   _____

© 2014 **BIOZONE** International
ISBN: 978-1-927173-84-8
Photocopying Prohibited

81 Energy Transfer Between Systems

Key Idea: The energy from sunlight is stored as glucose, which powers the production of ATP. ATP provides the energy for the various chemical reactions in living systems.

Summary of energy transformations in living systems

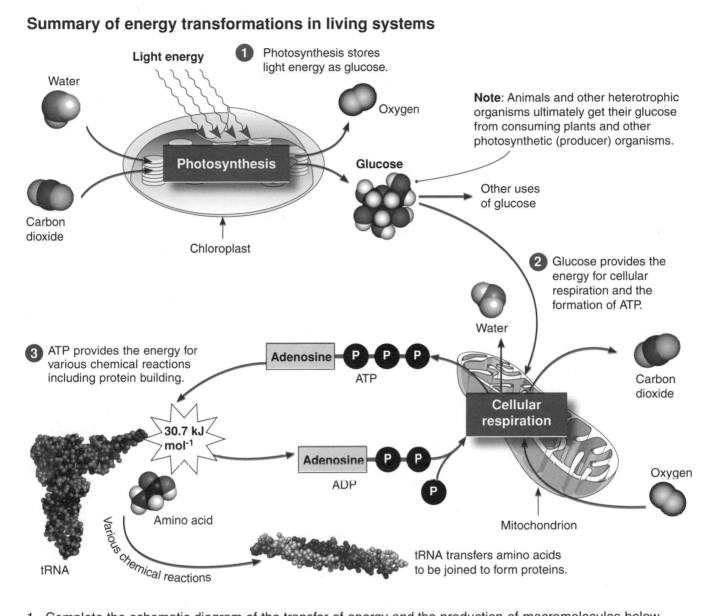

Light energy

Water

1 Photosynthesis stores light energy as glucose.

Oxygen

Note: Animals and other heterotrophic organisms ultimately get their glucose from consuming plants and other photosynthetic (producer) organisms.

Photosynthesis

Glucose

Carbon dioxide

Other uses of glucose

Chloroplast

2 Glucose provides the energy for cellular respiration and the formation of ATP.

Water

3 ATP provides the energy for various chemical reactions including protein building.

Adenosine P P P

ATP

Carbon dioxide

Cellular respiration

30.7 kJ mol⁻¹

Adenosine P P

ADP

P

Oxygen

Amino acid

Various chemical reactions

Mitochondrion

tRNA

tRNA transfers amino acids to be joined to form proteins.

1. Complete the schematic diagram of the transfer of energy and the production of macromolecules below using the following word list: *water, ADP, protein, carbon dioxide, amino acid, glucose, ATP.*

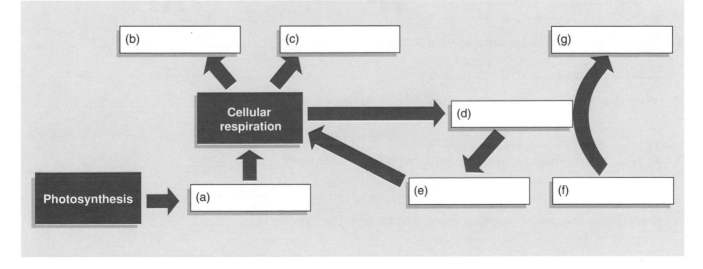

(b)

(c)

(g)

Cellular respiration

(d)

(e)

(f)

Photosynthesis

(a)

© 2014 **BIOZONE** International
ISBN: 978-1-927173-84-8
Photocopying Prohibited

KNOW

82 Cellular Respiration: Energy from Glucose

Key Idea: Cellular respiration is an aerobic process that converts the chemical energy in glucose into usable energy (in the form of ATP), carbon dioxide, and water.

▶ **Cellular respiration** is the process of extracting the energy stored in the chemical bonds in glucose and storing it in ATP molecules. The process includes many chemical reactions, some of which produce ATP molecules and some that prepare molecules for further chemical reactions.

▶ Cellular respiration can be divided into four major steps, each with its own set of chemical reactions. Every step, except the link reaction, produces ATP. The four steps are: glycolysis, the link reaction, the Krebs cycle, and the electron transport chain (ETC).

The overall equation for cellular respiration is:

Glucose + Oxygen $\longrightarrow$ Carbon dioxide + Water + Energy

$C_6H_{12}O_6 + 6O_2 \longrightarrow 6CO_2 + 6H_2O + Energy$

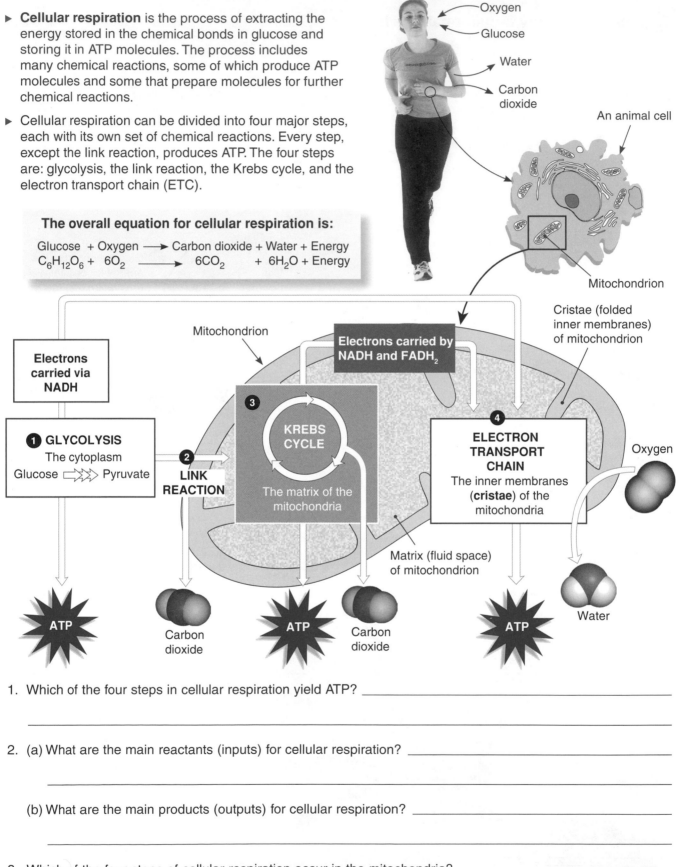

1. Which of the four steps in cellular respiration yield ATP? _____

_____

2. (a) What are the main reactants (inputs) for cellular respiration? _____

_____

(b) What are the main products (outputs) for cellular respiration? _____

_____

3. Which of the four steps of cellular respiration occur in the mitochondria? _____

_____

© 2014 **BIOZONE** International
ISBN: 978-1-927173-84-8
Photocopying Prohibited

How does cellular respiration provide energy?

▶ A molecule's energy is contained in the electrons within the molecule's chemical bonds. During a chemical reaction, energy (e.g. heat) can break the bonds of the reactants.

▶ When the reactants form products, the new bonds within the product will contain electrons with less energy, making the bonds more stable. The difference in energy is usually lost as heat. However, some of the energy can be captured to do work.

▶ Glucose contains 16 kJ of energy per gram (2870 kJ mol^{-1}). The step-wise breakdown of glucose through a series of chemical reactions yields ATP. In total, 38 ATP molecules can be produced from 1 glucose molecule.

A model for ATP production and energy transfer from glucose   **A model for ATP use in the muscles**

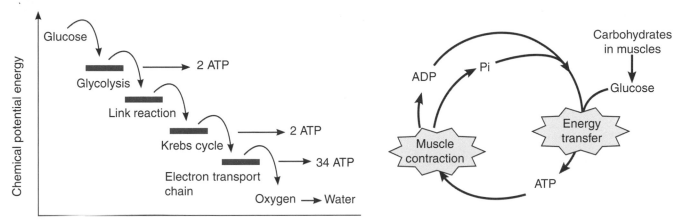

4. Explain how the energy in glucose is converted to useful energy in the body. Use the example of muscle contraction to help illustrate your ideas:

_____

_____

_____

_____

_____

5. (a) One mole of glucose contains 2870 kJ of energy. The hydrolysis of one mole of ATP releases 30.7 kJ of energy. Calculate the percentage of energy that is transformed to useful energy in the body. Show your working.

_____

_____

_____

  (b) Use your calculations above to explain why shivering keeps you warm and extreme muscular exertion causes you to get hot:

_____

_____

_____

_____

_____

_____

 © 2014 **BIOZONE** International
ISBN: 978-1-927173-84-8
Photocopying Prohibited

83 Chapter Review

Summarize what you know about this topic under the headings provided. You can draw diagrams or mind maps, or write short notes to organize your thoughts. Use the images and hints, included to help you:

The structure and role of ATP

HINT: How does the structure of ATP contribute to its function?

ATP

Photosynthesis

HINT: Where does photosynthesis take place? Include an equation to summarize the process. Why is photosynthesis important to all living organisms?

Chloroplast

© 2014 **BIOZONE** International
ISBN: 978-1-927173-84-8
Photocopying Prohibited

REVISE

Glucose

HINT: Include the role of glucose in living systems (e.g. how the components of glucose can be used to make macromolecules with different properties)

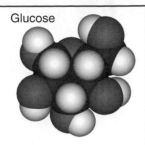

Glucose

Cellular respiration

HINT: Where does cellular respiration take place?
Include an equation to summarize the process.

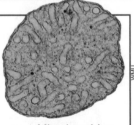

Mitochondrion

84 KEY TERMS: Did You Get It?

1. Test your vocabulary by matching each term to its definition, as identified by its preceding letter code.

ATP ....

cellular respiration ....

chloroplast ....

glucose ....

heat energy ....

mitochondria ....

photosynthesis ....

A These organelles are the cell's energy transformers, in which chemical energy is converted into ATP.

B The biochemical process that uses light energy to convert carbon dioxide and water into glucose molecules and oxygen.

C One of the main products of photosynthesis, this molecule consists of carbon, oxygen, and hydrogen.

D A set of biochemical reactions in which the chemical energy in glucose is converted to usable energy (as ATP) and waste products.

E An chlorophyll-containing organelle found in plants in which the reactions of photosynthesis take place.

F The cell's energy carrier.

G A by-product of many chemical reactions, this can be used to maintain body temperature.

2. Test your understanding of photosynthesis and cellular respiration by answer the questions below.

Plant cell

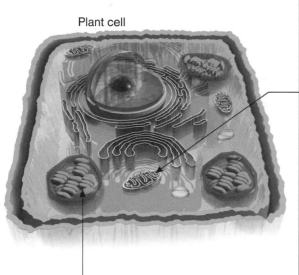

(a) Name this organelle: _____

Name the major process that occurs here:

_____

Write the word equation for this process:

_____

_____

Write the chemical equation for this process:

_____

_____

(b) Name this organelle: _____

Name the major process that occurs here: _____

Write the word equation for this process: _____

_____

Write the chemical equation for this process: _____

_____

TEST

© 2014 **BIOZONE** International
ISBN: 978-1-927173-84-8

Ecosystems: Interactions, Energy, and Dynamics

Concepts and Connections
Use arrows to make your own connections between related concepts in this section of this workbook

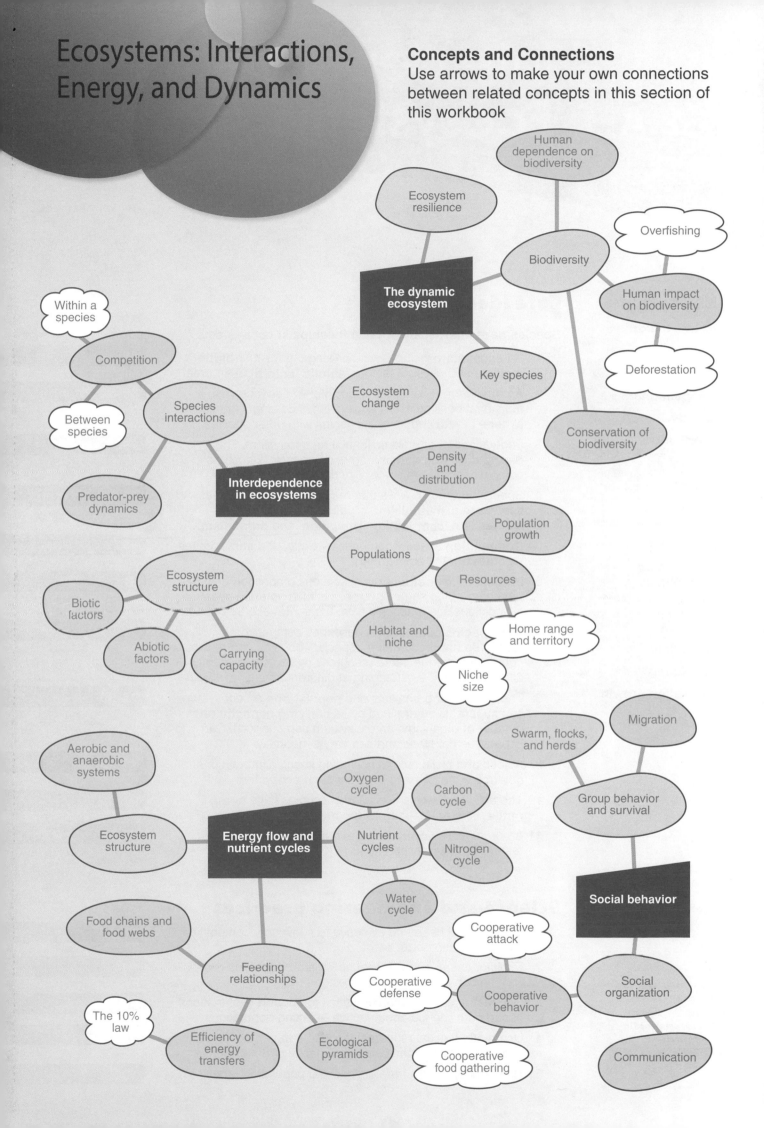

LS2.A Interdependence in Ecosystems

Key terms

- abiotic factor
- biotic factor
- carrying capacity
- competition
- density
- distribution
- ecosystem
- habitat
- home range
- interspecific competition
- intraspecific competition
- mutualism
- niche
- parasitism
- population
- population growth
- predation

Core ideas

Species have interdependent relationships in ecosystems

Activity number

☐ 1. An **ecosystem** includes all the living organisms (**biotic factors**) and physical factors (**abiotic factors**) in an area.   `85`

☐ 2. An organism's ecological niche (**niche**) describes its functional position in the ecosystem, including its habitat (where it lives) and its relationships with other species.   `86 87 97 98`

☐ 3. Populations of organisms inhabit an ecosystem. Their **distribution** and **density** depends on factors such as competition and the availability of resources.   `88`

☐ 4. Species interact in ways that may be beneficial to both species (e.g. **mutualism**), or harmful to at least one species (e.g. **competition**, **predation**, and **parasitism**).   `89 94`

☐ 5. **Competition** arises when species exploit the same limited resources in the environment.   `90`

☐ 6. **Intraspecific competition** describes competition between members of the same species. It is often intense because all individuals in a species have the same resource needs.   `91`

☐ 7. **Interspecific competition** describes competition for resources between different species. Species can reduce competition for the same resources by exploiting slightly different niches, e.g. foraging at different times.   `92 93`

☐ 8. The number of organisms and **populations** an ecosystem can support is determined by its **carrying capacity** (the number of organisms an ecosystem can support). Carrying capacity is not static and can vary seasonally.   `95 96`

☐ 9. Abiotic and biotic factors (including species interactions) determine the carrying capacity of an ecosystem.   `95 96`

☐ 10. The area an organism regularly occupies is its **home range**. Its size depends on the resources it contains.   `97 98`

☐ 11. An organism's **population growth** is limited by the resources of its environment.   `99 101`

Science and engineering practices

☐ 1. Use a model based on evidence to show how competition limits population size.   `91 92`

☐ 2. Explain, based on evidence, how different species reduce competition for limited resources.   `93`

☐ 3. Use mathematical representations to support explanations based on evidence about factors affecting populations.   `93 94`

☐ 4. Use mathematical representations to support explanations of factors affecting the carrying capacity of ecosystems.   `95 96`

☐ 5. Use a model to illustrate exponential population growth.   `100`

85 What is an Ecosystem?

Key Idea: An ecosystem is a natural unit encompassing all the living and non-living components in an area. These components are linked through nutrient cycles and energy flows.

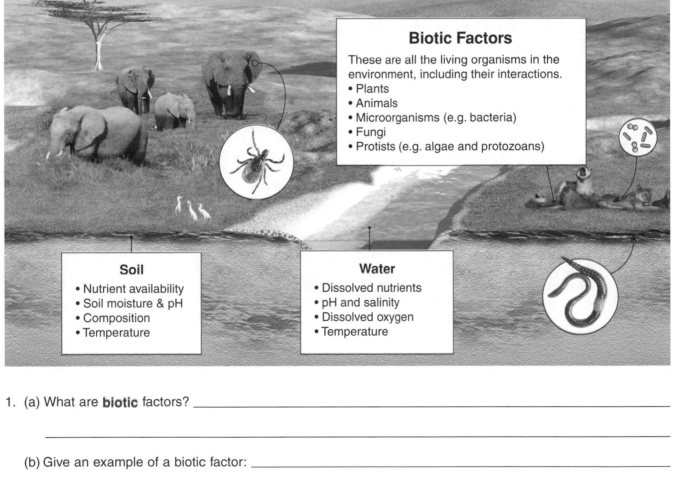

Atmosphere
• Wind speed and direction
• Humidity
• Light intensity and quality
• Precipitation
• Temperature

Ecosystems

Ecosystems are natural units made up of all the living organisms (biotic factors) and the physical (abiotic factors) in an area.

Abiotic factors (non-living chemical and physical factors) include the soil, water, atmosphere, temperature, and sunlight (SWATS). **Biotic factors** are all the living organisms (e.g. plants, animals, fungi, protists, and microorganisms).

Living organisms interact to influence each other. These interactions help to produce the characteristics of an ecosystem.

The components of an ecosystem are linked together (and to other ecosystems) through nutrient cycles and energy flows.

Biotic Factors
These are all the living organisms in the environment, including their interactions.
• Plants
• Animals
• Microorganisms (e.g. bacteria)
• Fungi
• Protists (e.g. algae and protozoans)

Soil
• Nutrient availability
• Soil moisture & pH
• Composition
• Temperature

Water
• Dissolved nutrients
• pH and salinity
• Dissolved oxygen
• Temperature

1. (a) What are **biotic** factors? _____

_____

(b) Give an example of a biotic factor: _____

2. (a) What is an **abiotic** factor? _____

_____

(b) Give an example of an abiotic factor: _____

3. How do biotic and abotic factors interact to form an ecosystem? _____

_____

86 Habitat and Niche

Key Idea: The environment in which an organism lives is called its habitat. An organism's niche describes its functional role within that environment.

Habitat: where you live. Niche: what you do.

The natural environment in which an organism lives is its **habitat**. It includes all the physical and biotic factors in that area. The **niche** (or ecological niche) of an organism, describes its functional position in that environment. It is a way of describing the life of an organism, and includes its interactions with other organisms and how it responds to changes in the availability of resources.

Each species has a **tolerance range** for factors in its environment (below). However, the members of a species population are all individually different and so vary in their ability to survive in a range of conditions.

Organisms are usually most abundant in their preferred niche, where the conditions for their survival and reproduction are optimal. Outside this optimal ecological space, in the marginal niche, the environment is less favorable for survival and there are many fewer individuals. For any species, there will also be environments that are unavailable to them because they are too far outside the tolerance range of all the individuals in the population.

Species tolerant of large environmental variations tend to be more widespread than organisms with a narrow tolerance range. The Atlantic blue crab (above) is widespread along the Atlantic coast from Nova Scotia to Argentina. Adults tolerate a wide range of water salinity ranging from almost fresh to highly saline. This species is an omnivore and eats anything from shellfish to carrion and animal waste.

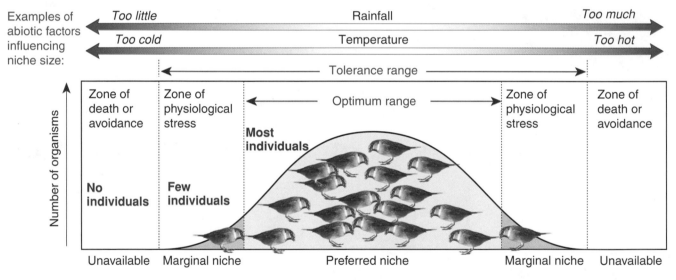

1. What is the difference between an organism's habitat and its niche? _____

_____

_____

_____

2. Suggest an advantage to being able to tolerate variations in a wide range of environmental factors?

_____

_____

3. In which part of an organism's range is competition for resources going to be most intense and why?

_____

_____

© 2014 **BIOZONE** International
ISBN: 978-1-927173-84-8
Photocopying Prohibited

87 Dingo Habitats

Key Idea: Organisms may need to exploit several different habitats to get the resources they need to survive.

Habitats provide resources

As we have seen (opposite) species may tolerate wide variations in a range of physical and biotic factors. As a result of this tolerance range, the habitat that is occupied by members of a species may be quite variable.

The important thing is that the habitat provides resources for the organisms that live there. These resources include water, food, shelter, and places to raise offspring.

Some habitats can be richer in resources than others and are usually described with reference to their main features. For example, riverine habitats (rivers and creeks containing water and thick vegetated cover) provide water, food, and cover (right).

Dingo habitats

Dingoes (right) are wild dogs found throughout Australia. The table on the far right gives information about five dingo packs at one location, including how much of their territory is made up of riverine areas. Kangaroos are the main prey for these dingoes.

Dingo pack name	Territory area (km²)	Pack size	Dingo density	% of total territory made up of riverine areas
Pack A	113	12	10.6	10
Pack B	94	12		14
Pack C	86	3		2
Pack D	63	6		12
Pack E	45	10		14

1. Calculate the density of each of the dingo packs using the equation below, and record it in the table above. The first one has been done for you.

> Density = pack size ÷ territory area × 100

2. (a) Plot a scatter graph of dingo density versus how much of their territory is made up of riverine areas for each pack.

   (b) Describe the relationship between dingo density and amount of riverine area:

   _____

   _____

   _____

   (c) Can you explain why this relationship might occur?

   _____

   _____

   _____

 © 2014 **BIOZONE** International
ISBN: 978-1-927173-84-8
Photocopying Prohibited

88 Density and Distribution of Populations

Key Idea: Population density is the number of organisms of one species in a specified area. Distribution describes how the organisms are distributed relative to each other.

▶ A population refers to all the organisms of the same species in a particular area. Populations can exist naturally at different densities and their individuals can be distributed in different ways within the environment.

Population density

Population **density** is the number of individuals of a species per unit area (for land organisms) or volume (for aquatic organisms). The characteristics of low density and high density populations are described below.

Low density

In low density populations, individuals are spaced well apart. There are only a few individuals per unit area (e.g. highly territorial, solitary mammal species, such as tigers).

High density

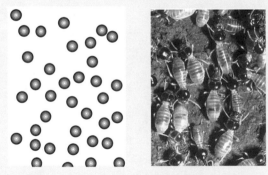

In high density populations, there are many individuals per unit area (e.g. colonial organisms, such as termites). This can be a natural feature of the population.

Population distribution

Population **distribution** describes how organisms are distributed relative to each other. The three distribution patterns (random, clumped, and uniform) are described below. In the examples, the circles represent individuals of the same species.

Random distribution

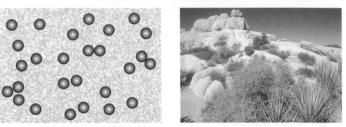

In random distributions, the spacing between individuals is irregular. Random distributions are uncommon in animals but are often seen in plants.

Clumped distribution

In clumped distributions, individuals are grouped in patches (sometimes around a resource). Clumped distributions occur in herding and highly social species.

Uniform distribution

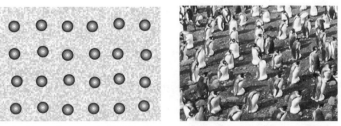

In uniform (regular) distribution, e.g. breeding colonies of birds, individuals are evenly spaced within the area.

1. What is the difference between population density and population distribution? _____

_____

_____

2. What type of distribution pattern would you expect to see where:

(a) Resources are not evenly spread out: _____   (c) Social insects: _____

(b) Resources are evenly spread out: _____   (d) Territorial animals: _____

© 2014 **BIOZONE** International
ISBN: 978-1-927173-84-8
Photocopying Prohibited

89 Species Interactions

Key Idea: Predation, competition, and parasitism are species interactions which have negative effects on at least one of the species involved.

Species interact in ways that limit the size of populations

▶ Within ecosystems, each species interacts with others in their community. In many of these interactions, at least one of the parties in the relationship is disadvantaged. Predators eat prey, parasites and pathogens exploit their hosts, and species compete for limited resources. These interactions (below) help to limit the number of organisms in a population, and prevent any one population from becoming too large.

▶ Not all relationships involve exploitation. Some species form relationships that are mutually beneficial, e.g. lichens are an organism formed by a mutualism between a fungus and a photosynthetic partner (alga or bacterium). Mutualisms such as these have been important in evolution of the eukaryotes.

Parasitism
Outcome:
Host is harmed (−)
Parasite benefits (+).
Example:
This tick benefits by gaining nutrients from the cat's blood it feeds on. The cat host is harmed (e.g. skin irritation and infection), but not killed.

Competition
Outcome:
All species (competitors) are harmed (−)
Example:
These plant species are all competing for the same resources (light and nutrients). All parties suffer because their access to resources is limited.

Predation
Outcome:
Prey is harmed (−)
Predator benefits (+).
Example:
This ladybug has captured and is eating an aphid. Often it is the prey numbers that limit the number of predators.

Herbivory
Outcome:
Plant is harmed (−)
Herbivore benefits (+).
Example:
This zebra eats grasses. Herbivory is a similar relationship of exploitation as predation, but the plant is not usually killed.

1. Use the information below to complete the table about the relationships between zebras and other savannah species. Fill in the species involved, the type of relationship, and its effect (+, −, 0). The first is done for you.

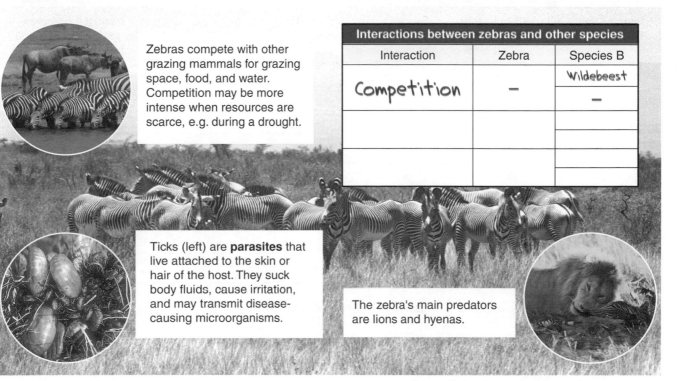

Zebras compete with other grazing mammals for grazing space, food, and water. Competition may be more intense when resources are scarce, e.g. during a drought.

Interactions between zebras and other species		
Interaction	Zebra	Species B
Competition	−	Wildebeest
		−

Ticks (left) are **parasites** that live attached to the skin or hair of the host. They suck body fluids, cause irritation, and may transmit disease-causing microorganisms.

The zebra's main predators are lions and hyenas.

© 2014 **BIOZONE** International
ISBN: 978-1-927173-84-8
Photocopying Prohibited

CONNECT **171** CONNECT **124** CONNECT **123** WEB **89** KNOW

90 Competition for Resources

Key Idea: Species interact with other living organisms in their environment. Competition occurs when species utilize the same limited resources.

▶ No organism exists in isolation. Each organism interacts with other organisms and with the physical (abiotic) components of the environment.

▶ **Competition** occurs when two or more organisms are competing for the same limited resource (e.g. food or space).

▶ Competition harms both competitors. The negative effects of competition limit population numbers because resources are limited and growth, reproduction, and survival are affected.

▶ Competition can occur between members of the same species (intraspecific competition), or between members of different species (interspecific competition).

A complex system of interactions occurs between the different species living on this coral reef in Hawaii. Population numbers will be limited by competition for limited resources, such as food and space on the reef.

Examples of limited resources

Space can be a limited resource
These sea anemones are competing for space in a tidal pool. Some species defend areas, called territories, which have resources they need.

Suitable mates can be hard to find
Within a species, individuals may compete for a mate. These male red deer are fighting to determine which one will mate will the females.

Food is usually a limited resource
In most natural systems, there is competition for food between individuals of the same species, and between different species with similar diets.

1. (a) What is competition? _____

   _____

   _____

   (b) Why does competition occur? _____

   _____

   _____

   (c) Why does competition have a negative effect on both competitors?_____

   _____

   _____

CONNECT CONNECT CONNECT

KNOW  95 97 98

 © 2014 **BIOZONE** International
ISBN: 978-1-927173-84-8
Photocopying Prohibited

91 Intraspecific Competition

Key Idea: Intraspecific competition describes competition between individuals of the same species for resources.

▸ Intraspecific competition occurs when individuals of the same species compete for the same limited resources. In addition to food, space, nutrients, and light, intraspecific competition also includes competition for mating partners and breeding sites.

▸ Intraspecific competition is more intense that competition between different species because the individuals are all competing for the same resources (e.g. same food and mates). It is an important factor in limiting the population size of many species.

These silkworm caterpillars are all competing for the same food. If there is inadequate food, none of the caterpillars may get enough, and they could all die.

Individuals may compete for the right to mate. Male elephant seals fight for territory and mates. Unsuccessful males may not mate at all.

In some animals, strict social orders limit competition for food. However, if food is very limited, only dominant individuals may receive enough to survive.

How does intraspecific competition limit population size?

Most resources are limited, and this is a major factor in determining how large a population can grow.

As population numbers increase, the demands on the resources are higher. The resources are used up more quickly and some individuals receive fewer resources than others. Populations respond by decreasing their numbers. This occurs by:

▸ Reduced survival (more individuals die).

▸ Reduced birth rates (fewer individuals are born).

If resources increase (e.g. food increases), population numbers can increase. The relationship between resources and population numbers is shown on the right.

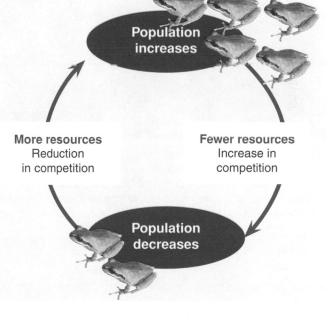

Population increases

More resources
Reduction in competition

Fewer resources
Increase in competition

Population decreases

1. What is intraspecific competition? _____

_____

2. (a) Explain the relationship between resource availability and intraspecific competition: _____

_____

_____

(b) What happens to population numbers as intraspecific competition increases? _____

_____

© 2014 **BIOZONE** International
ISBN: 978-1-927173-84-8
Photocopying Prohibited

CONNECT **138**   CONNECT **95**   WEB **91**   **KNOW**

92 Interspecific Competition

Key Idea: Interspecific competition describes competition between individuals of different species for resources. It can limit the number of one or both competing species.

Interspecific competition involves individuals of different species competing for the same limited resources.

▶ Interspecific competition is usually less intense than competition between members of the same species because competing species have different requirements for some of the resources (e.g. different mates, territories, or food preferences).

▶ Sometimes, humans may introduce a species with the same resource requirements as a native species. The resulting intraspecific competition can lead to the decline of the native species.

▶ Interspecific competition can be important in limiting a population's size, limiting the species present, or restricting habitat range.

Red squirrel   Gray squirrel

Competition between the European red squirrel and the introduced American gray squirrel in Britain is an example of how interspecific competition can limit species numbers.

Gray squirrels are larger and can compete more successfully than red squirrels for food and space. Since it was introduced into Britain 44 years ago, the gray squirrel has displaced many populations of native red squirrels. It is probably responsible for the major decline in red squirrel numbers.

Stork   Hyena   Vulture

The plant species in the forest community above will be competing for light, space, water, and nutrients. A tree that can grow taller than those around it will be able to absorb more sunlight, and grow more rapidly to a larger size than the plants in the shade below.

In some communities, many different species may be competing for the same resource. The example above shows several African carnivores competing for a dead animal carcass. The largest, most aggressive species will feed first (e.g. lions), and other species will compete for the remains.

1. (a) What is interspecific competition?_____

   _____

   (b) Why is interspecific competition often less intense than intraspecific competition?_____

   _____

   _____

2. Sometimes, humans may introduce a non-native species that exploits many of the same resources as a native. Suggest why the population numbers of the native species often decline after such introductions:

   _____

   _____

   _____

 © 2014 **BIOZONE** International
ISBN: 978-1-927173-84-8
Photocopying Prohibited

93 Reducing Competition Between Species

Key Idea: Competition between species for similar resources can be reduced if competing species have slightly different niches and exploit resources in different ways.

How species reduce competition

Species exploiting similar resources have adaptations (evolved features) to reduce competition. Each species exploits a smaller proportion of the entire spectrum of resources potentially available to it. For example:

▶ In a forest, each species may feed on a different part of a tree (e.g. trunk, branches, twigs, flowers, or leaves) or occupy different areas of vertical air-space (e.g. ground, understorey, sub-canopy, or canopy).

▶ Aquatic organisms may also inhabit different zones to reduce competition for resources. Some organisms will inhabit the bottom, and others may occupy surface waters.

▶ Competition may also be reduced by exploiting the same resources at a different time of the day or year (e.g. one species may feed at night and another may exploit the same resource in the morning).

Warblers of the genus *Setophaga* spend a lot of their time feeding in conifer trees. By feeding in different parts of the tree and on different food sources the warblers reduce competition. The blue areas shown on the tree below indicate areas where the warblers spent 50% or more of their time feeding.

Cape May warblers feed around new needles and buds near the top of tree. They hawk for flying insects and pick insects up from the tips of conifer branches. These warblers also feed on berry juice and nectar.

Blackburnian warblers forage in new needles and buds in upper branches of trees. They tend to move horizontally through tree feeding on insects or spiders.

Black-throated warbler forage in new needles and buds as well as older needles and branches in the upper to middle parts of tree. They feed mainly on insects, sometimes hovering-(gleaning), or catching insects in flight-(hawking). Berries will occasionally be consumed.

Bay-breasted warblers frequent older needles and lichen near middle branches. They feed on insects, particularly the spruce budworm. These birds will also feed on berries and nectar.

Myrtle warblers forage on lower trunks and middle branches. These birds are insectivorous, but will readily take wax-myrtle berries in winter, a habit which gives the species its name. They make short flights in search of bugs.

Photos:
Dan Pancamo: Bay-breasted warbler and black-throated warbler
Cephas: Myrtle warbler
Steve Maslowski, US Fish and Wildlife Service: Cape May warbler
Mdf: Blackburnian warbler

1. (a) How do the species of warbler avoid competition? _____

_____

(b) What evidence is there that their adaptations for doing this are largely behavioral? _____

_____

2. Cape May warblers and Blackburnian warblers appear to occupy the same habitat in the tree. Explain how they are able to avoid competition and coexist:

_____

_____

CONNECT CONNECT CONNECT

192 **183** **86** **KNOW**

Adaptations reduce competition in foraging bumblebees

Studies on bumblebee foraging have shown that when bumblebees forage in the presence of other bumblebee species they tend to spend the majority of their time on particular flower types. In many cases, the length of the corolla (the length of the flower petals) of flowers visited correlates with the length of the bumblebee's proboscis (mouthparts).

Bumblebee species in the mountains of Colorado (graph right) compete for nectar from flowers. Species with a long proboscis take nectar from flowers with long petals. Species with a short proboscis take nectar from flowers with short petals. This reduces competition for food between the bumblebee species.

Bumblebee species

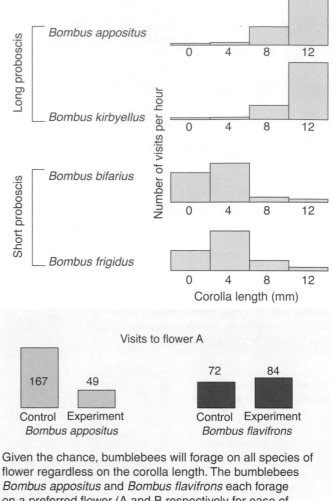

Corolla

Proboscis

Visits to flower A

167	49		72	84
Control	Experiment		Control	Experiment
Bombus appositus			*Bombus flavifrons*	

Given the chance, bumblebees will forage on all species of flower regardless on the corolla length. The bumblebees *Bombus appositus* and *Bombus flavifrons* each forage on a preferred flower (A and B respectively for ease of reference). In an experiment, *B. appositus* was prevented from visiting its preferred flower A. *B. flavifrons* responded by increasing its visits to that flower species.

3. (a) How do the *Bombus* species in Colorado reduce competition for flower resources? _____

_____

_____

(b) Are the differences between the Colorado species mainly structural, physiological, or behavioral? Explain:

_____

_____

4. What evidence is there that competition restricts bumblebee species to certain flower types:

_____

_____

_____

_____

_____

_____

© 2014 **BIOZONE** International
ISBN: 978-1-927173-84-8

94 Predator-Prey Relationships

Key Idea: Predators do not always regulate prey numbers, but predator numbers are often dependent upon prey numbers.

Do predators limit prey numbers?

▶ It was once thought that predators limited the numbers of their prey populations. However, we now know that this is often not the case. Prey species are more likely to be regulated by other factors such as climate and the availability of food.

▶ In contrast, predator populations are affected by the availability of prey, especially when there is little opportunity for prey switching (hunting another prey if the preferred one becomes scarce). This relationship is shown in the graph (right).

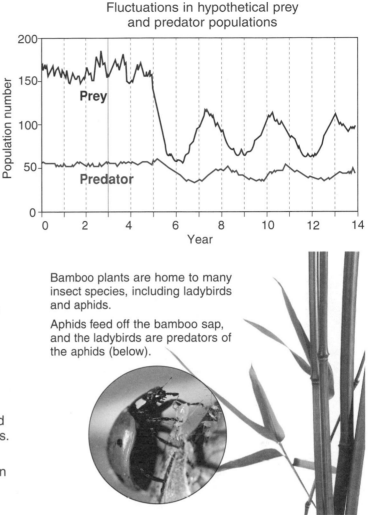

Fluctuations in hypothetical prey and predator populations

A case study in predator-prey numbers

In some areas of Northeast India, a number of woolly aphid species colonize and feed off bamboo plants. The aphids can damage the bamboo so much that it is no longer able to be utilized by the local people for construction and textile production.

Giant ladybird beetles (*Anisolemnia dilatata*) feed exclusively off the woolly aphids of bamboo plants. There is some interest in using them as biological control agents to reduce woolly aphid numbers, and limit the damage woolly aphids do to bamboo plants.

The graph below shows the relationship between the giant lady bird beetle and the woolly aphid when grown in controlled laboratory conditions.

Bamboo plants are home to many insect species, including ladybirds and aphids.

Aphids feed off the bamboo sap, and the ladybirds are predators of the aphids (below).

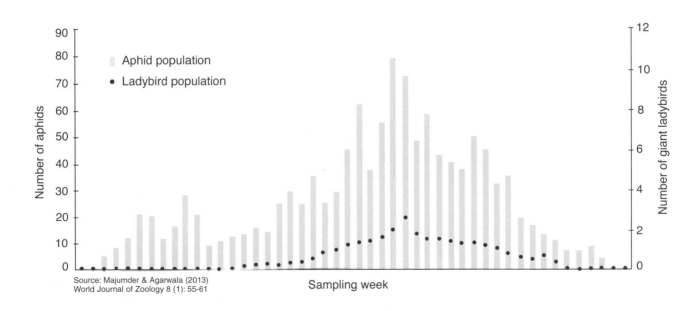

Source: Majumder & Agarwala (2013)
World Journal of Zoology 8 (1): 55-61

1. (a) On graph the above, mark the two points (using different colored pens) where the peak numbers of woolly aphids and giant ladybirds occurs.

CONNECT 108 CONNECT 96 WEB 94 **KNOW**

(b) Do the peak numbers for both species occur at the same time?_____

(c) Why do you think this is? _____

_____

_____

2. (a) Is the trend between the giant ladybirds woolly aphids positive or negative (circle one).

   (b) Explain your answer: _____

        _____

        _____

        _____

3. A census of a deer population on an island forest reserve indicated a population of 2000 animals in 1960. In 1961, ten wolves (natural predators of deer) were brought to the island in an attempt to control deer numbers. Over the next nine years, the numbers of deer and wolves were monitored. The results of these population surveys are presented in the table, right.

   (a) Plot a line graph for the tabulated results. Use one scale (on the left) for numbers of deer and another scale (on the right) for the number of wolves. Use different symbols or colors to distinguish the lines and include a key.

Field data notebook

Results of a population survey on an island

Time (year)	Wolf numbers	Deer numbers
1961	10	2000
1962	12	2300
1963	16	2500
1964	22	2360
1965	28	2244
1966	24	2094
1967	21	1968
1968	18	1916
1969	19	1952

(b) Study the line graph that you plotted and provide an explanation for the pattern in the data.

_____

_____

_____

© 2014 **BIOZONE** International
ISBN: 978-1-927173-84-8
Photocopying Prohibited

95 The Carrying Capacity of an Ecosystem

Key Idea: Carrying capacity is the maximum number of organisms a particular environment can support. Carrying capacity regulates population numbers.

Carrying capacity and population size

The **carrying capacity** is the maximum number of organisms of a given species a particular environment can support.

An ecosystem's carrying capacity, and therefore the maximum population size it can sustain, is limited by its resources. Factors affecting carrying capacity of an ecosystem can be biotic (e.g. food supply) or abiotic (e.g. water, climate, and available space).

The carrying capacity of an ecosystem is determined by the most limiting factor and can change over time (e.g. as a result of a change in food availability or a climate shift).

Below carrying capacity, population size increases because there are no limiting resources. As the population approaches carrying capacity (or passes above it) resources become limiting and environmental resistance increases, decreasing population growth.

Environmental resistance

Limiting factors:
Water, space, food

Stabilized population

Carrying capacity

Population size

Time

Factors affecting population size

Density dependent factors

The effect of these on population size is influenced by population density.

They include:
▶ Competition
▶ Predation
▶ Disease

Density dependent factors tend to be biotic and are less important when population density is low. They regulate population size by decreasing birth rates and increasing death rates.

Density independent factors

The effect of these on population size does not depend on population density.

They include catastrophic events such as:
▶ Volcanic eruptions, fire
▶ Drought, flood, tsunamis
▶ Earthquakes

Density independent factors tend to be abiotic.

They regulate population size by increasing death rates.

1. What is carrying capacity?_____

_____

2. How does carrying capacity limit population numbers? _____

_____

_____

3. What factors that influence carrying capacity? _____

_____

_____

96 A Case Study in Carrying Capacity

Key Idea: Predator-prey interactions are not always predictable because environmental factors influence the relationship.

When wolves were introduced to Coronation Island

Coronation Island is a small, 116 km^2 island off the coast of Alaska. In 1960, the Alaska Department of Fish and Game released two breeding pairs of wolves to the island. Their aim was to control the black-tailed deer that had been overgrazing the land. The results (below) were not what they expected. Introduction of the wolves initially appeared to have the desired effect. The wolves fed off the deer and successfully bred, and deer numbers fell. However, within a few years the deer numbers crashed. The wolves ran out of food (deer) and began eating each other, causing a drop in wolf numbers. Within eight years, only one wolf inhabited the island, and the deer were abundant. By 1983, there were no wolves on the island, and the deer numbers were high.

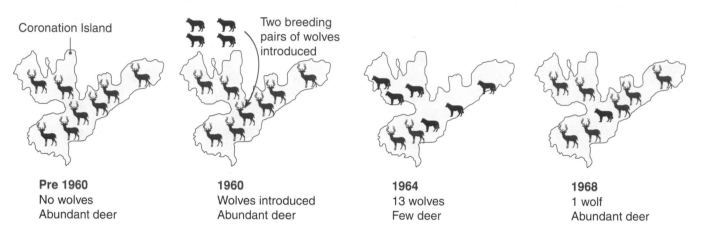

Pre 1960
No wolves
Abundant deer

1960
Wolves introduced
Abundant deer

1964
13 wolves
Few deer

1968
1 wolf
Abundant deer

What went wrong?

▶ The study showed Coronation Island was too small to sustain both the wolf and the deer populations.

▶ The deer could not easily find refuge from the wolves, so their numbers were quickly reduced.

▶ Reproductive rates in the deer may have been low because of poor quality forage following years of over-grazing. When wolves were introduced, predation and low reproductive rates, caused deer numbers to fall.

▶ The deer were the only food source for the wolves. When deer became scarce the wolves ate each other because there was no other prey available.

1. Why were wolves introduced to Coronation Island?

   _____

   _____

2. (a) What were some of the factors that caused the unexpected result?

   _____

   _____

   _____

   _____

   (b) What do these results tell you about the carrying capacity of Coronation Island? _____

   _____

   _____

   _____

 © 2014 **BIOZONE** International
ISBN: 978-1-927173-84-8
Photocopying Prohibited

97 Home Range Size in Dingoes

Key Idea: A home range is the area an animal normally inhabits. Home range size is influenced by the resources offered by the ecosystem.

Ecosystem and home range

The **home range** is the area where an animal normally lives and moves about in. An animal's home range can vary greatly in size. Animals that live in ecosystems rich in resources (e.g. good supply of food, water, shelter) tend to have smaller home ranges than animals that live in resource-poor ecosystems. This is because animals in a resource-poor ecosystem must cover a wider area to obtain the resources they need.

Dingo home ranges

Dingoes are found throughout Australia, in ecosystems as diverse as the tropical rainforests of the north, to the arid deserts of central Australia. The table (right) shows the home range sizes for dingo packs living in a variety of ecosystems. Some of the ecosystems in Australia in which dingoes are found are described below.

Dingo home range size in different ecosystems

Location (study site)	Ecosystem	Range (km²)
❶ Fortescue River, North-west Australia	Semi-arid, coastal plains and hills	77
❷ Simpson Desert, Central Australia	Arid, stony and sandy desert	67
❸ Kapalga, Kakadu N.P., North Australia	Tropical, coastal wet-lands and forests	39
❹ Harts Ranges, Central Australia	Semi-arid, river catch-ment and hills	25
❺ Kosciusko N.P., South-east Australia	Moist, cool forested mountains	21
❻ Georges Creek N.R., East Australia	Moist, cool forested tablelands (plateaux)	18
❼ Nadgee N.R., South-east Australia	Moist, cool coastal forests	10

Australian ecosystems

Temsabulta cc2.0

Arid: Little or no rain, and very dry. Very little, or no, vegetation grows. Often desert regions.

Semi arid: Rainfall is low, but sufficient to support some scrubby vegetation and grasses.

Cool forests: Moderate temperatures. Adequate water as rainfall or snow. Abundant vegetation.

Tropical forests: Warm regions with high rainfall. Abundant lush vegetation, including large trees.

1. Using the information on dingo home range size from the table above:

   (a) Name the two regions where home ranges were largest: _____

   (b) Name the two regions where home ranges were smallest: _____

   (c) Use the information provided on Australian ecosystems to explain how ecosystem type influences the home range of the dingo packs identified in (a) and (b):

   _____

   _____

   _____

   _____

   _____

98 Resources and Distribution

Key Idea: Within a home range, there are core areas or territories which animals vigorously defend because they contain the best resources.

Baboon home ranges in Nairobi Park

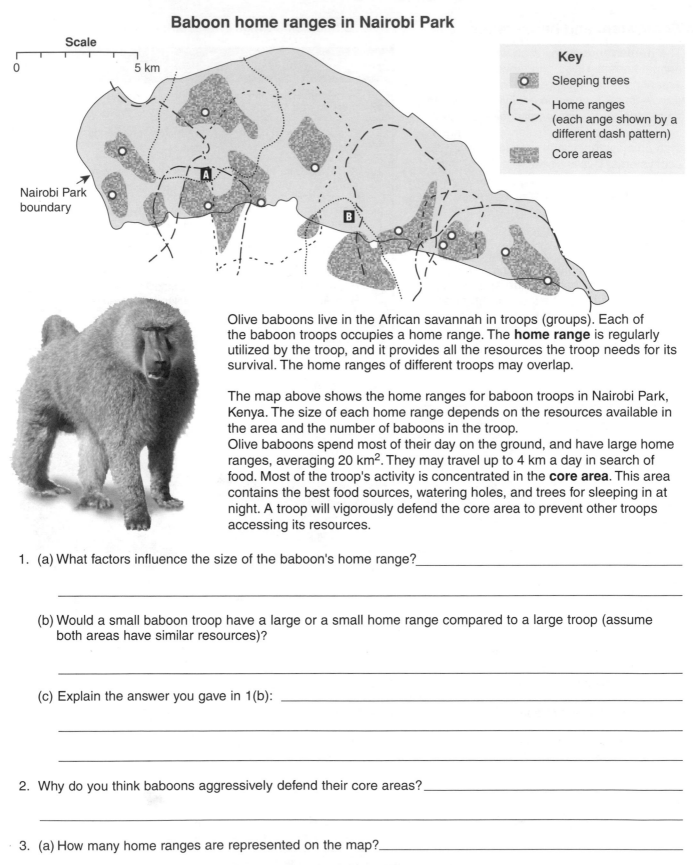

Scale

0 ——— 5 km

Nairobi Park boundary

Key

⊙ Sleeping trees

(⌐ ⌐) Home ranges (each ange shown by a different dash pattern)

▨ Core areas

Olive baboons live in the African savannah in troops (groups). Each of the baboon troops occupies a home range. The **home range** is regularly utilized by the troop, and it provides all the resources the troop needs for its survival. The home ranges of different troops may overlap.

The map above shows the home ranges for baboon troops in Nairobi Park, Kenya. The size of each home range depends on the resources available in the area and the number of baboons in the troop.
Olive baboons spend most of their day on the ground, and have large home ranges, averaging 20 km². They may travel up to 4 km a day in search of food. Most of the troop's activity is concentrated in the **core area**. This area contains the best food sources, watering holes, and trees for sleeping in at night. A troop will vigorously defend the core area to prevent other troops accessing its resources.

1. (a) What factors influence the size of the baboon's home range?_____

    _____

    (b) Would a small baboon troop have a large or a small home range compared to a large troop (assume both areas have similar resources)?

    _____

    (c) Explain the answer you gave in 1(b): _____

    _____

    _____

2. Why do you think baboons aggressively defend their core areas?_____

    _____

3. (a) How many home ranges are represented on the map?_____

    (b) How many home ranges overlap at the following points on the map?

    Point **A**: _____   Point **B**: _____

© 2014 **BIOZONE** International
ISBN: 978-1-927173-84-8
Photocopying Prohibited

99 Population Growth

Key Idea: Population growth can be exponential or logistic. The maximum sustainable population size is limited by the environment's carrying capacity.

Population growth

Population growth is the change in a population over time. It is regulated by the carrying capacity (the number of organisms an environment can support). There are two types of population growth:

▶ **Exponential growth** occurs when resources are unlimited (this rarely occurs). Growth is extremely rapid, but not sustainable. It produces a J-shaped growth curve.

▶ **Logistic growth** is characterized by a brief, early phase of exponential growth, followed by a slowing in growth as the population reaches carrying capacity. Logistic growth produces a S-shaped (sigmoidal) growth curve.

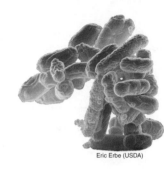

Exponential growth allows organisms to make the most of favorable environmental conditions, and rapidly expand population numbers. Bacteria (left) show exponential growth.

Eric Erbe (USDA)

Populations of large mammals, like this polar bear (right) show logistic growth. Their populations exist near the carrying capacity.

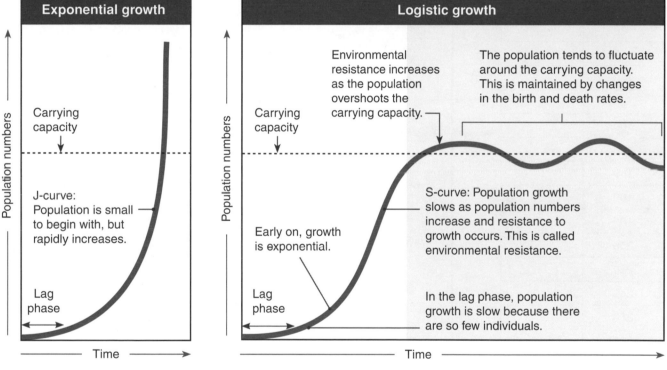

Exponential growth

Population numbers

Carrying capacity

J-curve: Population is small to begin with, but rapidly increases.

Lag phase

Time

Logistic growth

Population numbers

Carrying capacity

Environmental resistance increases as the population overshoots the carrying capacity.

The population tends to fluctuate around the carrying capacity. This is maintained by changes in the birth and death rates.

Early on, growth is exponential.

S-curve: Population growth slows as population numbers increase and resistance to growth occurs. This is called environmental resistance.

Lag phase

In the lag phase, population growth is slow because there are so few individuals.

Time

1. (a) Describe the features of exponential growth: _____

_____

(b) Why don't populations tend to continue to increase exponentially in an environment?_____

_____

2. (a) Describe the features of logistic growth: _____

_____

(b) What role does environmental resistance play in the growth of a population? _____

_____

© 2014 **BIOZONE** International
ISBN: 978-1-927173-84-8
Photocopying Prohibited

CONNECT CONNECT WEB

95 **2** **99** **KNOW**

100 Growth in a Bacterial Population

Key Idea: Population growth can be followed by plotting growth over time. Bacteria show exponential growth in an environment with unlimited resources.

In this activity, you will plot the growth of a hypothetical population of bacteria that reproduce every 20 minutes. The growth environment is maintained so that nutrients are not limiting.

1. Complete the table (below left) by doubling the number of bacteria for every 20 minute interval.

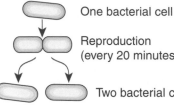

One bacterial cell

Reproduction (every 20 minutes)

Two bacterial cells

Time (mins)	Population size
0	1
20	2
40	4
60	8
80	
100	
120	
140	
160	
180	
200	
220	
240	
260	
280	
300	
320	
340	
360	

2. (a) Graph the results on the grid above.
   (b) Identify the **lag** and **exponential** phases of growth and mark them on the graph.

3. What shape is the curve you have plotted? _____

4. What would happen to the shape of the growth curve if no further nutrients were added? _____

_____

_____

 © 2014 **BIOZONE** International
ISBN: 978-1-927173-84-8
Photocopying Prohibited

101 A Case Study in Population Growth

Key Idea: Microbial growth is limited not only by the availability of resources, but also by the build up of toxic products.

The value of growing microbes

Many microbes (e.g. fungi and bacteria) produce commercially valuable products as they grow (e.g. medicines and food ingredients). The microbes must be kept in favorable growing conditions to maintain good microbial growth and keep production of these useful products high. For example, nutrients must be added, and toxic products removed from the growth solution.

The graph below shows a typical microbial growth curve for a population grown in a culture where no new nutrients are added, and toxic products are not removed.

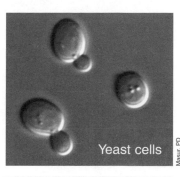

Yeast cells

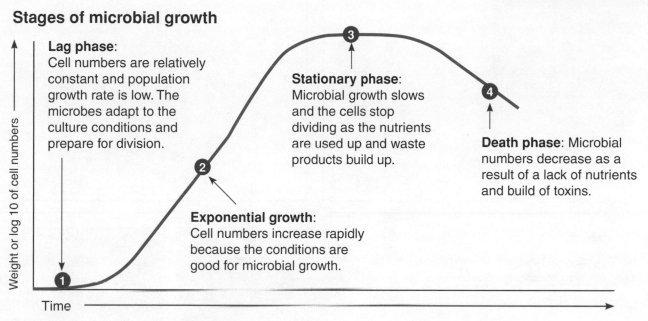

Stages of microbial growth

Lag phase:
Cell numbers are relatively constant and population growth rate is low. The microbes adapt to the culture conditions and prepare for division.

Stationary phase:
Microbial growth slows and the cells stop dividing as the nutrients are used up and waste products build up.

Death phase: Microbial numbers decrease as a result of a lack of nutrients and build of toxins.

Exponential growth:
Cell numbers increase rapidly because the conditions are good for microbial growth.

Weight or log 10 of cell numbers

Time

1. Describe what occurs during each of the four phases of microbial growth described above:

(a) Lag phase: _____

(b) Exponential phase: _____

(c) Stationary phase: _____

(d) Death phase: _____

2. (a) Why does microbial growth slow down and eventually stop?_____

(b) What would happen to the microbial growth curve if fresh nutrients were added and toxic products were removed just before the stationary phase?

CONNECT WEB

137 **101** **KNOW**

102 Chapter Review

Summarize what you know about this topic under the headings provided. You can draw diagrams or mind maps, or write short notes to organize your thoughts. Use the images and hints and guidelines included to help you:

Ecosystem structure

HINT: Include abiotic and biotic factors.

Species interactions

HINT: Describe the types of interactions between species. Is the interaction harmful or beneficial to the species involved?

Competition

HINT: Describe types of competition. How do species reduce competition?

REVISE

The carrying capacity of ecosystems

HINT: Include a definition of carrying capacity and describe factors that influence it.

Population growth

HINT: Describe exponential and logistic growth. How does carrying capacity limit population size?

Habitat and home range

HINT: Define home range and describe factors that influence habitat use and home range size.

103 KEY TERMS: Did You Get It?

1. Test your vocabulary by matching each term to its definition, as identified by its preceding letter code.

abiotic factor ....

biotic factor ....

carrying capacity ....

competition ....

density ....

ecosystem ....

habitat ....

home range ....

mutualism ....

niche ....

population ....

predation ....

A The number of individuals of a species per unit area (for land organisms) or volume (for aquatic organisms).

B The role of the organism in its environment, including its activities and interactions with other organisms.

C The maximum number of a specific organism that the environment can provide for.

D Exploitation in which one organism kills and eats another, usually of a different species.

E A term for any non-living part of the environment, e.g. rainfall, temperature.

F The area where an animal normally lives and moves about in.

G Any living component of the environment (or aspect of it) that has an effect on another organism in the environment.

H Community of interacting organisms and the environment (both biotic and abiotic) in which they both live and interact.

I An interaction between two species where both species benefit.

J The physical place or environment where an organism lives. It includes all the physical and biotic factors.

K The total number of individuals of a species within a set area.

L An interaction between organisms exploiting the same resource.

2. Study the graph of population growth for a hypothetical population below and answer the following questions:

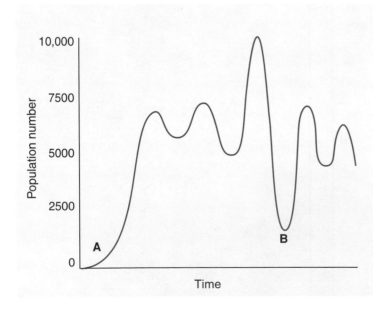

(a) Estimate the carrying capacity of the environment:

_____

(b) What happened at point **A**? _____

_____

_____

_____

(c) What happened at point **B**? _____

_____

_____

_____

_____

(d) What factors might have caused this? _____

_____

_____

VOCAB

LS2.B
PS3.D

Energy Flow and Nutrient Cycles

Key terms

anaerobic

aerobic

carbon cycle

cellular respiration

consumer

ecological pyramid

food chain

food web

hydrologic cycle

nitrogen cycle

oxygen cycle

photosynthesis

producer

trophic level

Core ideas

Energy flows through ecosystems

Activity number

☐ 1. **Photosynthesis** and **cellular respiration** provide most of the energy needed to carry out essential life processes. **104**

☐ 2. **Aerobic** respiration provides the useful energy in plants and animals. It requires oxygen to proceed. **105**

☐ 3. **Anaerobic** respiration is used by some bacteria to provide energy. It does not require oxygen to proceed. **105**

☐ 4. Organisms can be classified by how they obtain their energy (food). **Producers**, such as most plants and algae, make their own food. **Consumers** (e.g. animals) obtain their food from eating other organisms. **106 107**

☐ 5. A **food chain** provides a model of how energy, in the form of food, passes from one organism to another. Food chains can be connected to form **food webs**, which show all the feeding relationships within an ecosystem. **108 109 110**

☐ 6. Energy and matter are transferred through **trophic levels** (feeding levels). At each level, energy is lost from the system as heat, so ecosystems must receive a continuous input of light energy to sustain them. At each level in an ecosystem, matter and energy are conserved. **111**

☐ 7. **Ecological pyramids** provide a model the number of organisms, amount of energy, or biomass. **112**

Energy flows through ecosystems but matter is recycled

☐ 8. Chemical elements cycle through the ecosystem. Important nutrient cycles include the **hydrologic** (water), **carbon**, **oxygen**, and **nitrogen cycles**. **113-118**

☐ 9. Photosynthesis and cellular respiration have important roles in the cycling of carbon. **117**

Science and engineering practices

☐ 1. Construct an explanation based on evidence for how matter cycles and energy flows in aerobic and anaerobic systems. **104**

☐ 2. Use a model based on evidence to show how energy flows through an ecosystem. **109 110**

☐ 3. Use mathematical representations to show how energy flows and matter cycles through an ecosystem. **111 112 117**

☐ 4. Develop and use a model to illustrate the role of cellular respiration and photosynthesis in the cycling of carbon. **117**

BIOZONE APP
Student Review Series
Energy Flow and
Nutrient Cycles

104 Comparing Aerobic and Anaerobic Systems

Key Idea: Aerobic and anaerobic systems produce usable energy in different ways. Both systems are important in the cycling of matter through ecosystems.

Aerobic and anaerobic systems are important in producing **usable** energy in living systems. The reactions carried out during respiration move electrons from molecule to molecule. Each move releases energy, which is stored in chemical bonds and can be used to power essential chemical reactions in the cell.

Aerobic and anaerobic systems

An aerobic system: the breakdown of glucose

Aerobic systems, (e.g. cellular respiration, shown right) use oxygen as the final electron acceptor. Oxygen accepts electrons and joins with hydrogen forming water. Carbon dioxide is released earlier in the process.

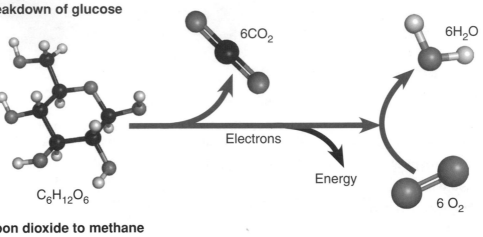

$C_6H_{12}O_6$

$6CO_2$

Electrons

Energy

$6H_2O$

$6 O_2$

An anaerobic system: carbon dioxide to methane

There are many different anaerobic systems used by living organisms to produce usable energy. Methanogenesis (right) is a form of anaerobic respiration found in methanogenic bacteria in the stomach of ruminants (e.g. cows) and many other animals.

CO_2

$4H_2$

$2H_2O$

Electrons

Energy

CH_4

Cycling matter

Nitrates are essential for plant growth. Some bacteria can convert nitrogen gas to nitrate. Other bacteria convert nitrates back into nitrogen. Both processes are anaerobic (do not need oxygen).

Aerobic respiration, the conversion of glucose and O_2 into CO_2 and water, is part of the carbon and oxygen cycles. The CO_2 and water are converted back into glucose and O_2 by the anaerobic process of photosynthesis.

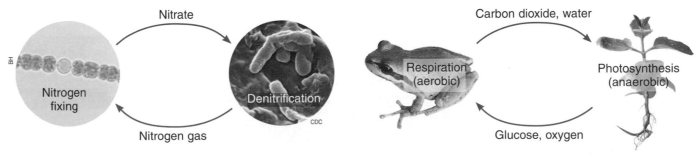

Nitrate

Nitrogen fixing

Denitrification

CDC

Nitrogen gas

Carbon dioxide, water

Respiration (aerobic)

Photosynthesis (anaerobic)

Glucose, oxygen

1. What is the main difference between an aerobic and anaerobic system? _____

_____

2. How are aerobic and anaerobic systems involved in nutrient cycling? _____

_____

_____

_____

© 2014 **BIOZONE** International
ISBN: 978-1-927173-84-8
Photocopying Prohibited

105 Energy in Ecosystems

Key Idea: Photosynthesis and cellular respiration provide most of the energy required for essential life processes.

Where does the energy for life processes come from?

Photosynthesis and **cellular respiration** provide most of the usable energy for life's essential processes. These processes are also involved in energy flow and nutrient cycling through an ecosystem.

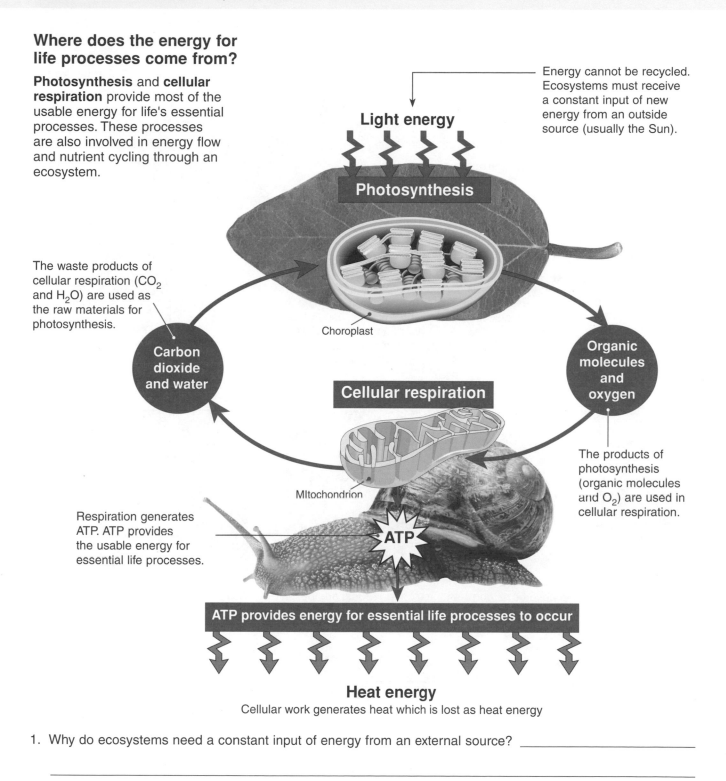

Energy cannot be recycled. Ecosystems must receive a constant input of new energy from an outside source (usually the Sun).

Light energy

Photosynthesis

Choroplast

The waste products of cellular respiration (CO_2 and H_2O) are used as the raw materials for photosynthesis.

Carbon dioxide and water

Cellular respiration

Organic molecules and oxygen

Mitochondrion

The products of photosynthesis (organic molecules and O_2) are used in cellular respiration.

Respiration generates ATP. ATP provides the usable energy for essential life processes.

ATP

ATP provides energy for essential life processes to occur

Heat energy
Cellular work generates heat which is lost as heat energy

1. Why do ecosystems need a constant input of energy from an external source? _____

_____

2. How do photosynthesis and cellular respiration interact to cycle matter through an ecosystem?

_____

_____

_____

© 2014 **BIOZONE** International
ISBN: 978-1-927173-84-8
Photocopying Prohibited

CONNECT CONNECT

80 **76** **KNOW**

106 Producers

Key Idea: Producers (autotrophs) make their own food. Most producers utilize the energy from the sun to do this, but some organisms use chemical energy.

What is a producer?

▶ A **producer** is an organism that can make its own food.

▶ Producers are also called **autotrophs**, which means self feeding.

▶ Plants, algae, and some bacteria are producers.

▶ Most producers are **photo**autotrophs, and use the energy in sunlight to make their food. The process by which they do this is called **photosynthesis**.

▶ Some producers are **chemo**autotrophs and use the chemical energy in inorganic molecules (e.g. hydrogen sulfide) to make their food.

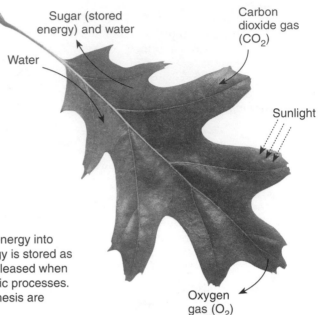

Sugar (stored energy) and water

Water

Carbon dioxide gas (CO_2)

Sunlight

Oxygen gas (O_2)

Photosynthesis transforms sunlight energy into chemical energy. The chemical energy is stored as sugar (glucose), and the energy is released when the sugar undergoes further metabolic processes. The inputs and outputs of photosynthesis are shown on the leaf diagram (right).

Photosynthesis by marine algae provides oxygen and absorbs carbon dioxide. Most algae are microscopic but some, like this kelp, are large.

On land, vascular plants (plants with transport tissues) are the main producers of food.

Producers make their own food, and they are also the ultimate source of food and energy for all consumers.

1. (a) What is a producer? _____

_____

(b) Name some organisms that are producers: _____

_____

2. Where do producers get their energy from? _____

_____

3. Why are producers so important in an ecosystem? _____

_____

_____

© 2014 **BIOZONE** International
ISBN: 978-1-927173-84-8
Photocopying Prohibited

107 Consumers

Key Idea: Consumers (heterotrophs) are organisms that cannot make their own food. They get their food by consuming other organisms.

What is a consumer?

Consumers (heterotrophs) are organisms that cannot make their own food and must get their food by consuming other organisms (by eating or extracellular digestion). Animals, fungi, and some bacteria are consumers. Consumers are categorized according to where they get their energy from (below).

Consumers need producers

Consumers rely on producers for survival, even if they do not consume them directly. Herbivores, such as the rabbit below, gain their energy by eating plants. Although higher level consumers, such as the eagle, may feed off herbivores, they still ultimately rely on plants to sustain them. Without the plants the rabbit would not survive, and the eagle could not eat it.

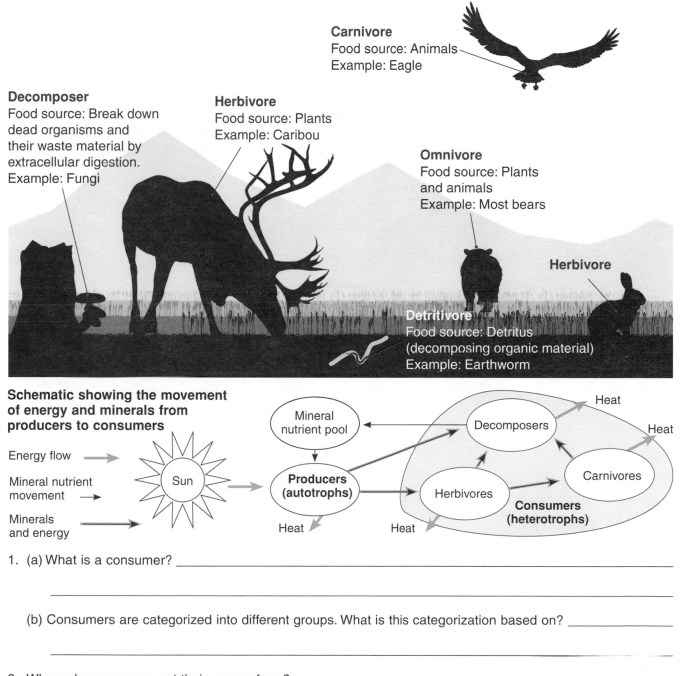

Carnivore
Food source: Animals
Example: Eagle

Decomposer
Food source: Break down dead organisms and their waste material by extracellular digestion.
Example: Fungi

Herbivore
Food source: Plants
Example: Caribou

Omnivore
Food source: Plants and animals
Example: Most bears

Herbivore

Detritivore
Food source: Detritus (decomposing organic material)
Example: Earthworm

Schematic showing the movement of energy and minerals from producers to consumers

Energy flow →

Mineral nutrient movement →

Minerals and energy →

Sun → Producers (autotrophs) → Herbivores → Carnivores

Mineral nutrient pool ← Decomposers

Consumers (heterotrophs)

Heat

1. (a) What is a consumer? _____

   _____

   (b) Consumers are categorized into different groups. What is this categorization based on? _____

   _____

2. Where do consumers get their energy from? _____

3. In the schematic system above, how are the movements of minerals and energy different? _____

   _____

108 Food Chains

Key Idea: A food chain is a model to illustrate the feeding relationships between organisms.

Food chains

Organisms in ecosystems interact in their feeding relationships. These interactions can be shown in a **food chain**, which is a simple model to illustrate how energy, in the form of food, passes from one organism to the next. Each organism in the chain is a food source for the next.

Trophic levels

The levels of a food chain are called **trophic** (feeding) **levels**. An organism is assigned to a trophic level based on its position in the food chain. Organisms may occupy different trophic levels in different food chains or during different stages of their life.

Arrows link the organisms in a food chain. The direction of the arrow shows the flow of energy through the trophic levels. At each link, energy is lost (as heat) from the system. This loss of energy limits how many links can be made. Most food chains begin with a producer, which is eaten by a primary consumer (herbivore). Second (and higher) level consumers eat other consumers, as shown below.

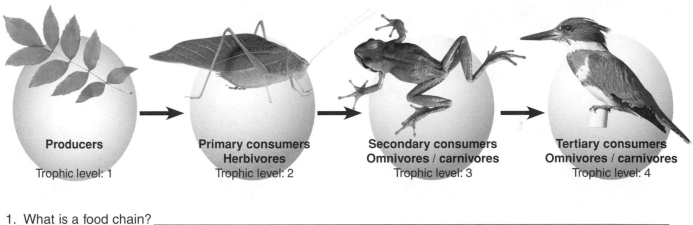

Producers	**Primary consumers** **Herbivores**	**Secondary consumers** **Omnivores / carnivores**	**Tertiary consumers** **Omnivores / carnivores**
Trophic level: 1	Trophic level: 2	Trophic level: 3	Trophic level: 4

1. What is a food chain? _____

_____

_____

2. (a) A simple food chain for a cropland ecosystem is pictured below. Label the organisms with their trophic level and trophic status (e.g. primary consumer).

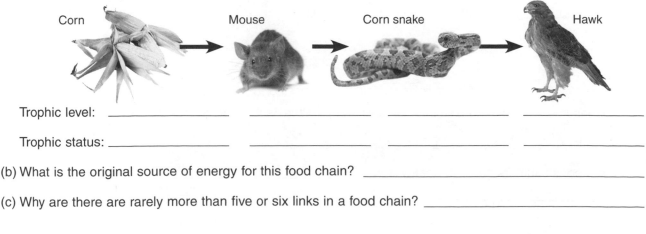

Corn   Mouse   Corn snake   Hawk

Trophic level: _____ _____ _____ _____

Trophic status: _____ _____ _____ _____

(b) What is the original source of energy for this food chain? _____

(c) Why are there are rarely more than five or six links in a food chain? _____

_____

© 2014 **BIOZONE** International
ISBN: 978-1-927173-84-8
Photocopying Prohibited

109 Food Webs

Key Idea: A food web consists of all the food chains in an ecosystem. Food webs show the complex feeding relationships between all the organisms in a community.

▶ If we show all the connections between all the food chains in an ecosystem, we can create a web of interactions called a **food web.** A food web is a model to illustrate the feeding relationships between all the organisms in a community.

▶ The complexity of a food web depends on the number of different food chains contributing to it. A simple ecosystem, with only a few organisms (and therefore only a few food chains) will have a simpler food web than an ecosystem that has many different food chains.

▶ A food web model, like the one shown below, can be used to show the linkages between different organisms in a community.

A simple food web for a lake ecosystem

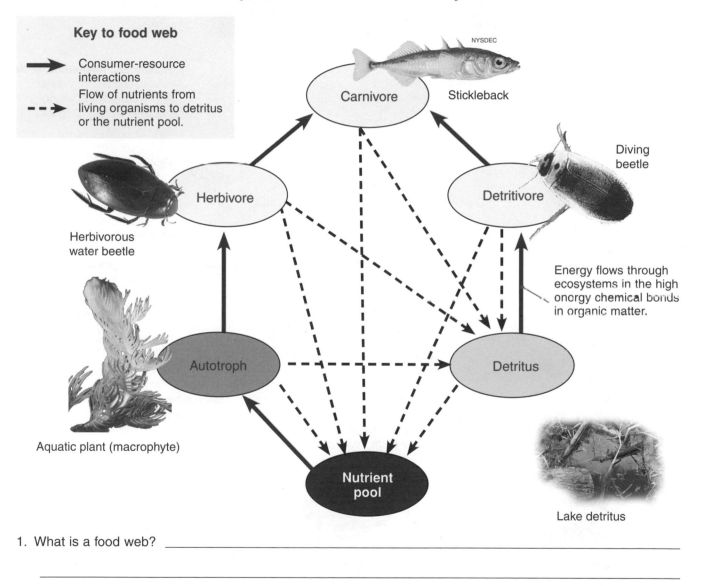

Key to food web

→ Consumer-resource interactions

- -▶ Flow of nutrients from living organisms to detritus or the nutrient pool.

NYSDEC

Carnivore   Stickleback

Herbivore   Detritivore   Diving beetle

Herbivorous water beetle

Autotroph   Detritus

Energy flows through ecosystems in the high energy chemical bonds in organic matter.

Aquatic plant (macrophyte)

Nutrient pool

Lake detritus

1. What is a food web? _____

_____

_____

2. Why would an ecosystem with only a few different types of organisms have a less complex food web than an ecosystem with many different types of organisms?

_____

_____

_____

CONNECT   WEB

111 109 **KNOW**

110 Constructing Food Webs

Key Idea: Knowing what the inhabitants of an ecosystem feed on allows food chains to be constructed. Food chains can be used to construct a food web.

The organisms below are typical of those found in many lakes. For simplicity, only a few organisms are represented here. Real lake communities have hundreds of different species interacting together. Your task is to assemble the organisms below into a food web in a way that shows how they are interconnected by their feeding relationships.

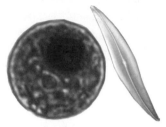

Autotrophic protists (algae)
Chlamydomonas (above left), and some diatoms (above right) photosynthesize.

Macrophytes
Aquatic green plants photosynthesize.

Detritus
Decaying organic matter.

Asplanchna (planktonic rotifer)
A large, carnivorous rotifer. Diet: Protozoa and young zooplankton (e.g. *Daphnia*).

Daphnia (zooplankton)
Small freshwater crustacean. Diet: Planktonic algae.

Leech
Fluid feeding predators. Diet: Small invertebrates, including rotifers, small pond snails, and worms.

Three-spined stickleback
Common in freshwater ponds and lakes. Diet: Small invertebrates such as *Daphnia* and insect larvae.

Diving beetle
Diet: Aquatic insect larvae and adult insects. The will also eat detritus collected from the bottom mud.

Common carp
Diet: Mainly feeds on bottom living insect larvae and snails, but will also eat some plant material (not algae).

Dragonfly larva
Large aquatic insect larvae. Diet: Small invertebrates including *Hydra*, *Daphnia*, insect larvae, and leeches.

Great pond snail
Diet: Omnivorous. Main diet is macrophytes but will eat decaying plant and animal material also.

Herbivorous water beetle
Diet: Adults feed on macrophytes. Young beetle larvae are carnivorous, feeding primarily on pond snails.

Protozan (e.g. *Paramecium*)
Diet: Mainly bacteria and microscopic green algae such as *Chlamydomonas*.

Pike
Diet: Smaller fish and amphibians. They are also opportunistic predators of rodents and small birds.

Mosquito larva
Diet: Planktonic algae.

Hydra
A small, carnivorous cnidarian. Diet: small *Daphnia* and insect larvae.

© 2014 **BIOZONE** International
ISBN: 978-1-927173-84-8
Photocopying Prohibited

1. From the information provided for the lake food web components on the previous page, construct ten different **food chains** to show the feeding relationships between the organisms. Some food chains may be shorter than others and most species will appear in more than one food chain. An example has been completed for you.

Example 1: Macrophyte ⟶ Herbivorous water beetle ⟶ Carp ⟶ Pike

(a) _____

(b) _____

(c) _____

(d) _____

(e) _____

(f) _____

(g) _____

(h) _____

(i) _____

(j) _____

2. Use the food chains you created above to help you to draw up a **food web** for this community in the box below. Use the information supplied on the previous page to draw arrows showing the flow of energy between species (only energy from the detritus is required).

Tertiary and higher level consumers

Pike   Carp

Tertiary consumers

Dragonfly larva   Three-spined stickleback

Hydra   Diving beetle   Leech

Secondary consumers

Mosquito larva   Asplanchna

Primary consumers

Daphnia   Paramecium   Herbivorous water beetle (adult)   Great pond snail

Producers

Autotrophic protists   Macrophytes

Detritus and bacteria

© 2014 **BIOZONE** International
ISBN: 978-1-927173-84-8

111 Energy Flow in an Ecosystem

Key Idea: Energy flows through an ecosystem from one trophic level to the next. Only 5-20% of energy is transferred from one trophic level to the next.

Energy is transferred from one trophic (feeding) level to the next. At each level, some of the energy is transformed to sustain essential life process, some is given out as heat, and some is transferred to the next level. These energy transfers can be represented diagrammatically using arrows of different sizes (see the diagrams below). Bigger arrows represent more energy.

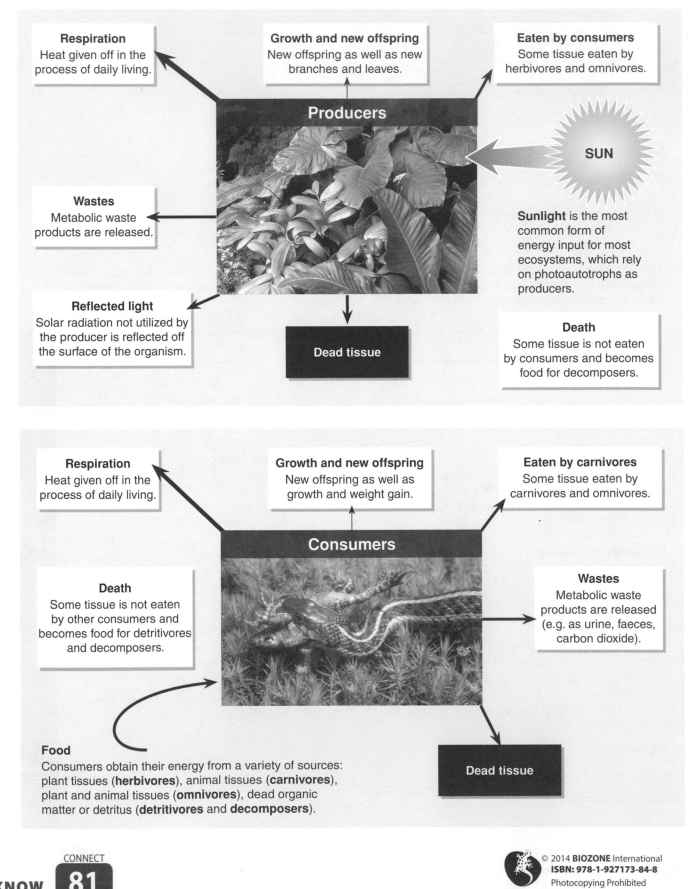

Respiration
Heat given off in the process of daily living.

Growth and new offspring
New offspring as well as new branches and leaves.

Eaten by consumers
Some tissue eaten by herbivores and omnivores.

Producers

SUN

Sunlight is the most common form of energy input for most ecosystems, which rely on photoautotrophs as producers.

Wastes
Metabolic waste products are released.

Reflected light
Solar radiation not utilized by the producer is reflected off the surface of the organism.

Dead tissue

Death
Some tissue is not eaten by consumers and becomes food for decomposers.

Respiration
Heat given off in the process of daily living.

Growth and new offspring
New offspring as well as growth and weight gain.

Eaten by carnivores
Some tissue eaten by carnivores and omnivores.

Consumers

Death
Some tissue is not eaten by other consumers and becomes food for detritivores and decomposers.

Wastes
Metabolic waste products are released (e.g. as urine, faeces, carbon dioxide).

Dead tissue

Food
Consumers obtain their energy from a variety of sources: plant tissues (**herbivores**), animal tissues (**carnivores**), plant and animal tissues (**omnivores**), dead organic matter or detritus (**detritivores** and **decomposers**).

© 2014 **BIOZONE** International
ISBN: 978-1-927173-84-8
Photocopying Prohibited

Conservation of energy and trophic efficiency

▶ The Law of Conservation of Energy states that energy cannot be created or destroyed, only transformed from one form (e.g. light energy) to another (e.g. chemical energy in the bonds of molecules). Each time energy is transferred (as food) from one trophic level to the next, some energy is given out as heat, usually during cellular respiration. This means the amount of energy available to the next trophic level is less than at the previous level.

▶ Potentially, we can account for the transfer of energy from its input (as solar radiation) to its release as heat from organisms, because energy is conserved. The percentage of energy transferred from one trophic level to the next is the **trophic efficiency**. It varies between 5% and 20% and measures the efficiency of energy transfer. An average figure of 10% trophic efficiency is often used. This is called the **ten percent rule** (below).

The ten percent rule

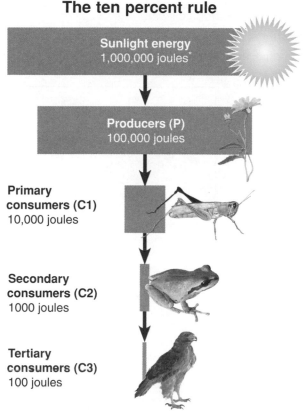

Sunlight energy
1,000,000 joules*

Producers (P)
100,000 joules

Primary consumers (C1)
10,000 joules

Secondary consumers (C2)
1000 joules

Tertiary consumers (C3)
100 joules

*Note: joules are units of energy

Calculating available energy

The energy available to each trophic level will equal the amount entering that trophic level, minus total losses from that level (energy lost as heat + energy lost to detritus).

Heat energy is lost from the ecosystem to the atmosphere. Other losses become part of the detritus and may be utilized by other organisms in the ecosystem.

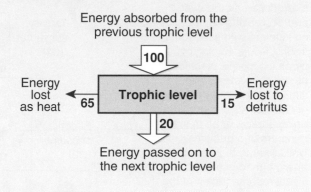

Energy absorbed from the previous trophic level

100

Energy lost as heat **65** | **Trophic level** | **15** Energy lost to detritus

20

Energy passed on to the next trophic level

1. Why is the energy available to a particular trophic level less than the energy in the previous trophic level?

_____

_____

_____

_____

_____

_____

_____

2. Why must ecosystems receive a continuous supply of energy from the Sun?

_____

_____

3. The diagram on the left shows the energy available to trophic levels in a hypothetical ecosystem. For each trophic level, calculate the amount of energy that can be passed onto the next level. Write you answer in the spaces provided (a-d).

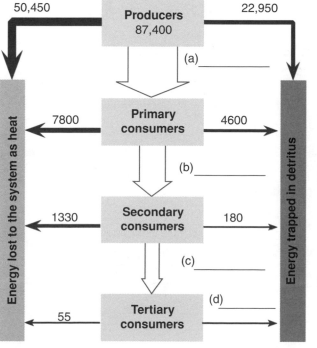

50,450 | **Producers** 87,400 | 22,950

(a) _____

Energy lost to the system as heat

7800 | **Primary consumers** | 4600

(b) _____

1330 | **Secondary consumers** | 180

Energy trapped in detritus

(c) _____

(d) _____

55 | **Tertiary consumers**

© 2014 **BIOZONE** International
ISBN: 978-1-927173-84-8
Photocopying Prohibited

112 Ecological Pyramids

Key Idea: Ecological pyramids are used to illustrate the number of organisms, amount of energy, or amount of biomass at each trophic level in an ecosystem.

Types of ecological pyramids

The energy, biomass, or numbers of organisms at each trophic level in any ecosystem can be represented by an ecological pyramid. The first trophic level is placed at the bottom of the pyramid and subsequent trophic levels are stacked on top in their 'feeding sequence'. Ecological pyramids provide a convenient model to illustrate the relationship between different trophic levels in an ecosystem.

▶ Pyramid of numbers shows the numbers of individual organisms at each trophic level.

▶ Pyramid of biomass measures the mass of the biological material at each trophic level.

▶ Pyramid of energy shows the energy contained within each trophic level. Pyramids of energy and biomass are usually quite similar in appearance.

This generalized ecological pyramid (right) shows a conventional pyramid shape, with a large number of producers at the base, and decreasing numbers of consumers with subsequent trophic levels. Ecological pyramids for this plankton-based ecosystem have a similar appearance regardless of whether we construct them using energy, or biomass, or numbers of organisms. Units refer to biomass or energy. Images provide a visual representation of the organisms present.

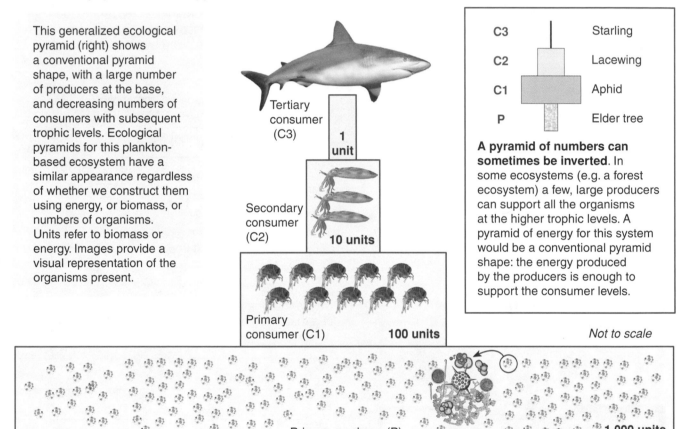

Tertiary consumer (C3) **1 unit**

Secondary consumer (C2) **10 units**

Primary consumer (C1) **100 units**

Primary producer (P) **1,000 units**

Marine amphipod image: uwe Kils-cc3.0

Not to scale

C3		Starling
C2		Lacewing
C1		Aphid
P		Elder tree

A pyramid of numbers can sometimes be inverted. In some ecosystems (e.g. a forest ecosystem) a few, large producers can support all the organisms at the higher trophic levels. A pyramid of energy for this system would be a conventional pyramid shape: the energy produced by the producers is enough to support the consumer levels.

1. What do each of the following types of ecological pyramids measure?

    (a) Number pyramid: _____

    (b) Biomass pyramid: _____

    (c) Energy pyramid: _____

2. (a) Why are some ecological pyramids not a typical pyramid shape? _____

    _____

    (b) Would a pyramid of energy ever have an inverted (upturned) shape? Explain your answer:

    _____

    _____

    _____

 © 2014 **BIOZONE** International
ISBN: 978-1-927173-84-8
Photocopying Prohibited

113 Cycles of Matter

Key Idea: Matter cycles through the biotic and abiotic compartments of Earth's ecosystems. These cycles are called nutrient cycles or biogeochemical cycles.

Nutrients cycle through ecosystems

▶ Nutrient cycles move and transfer chemical elements (e.g. carbon, hydrogen, nitrogen, and oxygen) through an ecosystem. Because these elements are part of many essential nutrients, their cycling is called a **nutrient cycle**, or a **biogeochemical cycle**. The term biogeochemical means that **bio**logical, **geo**logical, and **chemical** processes are involved in nutrient cycling.

▶ In a nutrient cycle, the nutrient passes through the biotic (living) and abiotic (physical) components of an ecosystem (see diagram below). Matter is conserved throughout all these transformations, although it may pass from one ecosystem to another.

Flow through a generalized biogeochemical cycle

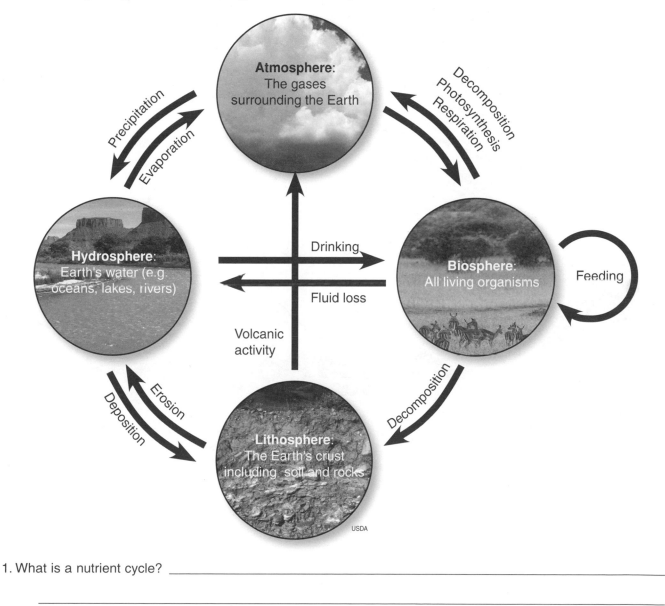

1. What is a nutrient cycle? _____

_____

2. Why do you think it is important that matter is cycled through an ecosystem? _____

_____

_____

CONNECT
104

WEB
113

KNOW

114 The Hydrologic Cycle

Key Idea: The hydrologic cycle results from the cycling of water from the oceans to the land and back.

▶ About 97% of the water on Earth is stored in the oceans, which contain more than 1.3 billion cubic kilometres of water. Less than 1% of Earth's water is freely available fresh water (in lakes and streams).

▶ Water evaporates from the oceans and lakes into the atmosphere and falls as rain (or snow and hail). Rain that falls on the land is transported back to the oceans by rivers and streams, or is returned to the atmosphere by evaporation or transpiration.

▶ Water can cycle very quickly if it remains near the Earth's surface, but in some circumstances it can remain locked away for hundreds or even thousands of years (e.g. in deep ice layers at the poles or in ground water).

▶ Humans intervene in the water cycle by using water for their own needs. Irrigation from rivers and lakes changes evaporation patterns, lowers lake levels, and reduces river flows.

Water is only substance on Earth that can be found naturally as a solid, liquid, or gas. It has the unique property of being less dense as a solid than a liquid, causing water to freeze from the top down (and float), and it has an unexpectedly high boiling point compared to other similar molecules.

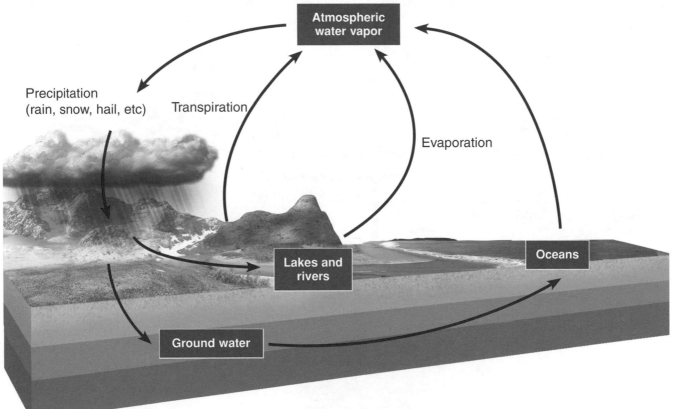

1. The main storage reservoir for water on Earth is: _____

2. Name the two processes by which water moves from the land or oceans to the atmosphere: _____

_____

3. Identify a feature of water that allows it cycle as described above: _____

_____

_____

_____

 © 2014 **BIOZONE** International
ISBN: 978-1-927173-84-8
Photocopying Prohibited

115 The Carbon Cycle

Key Idea: All life is carbon-based. Carbon cycles between the atmosphere, biosphere, geosphere, and hydrosphere. Photosynthesis and respiration are central to this.

▸ Carbon is the essential element of life. Its unique properties allow it to form an almost infinite number of different molecules. In living systems, the most important of these are carbohydrates, fats, nucleic acids, and proteins.

▸ Carbon in the atmosphere is found as carbon dioxide (CO_2). In rocks, it is most commonly found as either coal (mostly carbon) or limestone (calcium carbonate).

▸ The most important processes in the carbon cycle are photosynthesis and respiration.

▸ Photosynthesis removes carbon from the atmosphere and converts it to organic molecules. This organic carbon may eventually be returned to the atmosphere through respiration.

▸ Carbon cycles at different rates depending on where it is. On average, carbon remains in the atmosphere as CO_2 for about 5 years, in plants and animals for about 10 years, and in oceans for about 400 years. Carbon can remain in rocks (e.g. coal) for millions of years.

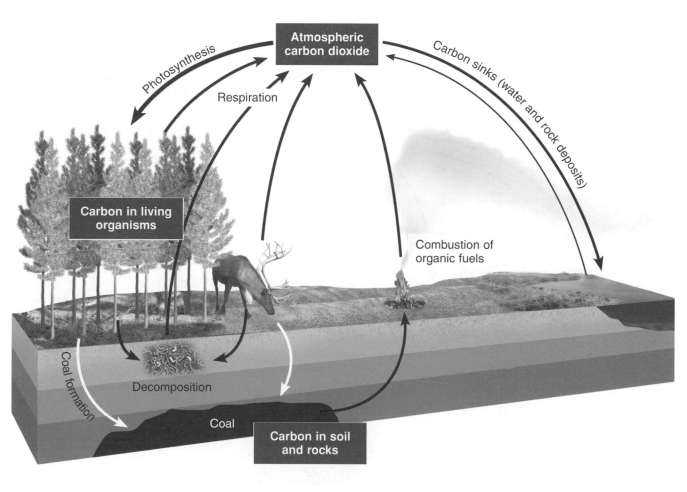

1. (a) In what form is carbon found in the atmosphere? _____

   (b) In what three important molecules is carbon found in living systems? _____

   (c) In what two forms is carbon found in rocks? _____

2. (a) Name two processes that remove carbon from the atmosphere: _____

   (b) Name two processes that add carbon to the atmosphere: _____

3. What is the effect of deforestation and burning of coal and oil on carbon cycling?_____

   _____

   _____

116 The Oxygen Cycle

Key Idea: The oxygen cycle describes the movement of oxygen (O_2) through an ecosystem. The oxygen cycle is closely linked to the carbon cycle.

The importance of the oxygen cycle

The **oxygen cycle** describes the movement of oxygen between the biotic and abiotic components of ecosystems. Photosynthesis is the main source of oxygen in ecosystems.

Oxygen is involved to some degree in all the other biogeochemical cycles, but is closely linked to the carbon cycle in particular. This is because most producers utilize carbon dioxide in photosynthesis, and produce oxygen as a waste product. The oxygen is used in cellular respiration and carbon dioxide is produced as a waste product. This is the oxygen-carbon dioxide cycle (simplified in the diagram, right).

The link between the oxygen and carbon cycles

The oxygen cycle

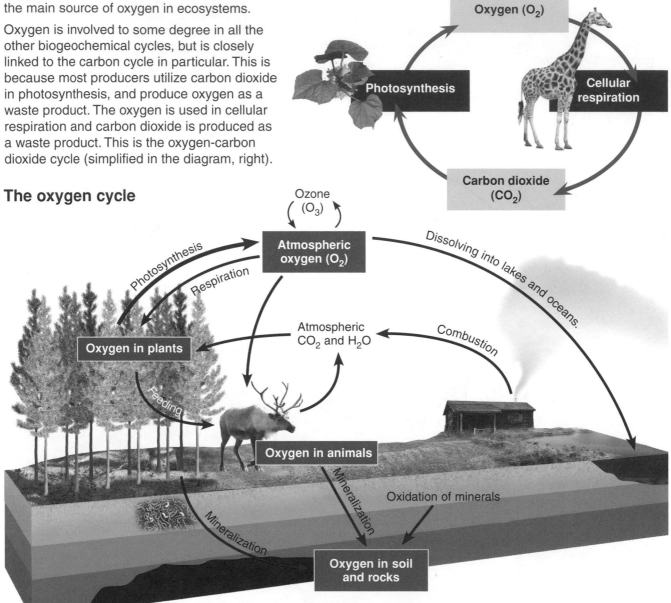

1. What is the main source of oxygen for the oxygen cycle? _____

_____

2. Why are the oxygen cycle and carbon cycle so interdependent?_____

_____

_____

_____

_____

© 2014 **BIOZONE** International
ISBN: 978-1-927173-84-8
Photocopying Prohibited

117 Role of Photosynthesis in Carbon Cycling

Key Idea: Photosynthesis removes carbon from the atmosphere, and adds it to the biosphere. Respiration removes carbon from the biosphere and adds it the atmosphere.

Photosynthesis and carbon

▶ Photosynthesis removes carbon from the atmosphere by fixing the carbon in CO_2 into carbohydrate molecules. Plants use the carbohydrates (e.g. glucose) to build structures such as wood.

▶ Some carbon may be returned to the atmosphere during respiration (either from the plant or from animals). If the amount or rate of carbon fixation is greater than that released during respiration then carbon will build up in the biosphere and be reduced in the atmosphere (diagram, right).

Respiration and carbon

▶ Cellular respiration releases carbon into the atmosphere as carbon dioxide as a result of the breakdown of glucose.

▶ If the rate of carbon release is greater than that fixed by photosynthesis then over time carbon may accumulate in the atmosphere (diagram bottom right). Before the Industrial Revolution, many thousands of gigatonnes (Gt) of carbon were contained in the biosphere of in the Earth's crust (e.g. coal).

▶ Deforestation and the burning of fossil fuels have increased the amount of carbon in the atmosphere.

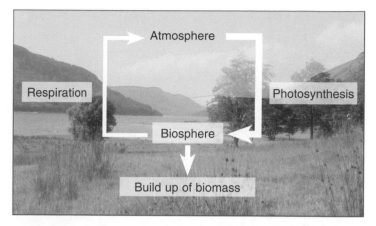

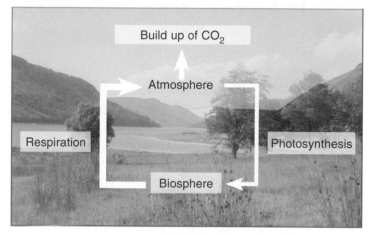

Carbon cycling simulation

Plants move about 120 Gt of carbon from the atmosphere to the biosphere a year. Respiration accounts for about 60 Gt of carbon a year. A simulation was carried out to study the effect of varying the rates of respiration and photosynthesis on carbon deposition in the biosphere or atmosphere. To keep the simulation simple, only the effects to the atmosphere and biosphere were simulated. Effects such as ocean deposition and deforestation were not studied. The results are shown in the tables right and below.

Table 1: Rate of photosynthesis equals the rate of cellular respiration.

Years	Gt carbon in biosphere	Gt carbon in atmosphere
0	610	600
20	608	600
40	608	600
60	609	598
80	612	598
100	610	596

Table 2: Rate of photosynthesis increases by 1 Gt per year.

Years	Gt carbon in biosphere	Gt carbon in atmosphere
0	610	600
20	632	580
40	651	558
60	671	538
80	691	518
100	710	498

Table 3: Rate of cellular respiration increases by 1 Gt per year.

Years	Gt carbon in biosphere	Gt carbon in atmosphere
0	610	600
20	590	619
40	570	641
60	548	664
80	528	686
100	509	703

CONNECT **82** CONNECT **81** CONNECT **77** KNOW

1. Plot the data for tables 1,2, and 3 on the grid provided (above). Include a key and appropriate titles and axes.

2. (a) What is the effect of increasing the rate of photosynthesis on atmospheric carbon? _____

   _____

   (b) i. What is the effect of increasing the rate of photosynthesis on biospheric carbon?_____

      _____

      ii. How does this effect occur? _____

      _____

      _____

3. What is the effect of increasing the rate of cellular respiration on atmospheric and biospheric carbon?

   _____

   _____

4. In the real world, respiration is not necessarily increasing in comparison to photosynthesis, but many human activities cause the same effect.

   (a) Name two human activities that have the same effect on atmospheric carbon as increasing the rate of cellular respiration:

   _____

   (b) What effect does this extra atmospheric carbon have on the global climate? _____

   _____

© 2014 **BIOZONE** International
ISBN: 978-1-927173-84-8
Photocopying Prohibited

118 The Nitrogen Cycle

Key Idea: Nitrogen is essential for building proteins. Nitrogen gas is converted to nitrates, which are taken up by plants. Animals gain nitrogen by feeding off plants or animals.

▶ Nearly eighty percent of the Earth's atmosphere is made of nitrogen gas. As a gas nitrogen is very stable and unreactive, effectively having no interaction with living systems. However, nitrogen is extremely important in the formation of amino acids, which are the building blocks of proteins.

▶ Nitrogen may enter the biosphere during lightning storms. Lightning produces extremely high temperatures in the air (around 30,000 °C). At such high temperatures, nitrogen reacts with oxygen in the air to form ammonia and nitrates which dissolve in water and are washed into the soil.

▶ Some bacteria can fix nitrogen directly from the air. Some of these bacteria are associated with plants (especially legumes) and produce ammonia (NH_3). This can be converted to nitrates (NO_3^-) by other bacteria. Other bacteria produce nitrites (NO_2^-).

▶ Nitrates are absorbed and used by plants to make amino acids. Animals gain their nitrogen by feeding on plants (or on herbivores).

▶ Nitrogen is returned to the atmosphere by denitrifying bacteria which convert nitrates back into nitrogen gas.

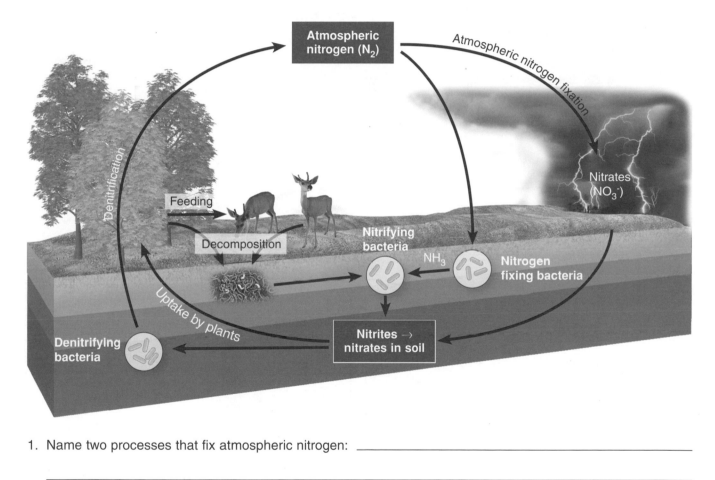

1. Name two processes that fix atmospheric nitrogen: _____

   _____

2. What process returns nitrogen to the atmosphere?_____

3. What essential organic molecule does nitrogen help form? _____

4. Explain why farmers often plant legumes between cropping seasons, then plow them into the soil rather than harvest them?

   _____

   _____

   _____

5. Where do animals get their nitrogen from?_____

© 2014 **BIOZONE** International
ISBN: 978-1-927173-84-8
Photocopying Prohibited

119 Chapter Review

Summarize what you know about this topic under the headings provided. You can draw diagrams or mind maps, or write short notes to organize your thoughts. Use the images, hints, and guidelines included to help you:

Organisms are classified on how they produce their energy

HINT: Define the terms producers and consumers and say how each obtains its energy.

Food chains and food webs

HINT: Define trophic level and describe the importance of producers in food chains.

Energy flow through ecosystems

HINT: Include reference to the 10% rule.

© 2014 **BIOZONE** International
ISBN: 978-1-927173-84-8
Photocopying Prohibited

REVISE

Cycling of matter in ecosystems

HINT: Include reference to the carbon, oxygen, nitrogen, and hydrologic cycles. State the importance of photosynthesis and cellular respiration in the carbon cycle.

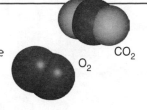

120 KEY TERMS: Crossword

Complete the crossword below, which will test your understanding of key terms in this chapter.

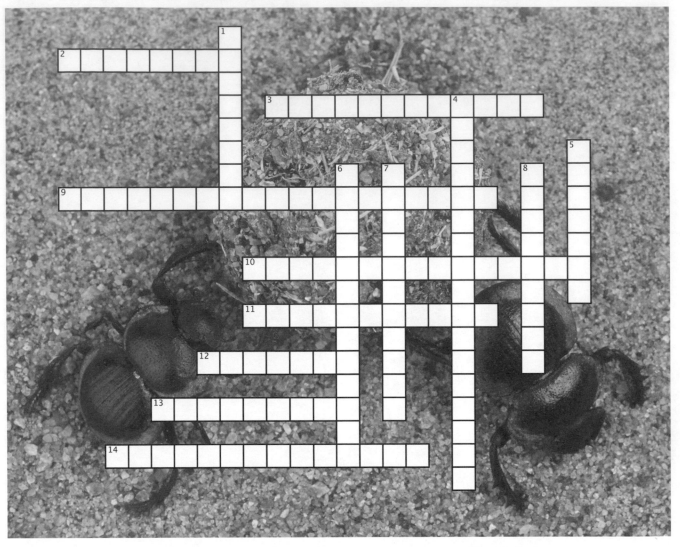

Clues Across

2. The term to describe an organism that obtains its carbon and energy from other organisms.

3. Any of the feeding levels that energy passes through as it proceeds through the ecosystem is termed this (2 words; 7, 5).

9. The catabolic process in which the chemical energy in complex organic molecules is coupled to ATP production (2 words: 8, 11).

10. The cycling of water through the environment, from lakes, seas and oceans to clouds, rain, through organisms, to ground water and rivers and back (2 words: 10, 5).

11. The cycle describing the movement of oxygen through the ecosystem (2 words: 6, 5).

12. A metabolic process that requires oxygen in order for it to proceed.

13. A metabolic process that does not require oxygen to proceed.

14. Process in which carbon dioxide, water and sunlight are used to produce glucose and oxygen.

Clues Down

1. Organism that is capable of making its own food, e.g. green plants.

4. Diagram illustrating changes in biomass, numbers, or amount of energy in organisms at each trophic level of organization in an ecosystem (2 words: 10, 7).

5. A complex series of interactions showing the feeding relationships between organisms in an ecosystem. (2 words: 4, 3).

6. The processes by which nitrogen circulates between the atmosphere and the biosphere (2 words: 8, 5).

7. The cycling of carbon (as various carbon based compounds) through the environment (2 words: 6, 5).

8. The name of a sequence of steps describing how an organism derives energy from the ones before it (2 words; 4, 5).

 © 2014 **BIOZONE** International
ISBN: 978-1-927173-84-8
Photocopying Prohibited

VOCAB

LS2.C ETS1.B
The Dynamic Ecosystem

Key terms

biodiversity

conservation

ecosystem

deforestation

global warming

keystone species

overfishing

resilience

Core ideas

Activity number

Ecosystems are dynamic, open systems

☐ 1. The physical and biotic environments contribute to the characteristics of different **ecosystems**.

121

☐ 2. In stable conditions, the numbers and types of organisms within the ecosystem are relatively stable.

121

☐ 3. Organisms with pivotal roles in the functioning of an ecosystem are called **keystone species**. Their removal may cause a fundamental change in a ecosystem.

122

☐ 4. The ecosystem is resilient. An ecosystem generally returns to its original state after modest fluctuations and disturbances. The **resilience** of an ecosystem depends on its biodiversity, health, and the frequency of disturbances to that ecosystem.

123 124

☐ 5. Extreme fluctuations in conditions or a population's size may challenge normal ecosystem functioning (e.g. by disturbing energy flows, species interactions, or nutrient cycling) and may cause irreversible changes.

125

☐ 6. **Biodiversity** is the biological variety in an ecosystem. Biodiversity is an important factor in ecosystem stability and resilience.

126

Anthropogenic changes can threaten species survival

☐ 7. Human (anthropogenic) activity can alter ecosystems and diminish biodiversity.

127

☐ 8. Human actions that threaten biodiversity include anthropogenic causes of **global warming**, **overfishing**, and **deforestation**.

128 129 130

☐ 9. **Conservation** methods are used to preserve species and maintain biodiversity.

131

☐ 10. Solutions to problems created by human activities must consider practical constraints, such as costs, and social, cultural, and environmental impacts.

131 132

Science and engineering practices

☐ 1. Analyze examples of ecosystem resilience using second-hand data provided.

124

☐ 2. Use models based on evidence to show the effects of change (including human-induced change) on ecosystems.

125 128-130

☐ 3. Develop and evaluate solutions for sustaining biodiversity while allowing essential human use of resources.

132

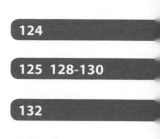

BIOZONE APP
Student Review Series
The Dynamic Ecosystem

121 Ecosystem Dynamics

Key Idea: Ecosystems are dynamic systems, responding to short-term and cyclical changes, but remaining relatively stable in the long term.

What is an ecosystem?

▶ An **ecosystem** consists of a community of organisms and their physical environment. For example a forest ecosystem consists of all the organisms within the defined area of the forest, along with the physical factors in the forest such as the temperature and the amount of wind or rain.

The dynamic ecosystem

▶ Ecosystems are dynamic in that they are constantly changing. Many ecosystem components, including the seasons, predator-prey cycles, and disease cycles, are cyclical. Some cycles may be short term, such as the change of seasons, or long term, such as the growth and retreat of deserts.

▶ Although ecosystems may change constantly over the short term, they may be relatively static over longer periods. For example, some tropical areas have wet and dry seasons, but over hundreds of years the ecosystem as a whole remains unchanged.

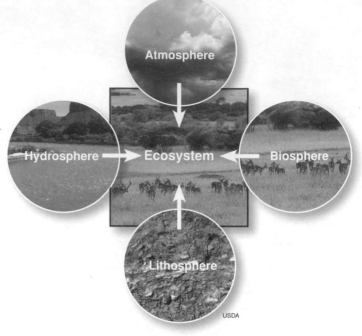

USDA

The type of ecosystem in a particular area is a result of the interactions between biological (biotic) and physical (abiotic) factors.

An ecosystem may remain stable for many hundreds or thousands of years provided that the components interacting within it remain stable.

Small scale changes usually have little effect on an ecosystem. Fire or flood may destroy some parts, but enough is left for the ecosystem to return to is original state relatively quickly.

Large scale disturbances such as volcanic eruptions, sea level rise or large scale open cast mining remove all components of the ecosystem, changing it forever.

1. What is meant by the term dynamic ecosystem? _____

   _____

2. (a) Describe two small scale events that an ecosystem may recover from: _____

   _____

   (b) Describe two large scale events that an ecosystem may not recover from: _____

   _____

   _____

CONNECT

KNOW 51

© 2014 **BIOZONE** International
ISBN: **978-1-927173-84-8**
Photocopying Prohibited

122 Keystone Species

Key Idea: Keystone species play a crucial role in ecosystems. Their actions are key to maintaining the dynamic equilibrium of an ecosystem.

Keystone species

▶ Some species have a disproportionate effect on the stability of an ecosystem. These species are called **keystone species** (or key species). The term keystone species comes from the analogy of the keystone in a true arch. If the keystone is removed, the arch collapses.

▶ The role of the keystone species varies from ecosystem to ecosystem, but if they are lost the ecosystem can rapidly change, or collapse completely. The pivotal role of keystone species is a result of their influence in some aspect of ecosystem functioning, e.g. as predators, prey, or processors of biological material.

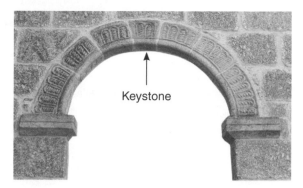

An archway is supported by a series of stones, the central one being the keystone. Although this stone is under less pressure than any other stone in the arch, the arch collapses if it is removed.

Keystone species make a difference

▶ The idea of the keystone species way first hypothesized in 1969 by Robert Paine. He studied an area of rocky seashore, noting that diversity seemed to be correlated with the number of predators present. To test this he removed the starfish from an 8 m by 2 m area of seashore. Initially the barnacle population increased rapidly before collapsing and being replaced by mussels and gooseneck barnacles. Eventually the mussels crowded out the gooseneck barnacles and the algae that covered the rocks. Limpets that fed on the algae were lost. The number of species in the study area dropped from 15 to 8.

Ochre starfish - Paine removed these in his famous study.

Elephants play a key role in maintaining the savannas by pulling down even very large trees for food. This activity maintains the grasslands.

The burrowing of prairie dogs increases soil fertility and channels water into underground stores. Their grazing promotes grass growth and diversity.

Mountain lions prey on a wide range of herbivores and to some extent dictate their home ranges and the distribution of scavenger species.

1. Define the term **keystone species**: _____

   _____

   _____

2. Prairie dogs colonies are often destroyed by ranchers who believe they compete with cattle for food. How might this lead to the collapse of prairie ecosystems?

   _____

   _____

   _____

© 2014 **BIOZONE** International
ISBN: 978-1-927173-84-8
Photocopying Prohibited

CONNECT
WEB

89 **122** **KNOW**

123 The Resilient Ecosystem

Key Idea: The resilience of a ecosystem depends on its biodiversity, health, and the frequency with which it is disturbed.

Factors affecting ecosystem resilience

Resilience is the ability of the ecosystem to recover after disturbance and is affected by three important factors: diversity, ecosystem health, and frequency of disturbance (below). Some ecosystems are naturally more resilient than others.

▶ **Ecosystem biodiversity**
The greater the diversity of an ecosystem the greater the chance that all the roles (niches) in an ecosystem will be occupied, making it harder for invasive species to establish and easier for the ecosystem to recover after a disturbance.

▶ **Ecosystem health**
Intact ecosystems are more likely to be resilient than ecosystems suffering from species loss or disease.

▶ **Disturbance frequency**
Single disturbances to an ecosystem can be survived, but frequent disturbances make it more difficult for an ecosystem to recover. Some ecosystems depend on frequent natural disturbances for their maintenance, e.g. grasslands rely on natural fires to prevent shrubs and trees from establishing. The keystone grass species have evolved to survive frequent fires.

▶ A study of coral and algae cover at two locations in Australia's Great Barrier Reef (right) showed how ecosystems recover after a disturbance. At Low Isles, frequent disturbances (e.g. from cyclones) made it difficult for corals to reestablish, while at Middle Reef, infrequent disturbances made it possible to coral to reestablish its dominant position in the ecosystem.

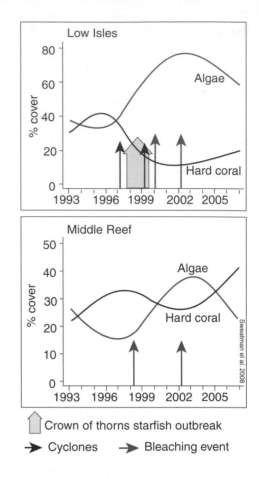

Crown of thorns starfish outbreak
→ Cyclones → Bleaching event

Sweatman et al. 2008

Stability

The stability of an ecosystem can be illustrated by a ball in a tilted bowl. The system is resilient under a certain set of conditions. If the ball is given a slight push (the disturbance is small) it will eventually return to its original state (**line A**). However if the ball is pushed too hard (the disturbance is too large) then the ball will roll out of the bowl and the original state with never be restored (**line B**).

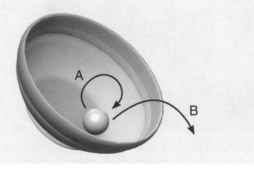

1. Define **ecosystem resilience**: _____

_____

2. Why did the coral at Middle Reef remain abundant from 1993 to 2005 while the coral at Low Isles did not?

_____

_____

_____

3. A monoculture (single species system) of cabbages is an unstable ecosystem. What is required for a cabbage monoculture to remain unchanged?

_____

 © 2014 **BIOZONE** International
ISBN: 978-1-927173-84-8
Photocopying Prohibited

124 A Case Study in Ecosystem Resilience

Key Idea: Ecosystems fluctuate between extremes. Resilient ecosystems are able to recover from moderate fluctuations.

Spruce budworm and balsam fir

A case study of ecosystem resilience is provided by the spruce-fir forest community in northern North America. Organisms in the community include the spruce budworm, and balsam fir, spruce, and birch trees. The community fluctuates between two extremes:

▶ Between spruce budworm outbreaks the environment favors the balsam fir.

▶ During budworm outbreaks the environment favors the spruce and birch species.

Balsam fir   Spruce budworm

❶ Under certain environmental conditions, the spruce budworm population grows so rapidly it overwhelms the ability of predators and parasites to control it.

❷ The budworm feeds on balsam fir (despite their name), killing many trees. The spruce and birch trees are left as the major species.

❸ The population of budworm eventually collapses because of a lack of food.

❹ Balsam fir saplings grow back in thick stands, eventually out-competing the spruce and birch. Evidence suggests these cycles have been occurring for possibly thousands of years.

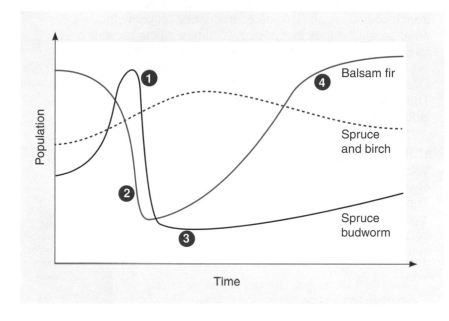

1. (a) Why could predators not control the budworm population? _____

   _____

   _____

   (b) What was the cause of the budworm population collapse after its initial rise?_____

   _____

2. Under what conditions does the balsam fir out-compete the spruce and birch? _____

   _____

3. In what way is the system resilient in the long term? _____

   _____

   _____

   _____

   _____

   _____

© 2014 **BIOZONE** International
ISBN: 978-1-927173-84-8

KNOW

125 Ecosystem Changes

Key Idea: Sometimes the disturbances to an ecosystem are so extreme that the ecosystem never returns to its original state.

Ecosystems are dynamic, constantly fluctuating between particular conditions. However, large scale changes can occasionally occur to completely change the ecosystem. These include climate change, volcanic eruptions, or large scale fires.

Human influenced changes

▶ Dolly Sods is a rocky high plateau area in the Allegheny Mountains of eastern West Virginia, USA. Originally the area was covered with spruce, hemlock, and black cherry. During the 1880s, logging began in the area and virtually all of the commercially viable trees were cut down. The logging caused the underlying humus and peat to dry out. Sparks from locomotives and campfires frequently set fire to this dry peat, producing fires that destroyed almost all the remaining forest. In some areas, the fires were so intense that they burnt everything right down to the bedrock, destroying seed banks. One fire during the 1930s destroyed over 100 km² of forest. The forests have never recovered. What was once a forested landscape is now mostly open meadow.

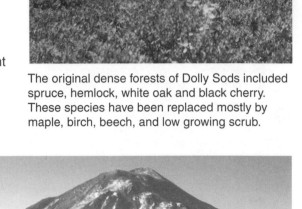

The original dense forests of Dolly Sods included spruce, hemlock, white oak and black cherry. These species have been replaced mostly by maple, birch, beech, and low growing scrub.

Natural changes

▶ Volcanic eruptions can cause extreme and sudden changes to the local (or even global) ecosystems. The eruption of Mount St. Helens in 1980 provides a good example of how the natural event of a volcanic eruption can cause extreme and long lasting changes to an ecosystem.

Before the 1980 eruption, Mount St. Helens had an almost perfect and classic conical structure. The forests surrounding it were predominantly conifer, including Douglas-fir, western red cedar, and western white pine.

Mount St. Helens, one day before the eruption.

Schematic of eruption and recovery

The eruption covered about 600 km² (dark blue) in ash (up to 180 m deep in some areas) and blasted flat 370 km² of forest. The schematic shows the general area of bare land per decade since the eruption.

■ 1980   ■ 1990
■ 2000   ■ 2010

Coldwater Lake was formed when Coldwater Creek was blocked by eruption debris.

Spirit Lake was completely emptied in the initial eruption. All life except bacteria was extinguished. The lake has since begun to recover.

The forests on the northern flank of the mountain have vanished, replaced by pumice plains thinly covered in low growing vegetation.

North Fork Toutle River originated from Spirit Lake before the eruption. It now originates from the mountain's crater. The river itself is now laden with sediment due to erosion.

N

Eruption crater
Original summit
Mt St Helens

© 2014 **BIOZONE** International
ISBN: 978-1-927173-84-8
Photocopying Prohibited

1. (a) Identify the large scale ecosystem change that occurred at Dolly Sods: _____

    _____

    (b) What caused this change in the ecosystem? _____

    _____

    (c) Explain why the ecosystem has not been able to recover quickly after the change: _____

    _____

    _____

    _____

2. Describe the major change in the ecosystem on Mount St Helens' northern flank after the eruption:

    _____

    _____

3. Study the eruption schematic on the previous page. Why has the recovery of the ecosystem area shown in light blue (2010) been so much slower than elsewhere?

    _____

    _____

    _____

4. Describe the large scale change than occurred at Coldwater Creek after the eruption. What effect would this have had on the local ecosystem?

    _____

    _____

    _____

5. Describe the large scale change that occurred at Spirit Lake. Explain why this would cause an almost complete change in the lake ecosystem:

    _____

    _____

    _____

    _____

6. (a) What two major changes have occurred in the North Fork Toutle River? _____

    _____

    _____

    (b) How might this affect this riverine ecosystem? _____

    _____

    _____

    _____

    _____

 © 2014 **BIOZONE** International
ISBN: 978-1-927173-84-8
Photocopying Prohibited

126 Biodiversity

Key Idea: Biodiversity describes the biotic variation in an ecosystem. The perception of an ecosystem's diversity may differ depending on how the biodiversity is measured.

What is biodiversity?

Biodiversity is the amount of biotic variation within a given group. It could be the number of species in a particular area or the amount of genetic diversity in a species. Ecosystem biodiversity is often used as a measure of the different number of species within a specific ecosystem.

Biodiversity tends to be clustered in certain parts of the world ,called hotspots, where there are very large numbers of different species. Tropical forests and coral reefs (above) are some of the most diverse ecosystems on Earth.

Biodiversity of Earth	
Type of organism	**Estimated number of species**
Protozoa	36,400
Brown algae, diatoms	27,500
Invertebrates	7.6 million
Plants	298,000
Fungi	611,000
Vertebrates	60,000

The latest estimate of the number of eukaryotic species on Earth is 8.7 million, of which only 1.2 million have been formally described. Prokaryotic species are so variable and prolific the number of species is virtually inestimable and could be hundreds of millions.

Measuring biodiversity

Biodiversity is measured for a variety of reasons, e.g. to assess the success of conservation work or to measure the impact of human activity. One measure of biodiversity is to simply count all the species present (the species richness) although this may give an imprecise impression of the ecosystem's biodiversity. Species evenness gives a measure of relative abundance of species, i.e. how close in numbers each species in an environment are. It describes how the biodiversity is apportioned.

Two ecosystems with quite different biodiversity.

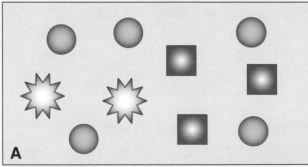

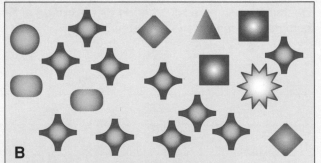

Species richness is a simple method of estimating biodiversity in which the number of species is counted. This does not show if one species is more abundant than others. **Species evenness** is a method in which the proportions of species in an ecosystem are estimated.

Diversity Indices use mathematical formula based on the species abundance and the number of each species to describe the biodiversity of an ecosystem. Many indices produce a number between 0 and 1 to describe biodiversity, 1 being high diversity and 0 being low diversity.

1. Calculate the number and percentage of eukaryotic species that are still to be formally described.

   _____

2. Describe in words the species richness and species evenness of ecosystem A and B above:

   A _____

   B _____

WEB   CONNECT   CONNECT

KNOW   **126**   **122**   **127**

© 2014 **BIOZONE** International
ISBN: 978-1-927173-84-8
Photocopying Prohibited

127 Humans Depend on Biodiversity

Key Idea: Humans rely on natural resources for their health, well being, and livelihood. The biodiversity of an ecosystem affects its ability to provide these services.

The importance of biodiversity

▶ Biodiversity is important to ecosystem functioning. A properly functioning ecosystem is able to provide **ecosystem services**, e.g. production of clean water and air. These have a direct impact on human health. The UN has identified four categories of ecosystem services: supporting, provisioning, regulating, and cultural.

Biodiversity supports human wellbeing

Ecosystem services

Constituents of human wellbeing

Provisioning:
Food
Freshwater
Resources for building
Fuel

Supporting systems:
Nutrient cycling
Soil formation
Primary production

Regulating:
Climate
Flood
Disease
Water purification

Cultural:
Aesthetic appeal
Education
Recreation

Security:
- Secure resource access
- Protection from disasters (flood, famine, epidemics)

Materials for living:
- Adequate livelihoods
- Sufficient food
- Shelter
- Access to goods

Health:
- Access to clean water and air
- Avoidance of disease (e.g. mosquitoes carrying malaria)

Social relations:
- Social cohesion
- Ability and willingness to help others

Freedom of choice and action
Providing adequate food, safety, and social support frees individuals to achieve their particular goals and aspirations.

Adapted from the Millennium Ecosystem Assessment

Intensity of linkage: ——→ Weak ——→ Moderate ——→ Strong

Disease resistance in sorghum

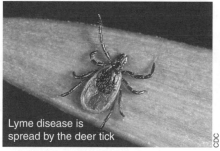

Lyme disease is spread by the deer tick — CDC

Landslide — NOAA

Biodiversity is important in crop development, e.g promoting disease resistance. Many medical breakthroughs have come from understanding the biology of wild plants and animals.

High biodiversity creates buffers between humans and infectious diseases (e.g. Lyme disease) and increases the efficiency of processes such as water purification.

Biodiversity and ecosystem health are essential for reducing the effects of human activities (e.g. pollution) and the effects of environmental disasters (e.g. eruptions and landslides).

1. Explain what is meant by **ecosystem services**: _____

_____

2. What is the role of biodiversity in the ability of an ecosystem to provide ecosystem services?

_____

_____

_____

© 2014 **BIOZONE** International
ISBN: 978-1-927173-84-8
Photocopying Prohibited

CONNECT
122

WEB
127 KNOW

Human effects on biodiversity

► Human activities have had major effects on the biodiversity of Earth. Nearly 40% of the Earth's land surface is devoted to agricultural use. In these areas, the original biodiversity, a polyculture of plants and animals, has been severely reduced. Many of these areas are effectively monocultures, where just one type of plant is grown. The graph right shows that as the land is more intensively used, the populations and variety of plants and animals fall.

Average remaining % of population under each land use

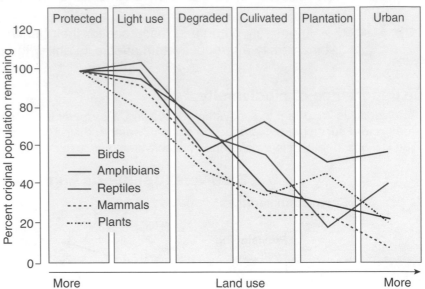

Climate change from burning fossil fuels will have a major effect on biodiversity. Warming climates and rises in sea levels will force organisms into new environments. Those that can't adapt may become extinct.

Overfishing has severely reduced fish stocks throughout the world. In some areas, the dominant species of fish have virtually disappeared and have been replaced by other species, disrupting marine ecosystems.

Demand for farmland, timber, and plantation resources have resulted in the clearance of many forests. Thousands of square kilometers of forest a year are destroyed by logging or fire, reducing local biodiversity.

3. How do ecosystem services contribute to human well being? _____

_____

_____

4. Describe the effect of land development on biodiversity: _____

_____

_____

_____

5. How have the following human activities affected biodiversity?

(a) Burning fossil fuels: _____

_____

(b) Overfishing: _____

_____

(c) Deforestation: _____

_____

© 2014 **BIOZONE** International
ISBN: 978-1-927173-84-8
Photocopying Prohibited

128 Biodiversity and Global Warming

Key Idea: Changing climates may cause changes to biodiversity by expanding or shrinking the current range of suitable habitat for plant and animal species.

▶ Global warming is the warming of the Earth's climate, mainly due to human activities such as deforestation and the use of fossil fuels.

▶ Habitats are closely linked to climate, e.g. desert habitats occur where there is very low rainfall and very high temperatures. Global warming will change habitats throughout the world and affect global and local biodiversity. In general, plants and animals adapted to colder climates will reduce their range and those adapted to warmer climates will increase their range.

▶ Observations of animal and plant populations over many decades have noted the effect of climate change. From these observations is it possible to predict the likely impact of sustained climate change.

Effects of climate variation on plant and animal populations in the United States

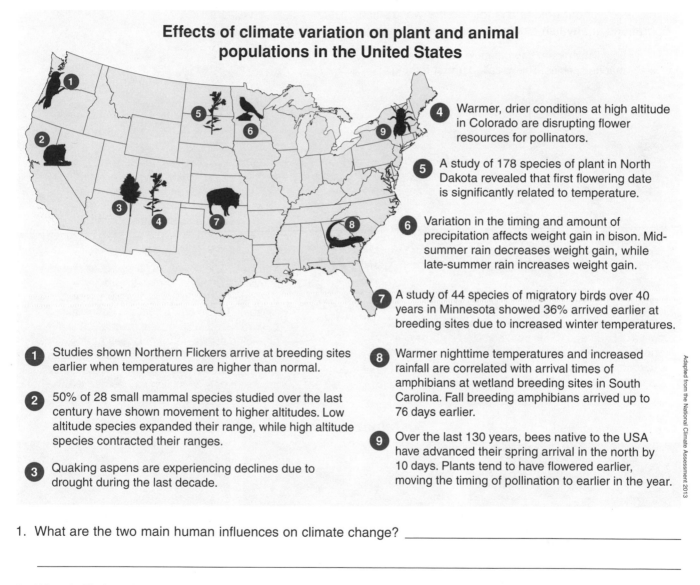

4 Warmer, drier conditions at high altitude in Colorado are disrupting flower resources for pollinators.

5 A study of 178 species of plant in North Dakota revealed that first flowering date is significantly related to temperature.

6 Variation in the timing and amount of precipitation affects weight gain in bison. Mid-summer rain decreases weight gain, while late-summer rain increases weight gain.

7 A study of 44 species of migratory birds over 40 years in Minnesota showed 36% arrived earlier at breeding sites due to increased winter temperatures.

1 Studies shown Northern Flickers arrive at breeding sites earlier when temperatures are higher than normal.

2 50% of 28 small mammal species studied over the last century have shown movement to higher altitudes. Low altitude species expanded their range, while high altitude species contracted their ranges.

3 Quaking aspens are experiencing declines due to drought during the last decade.

8 Warmer nighttime temperatures and increased rainfall are correlated with arrival times of amphibians at wetland breeding sites in South Carolina. Fall breeding amphibians arrived up to 76 days earlier.

9 Over the last 130 years, bees native to the USA have advanced their spring arrival in the north by 10 days. Plants tend to have flowered earlier, moving the timing of pollination to earlier in the year.

Adapted from the National Climate Assessment 2013

1. What are the two main human influences on climate change? _____

_____

2. What is likely to happen to polar and high altitude habitats as the climate warms? _____

_____

3. Explain how global warming is likely to affect a plant or animal behavior that is correlated to climate:

_____

_____

_____

CONNECT WEB
125 128 KNOW

129 Overfishing

Key Idea: Overfishing has caused the collapse of many fisheries. Unsustainable fishing practices continue throughout the world's oceans.

▶ Fishing is an ancient human tradition. It provides food, and is economically, socially, and culturally important. Today, it is a worldwide resource extraction industry. Decades of overfishing in all of the world's oceans has pushed commercially important species (such as cod, right) into steep decline.

▶ According to the United Nation's Food and Agriculture Organization (FAO) almost half the ocean's commercially targeted marine fish stocks are either heavily or over-exploited. Without drastic changes to the world's fishing operations, many fish stocks will soon be effectively lost.

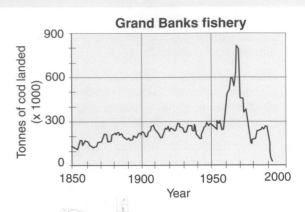
Grand Banks fishery

Lost fishing gear can entangle all kinds of marine species. This is called **ghost fishing**.

Overfishing has resulted in many fish stocks at historic lows and fishing effort (the effort expended to catch fish) at unprecedented highs.

Huge fishing trawlers are capable of taking enormous amounts of fish at once. Captures of 400 tonnes at once are common.

Bottom trawls and dredges cause large scale physical damage to the seafloor. Non-commercial, bottom-dwelling species in the path of the net can be uprooted, damaged, or killed. An area of 8 million km² is bottom trawled annually.

The limited selectivity of fishing gear results in millions of marine organisms being discarded for economic, legal, or personal reasons. These organisms are defined as **by-catch** and include fish, invertebrates, protected marine mammals, sea turtles, and sea birds. Many of the discarded organisms die. Estimates of the worldwide by-catch is approximately 30 million tonnes per year.

Sustainable fisheries

Creating marine reserves provides areas where fish species can spawn and rebuild the fish stock.

Giant river prawns are farmed in 37 countries with around 300,000 tonnes being produced per year.

Marine species such as mussels can be farmed offshore using stockings hung from ropes to anchor the mussels.

1. Why is repeated overfishing of a fish species not sustainable? _____

_____

_____

2. Why is returning by-catch to the sea not always as useful as it might appear? _____

_____

© 2014 **BIOZONE** International
ISBN: 978-1-927173-84-8
Photocopying Prohibited

130 Deforestation

Key Idea: Deforestation is the permanent removal of forest from an area. Causes of deforestation include land development for farming and plantations.

Deforestation

▶ At the end of the last glacial period, about 10,000 years ago, forests covered an estimated 6 billion hectares, about 45% of the Earth's land surface. Forests currently still cover about 4 billion hectares of land (31% of Earth's surface). Over the last 5,000 years, the loss of forest cover is estimated at 1.8 billion hectares. 5.2 million hectares has been lost in the last 10 years alone. Areas where humans have historically lived longer (e.g. Europe) have suffered the most. Approximately 60% of Europe's original forest has been lost. Intensive clearance of forests during settlement of recently discovered lands has extensively altered the landscape in places such as New Zealand (where 75% of the original forests were lost in little over a few hundred years).

Causes of deforestation

▶ **Deforestation** is caused by many interrelated reasons, which often center around socioeconomic situation. In many tropical regions, the vast majority of deforestation is the result of subsistence farming. Poverty and a lack of secure land can be partially solved by clearing forests and producing family plots. However huge areas of forests have been cleared for agriculture, both ranching and production of palm oil plantations. These produce revenue to governments through taxes and permits, producing a incentive to clear more forest. Just 14% of deforestation is attributable to commercial logging (although combined with illegal logging it may be higher).

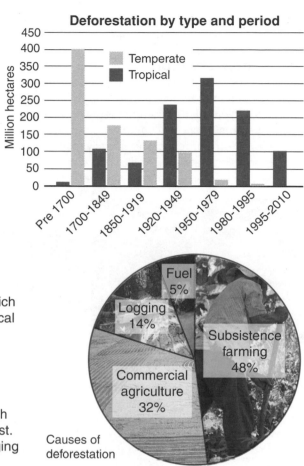

Deforestation by type and period

Million hectares

(Legend: Temperate, Tropical)

Periods: Pre 1700, 1700-1849, 1850-1919, 1920-1949, 1950-1979, 1980-1995, 1995-2010

Fuel 5%
Logging 14%
Subsistence farming 48%
Commercial agriculture 32%

Causes of deforestation

Deforestation and biodiversity

Deforestation can have a major effect on biodiversity, especially in tropical rainforests where biodiversity is naturally high, as it results in loss of habitat and makes it easier for invasive species to become established, further degrading habitats. However, the loss of biodiversity is not the same from forest to forest.

Loss of forest in a temperate woodland (left) may not reduce biodiversity as much as it might in a tropical rainforest (right) where species richness is much higher.

1. Describe the trend in temperate and tropical deforestation over the last 300 years. _____

_____

2. What are some causes of deforestation? _____

_____

_____

3. How does the type of forest lost by deforestation influence how much biodiversity is lost? _____

_____

_____

CONNECT WEB
 **127** **130** KNOW

131 Sustaining Biodiversity

Key Idea: Conservation aims to maintain or improve biodiversity. This can be done by managing the organisms in their environment or in captivity.

▶ Conservation aims to maintain the biodiversity of a particular ecosystem. This is achieved by managing species in their habitats (*in-situ*) or by conservation programs away from the natural environment (*ex-situ*).

In-situ conservation

▶ *In-situ* conservation describes the conservation of organisms in their own environment. Methods focus on ecological restoration and legislation to protect ecosystems of special value. Ecological restoration is a long term process and usually involves collaboration between institutions with scientific expertise and the local communities involved.

In the US, the **Endangered Species Act** (**ESA**) is designed to protect species and the ecosystems on which they depend. The act is administered by the US Fish and Wildlife Service and the National Oceanic and Atmospheric Administration. As a result of this and other efforts, species such as the snowy egret, white tailed deer, and wild turkey, which were all critically endangered, are common once again.

Mainland islands (pioneered in New Zealand) were developed to protect and restore ecosystems of ecological importance and to rebuild native populations of animals and plants. Methods include predator-proof fencing around large areas of forest, pest control, and translocation of species to spread the risk to recovering species. Volunteer programs are an important part of these developments.

Ex-situ conservation

▶ *Ex-situ* conservation methods operate away from the natural environment and are particularly useful where species are critically endangered. Zoos, aquaria, and botanical gardens are the most conventional vehicles for *ex-situ* conservation. They house and protect specimens for breeding and, where necessary and possible, they reintroduce them into the wild to restore natural populations.

Captive breeding programs can be very difficult and expensive to implement, and may not be appropriate for some species. In some cases they can be extremely successful. Here a puppet 'mother' shelters a takahe chick. Takahe, a native New Zealand bird, were brought back from the brink of extinction through a successful captive breeding program.

Zoos have a major role in public education, captive breeding programs, and as custodians of rare species. They raise awareness of the threats facing species in their natural environments and engender public empathy for conservation work. Modern zoos tend to concentrate on particular species and are part of breeding and education programs.

1. Explain the difference between *in-situ* and *ex-situ* conservation: _____

_____

_____

2. How does conservation help to sustain, or even increase, biodiversity? _____

_____

_____

_____

CONNECT 195

KNOW

© 2014 **BIOZONE** International
ISBN: 978-1-927173-84-8

132 Modeling a Solution

Key Idea: Conservation efforts are often a compromise between environmental, economic, and cultural needs.

Deciding on a course of action for preserving biodiversity is not always simple. Environmental, cultural, and economic impacts must be taken into account, and compromises must often be made. The map below shows a hypothetical area of 9,300 ha (93 km²) in which two separate populations of an endangered bird species exist within a forested area of public land. A proposal to turn part of the area into an wildlife reserve has been put forward by local conservation groups. However, the area is known to have large deposits of economically viable minerals and is frequented by trampers. Hunters also spend time in the area because part of it has an established population of introduced game animals. The proposal would allow a single area of up to 1,500 ha (15 km²) to be reserved exclusively for conservation efforts.

1. Study the map below and draw on to the map where you would place the proposed reserve, taking into account economic, cultural, and environmental values. On a separate sheet, write a report justifying your decision as to where you placed the proposed reserve.

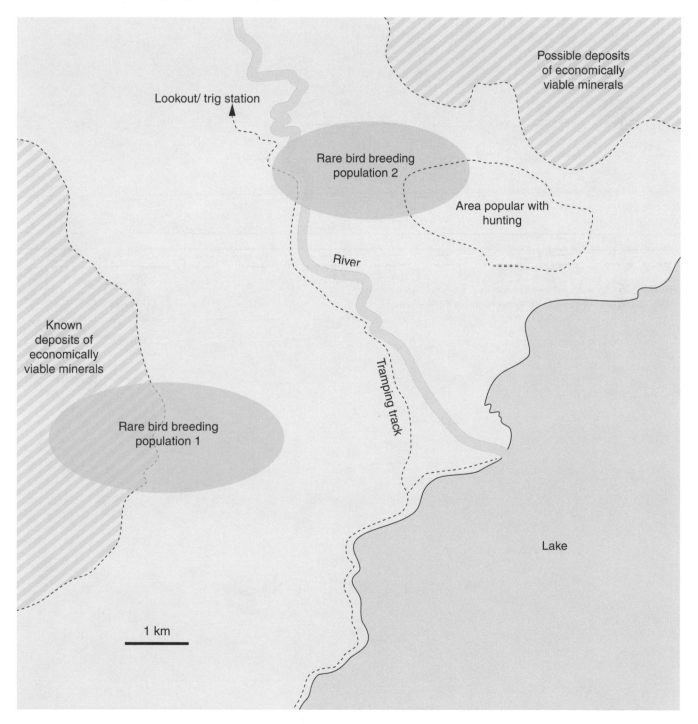

 © 2014 **BIOZONE** International
ISBN: 978-1-927173-84-8
Photocopying Prohibited

133 Chapter Review

Summarize what you know about this topic so far under the headings provided. You can draw diagrams or mind maps, or write short notes to organize your thoughts. Use the images, hints, and guidelines included to help you:

The dynamic ecosystem

HINT: What makes up an ecosystem?

Ecosystem change and resilience

HINT: How do ecosystems change and to what extent do the organisms in them help them recover from change?

REVISE

Biodiversity

HINT: What are the benefits of ecosystem biodiversity?

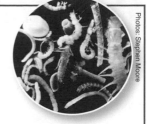

Photos: Stephen Moore

Humans and biodiversity

HINT: How do humans depend on biodiversity and how do they affect biodiversity?

134 KEY TERMS: Did You Get It?

1. Test your vocabulary by matching each term to its definition, as identified by its preceding letter code.

biodiversity	**A** A species that is influential in the stability of an ecosystem and has a disproportionate effect on ecosystem function due to their pivotal role.
conservation	**B** The removal of forests for fuel, building material, or grazing land.
deforestation	**C** The active management of natural populations in order to rebuild numbers and ensure species survival.
ecosystem	**D** A term describing the variation of life at all levels of biological organization.
global warming	**E** The process of the Earth's surface steadily increasing in temperature (and its projected continuation). Usually attributed to the rise in greenhouse gases produced by fossil fuels and industrial processes.
keystone species	**F** Community of interacting organisms and the physical environment in which they both live and interact.
overfishing	**G** The ability of an ecosystem to recover from external disturbances.
resilience	**H** The removal of fish stocks above their natural rate of regeneration.

2. (a) Identify the two groups of organisms (right) as high diversity or low diversity:

   Group A. _____

   Group B. _____

   (b) Which group appears to have the highest species evenness:

   _____

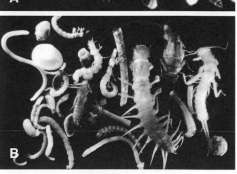

Photos: Stephen Moore

3. Circle the correct statement below:

   A. The higher an ecosystem's resilience, the more difficult it is for it to return to its original state after a disturbance.
   B. Natural ecosystem change is always less disruptive than human induced changes.
   C. When keystone species are removed from their environment, biodiversity tends to decrease.
   D. Captive breeding is a key component of *in-situ* conservation.

4. Explain how conservation can help sustain biodiversity and how this can affect human well-being:

   _____

   _____

   _____

   _____

   _____

   _____

   _____

 © 2014 **BIOZONE** International
ISBN: 978-1-927173-84-8
Photocopying Prohibited

Social Behavior

LS2.D

Key terms

altruism

cooperative behavior

flocking

herding

kin selection

migration

schooling

social group

Core ideas

Social interactions and group behavior increase the chances of survival

Activity number

☐ 1. The organization of a group into a social structure improves the chances of survival and reproduction for members of the group.

135

☐ 2. Animals are classed as solitary, grouped together (but without any social order), or grouped together in close **social groups**. Social organization has both advantages and disadvantages.

135

☐ 3. Group behavior includes remaining as a group (e.g. **schooling**, **herding**, **flocking**), group attack and group defense.

136 137

☐ 4. Schooling, herding, and flocking provide benefits to the group. These include increased protection from predators and improved foraging and efficiency of locomotion (e.g. during long distance flight).

136 137

☐ 5. Social animals organize themselves in a way that divides resources and roles between members of the social group. Social behavior evolves because, on average, it improves the survival of individuals.

138 139

☐ 6. **Cooperative behavior** involves two or more individuals working together to achieve a common goal. **Altruism** and **kin selection** are examples of cooperative behavior.

140

☐ 7. Working together in defense and attack can help increase or maintain resources for the group, and increase the survival chances of individuals.

141

☐ 8. Cooperative behavior in gathering food increases the chances of foraging success and improves efficiencies.

142

Science and engineering practices

☐ 1. Analyze data to explain how group behavior is beneficial to survival during migration.

137

☐ 2. Use a model based on evidence to demonstrate how cooperative behavior (help given) can depend on relatedness of the individuals involved.

139

☐ 3. Construct and revise an explanation based on evidence for the benefits of cooperation in hunting or foraging.

142

135 Social Groupings

Key Idea: Animals may be solitary, form loosely associated groups, or form complex groups with clear social structures. Each behavior has its advantages and disadvantages.

▶ No animal lives completely alone. At some stage in their lives all animals must interact with others of their species (e.g. to reproduce or through competitive interactions for food or resources).

▶ Generally animals are classed as solitary, grouped together (but without any social order), or grouped together in close social groups.

Solitary animal

Non-social groups

Social groups

Solitary animals spend the majority of their lives alone, often in defended territories. They may only seek out others of their species for breeding. Offspring are often driven away shortly after they become independent.

Solitary life is often an advantage when resources are scarce or scattered over a large area. Solitary animals include many of the cat family e.g. tiger (above), bears, and various invertebrates.

Many animals form loose associations but do not interact socially. Each animal is acting directly for its own benefit with little or no direct cooperation between them. Schools of fish, flocks of birds and many herding mammals exhibit this non-social grouping.

Non-social groups provide protection from predators by reducing the possibility of being preyed upon individually. There may also be benefits during feeding and moving.

Primates form complex social structures which are usually based around a family group. Some animals that form social groups also form dominance hierarchies.

Dominance hierarchies help distribute resources and maintain social structure. In some species members of the group are divided into castes with specialized roles (e.g. ants and bees). Some produce offspring or help raise young, others may be workers or help with defense of the colony.

Advantages of large social groupings

1. Protection from physical factors and predators.
2. Assembly for mate selection.
3. Locating and obtaining food.
4. Defense of resources against other groups.
5. Division of labor amongst specialists.
6. Richer learning environment.
7. Population regulation (e.g. breeding restricted to a dominant pair).

Possible disadvantages of large social groups

1. Increased competition between group members for resources as group size increases.
2. Increased chance of the spread of diseases and parasites.
3. Interference with reproduction, e.g. infanticide by non-parents or cheating in parental care (so that non-parents may unknowingly raise another's offspring).

1. Give one advantage and one disadvantage of solitary living: _____

_____

2. Explain why group behavior, such as schooling, is more about individual advantage than group advantage:

_____

_____

_____

3. Give one advantage and one disadvantage of a dominance hierarchy: _____

_____

_____

© 2014 **BIOZONE** International
ISBN: 978-1-927173-84-8
Photocopying Prohibited

136 Schooling, Flocking, Herding

Key Idea: Being part of a group enhances survival by providing protection from predators and by reducing energy expenditure during movement.

Dynamics in a flock, school, or herd

▶ **Schooling** by fish, **flocking** by birds, and **herding** by grazing mammals are essentially all the same behavior. Each individual is behaving in a way that helps its own survival regardless of the others within that group. Within the group, the application of a few simple rules results in apparently complex behavior.

▶ In a school, flock, or herd, three simple rules tend to apply:
- Move towards the group or others in the group.
- Avoid collision with others in the group or external objects.
- Align your movement with the movement of the others in the group.

▶ If every individual moves according to these rules, the school, flock, or herd will stay as a dynamic cohesive unit, changing according to the movement of others and the cues from the environment.

Flock of auklets (a small seabird)
D. Dibenski

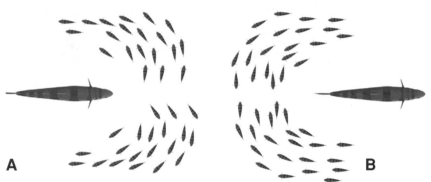

A **B**

Schooling dynamics: In schooling fish or flocking birds, every individual behaves according to a set of rules. In diagram A, each fish moves away from the predator while remaining close to each other. The school splits to avoid the predator (A), before moving close together again behind the predator (B).

In a **flash expansion**, each individual moves directly away from the predator. Collisions have never been observed, suggesting each fish is able to sense the direction of movement of the fish next to it.

Why do fish school?

▶ Schooling in predatory fish may enhance the ability of any individual to catch its prey. If the prey avoids one predator, it may get caught by the next predator. Prey fish may school for defensive reasons.

Advantages of schooling

▶ **Avoidance of predators**
- Confusion caused by the movement of the school.
- Protection by reduced probability of individual capture.
- Predator satiation (more than the predator could eat)
- Better predator detection (the many eyes effect)

▶ **Better hydrodynamics** within the school, so less energy is expended in swimming.

School of jacks

1. How do the rules for flocking and schooling help to maintain a cohesive group?

_____

_____

_____

_____

CONNECT CONNECT WEB
183 **88** **136** **KNOW**

Flocking in birds

▶ Flocking in birds follows similar rules to schooling. In flight, each individual maintains a constant distance from others and keeps flying in the average direction of the group. Flocks can be very large, with thousands of birds flying together as a loosely organized unit, such as starlings flocking in evenings or queleas, flocking over feeding or watering sites (photo A, right).

A: Flocking queleas (small weaver birds)

Alastair Rae cc2.0

▶ Migrating flocks generally fly in a V formation (photo B, right). The V formation provides the best aerodynamics for all in the flock except the leader. Each bird gains lift from the movement of the air caused by the bird ahead of it.

B: Migrating geese

Research on great white pelicans has shown that the V formation helps the birds conserve energy. As the wing moves down, air rushes from underneath the wing to above it, causing an upward moving vortex behind the wing. This provides lift to the bird flying behind and to one side, requiring it to use less effort to maintain lift. Energy savings come from increased gliding.

The wingtip vortex provides lift for the trailing bird.

Herding in mammals

▶ Herding is common in hoofed mammals, especially those on grass plains like the African savanna. A herd provides protection because while one animal has its head down feeding, another will have its head up looking for predators. In this way, each individual benefits from a continual supply of lookouts (the many eyes principle). During an attack, individuals move closer to the center of the herd, away from the predator, as those on the outside are more frequently captured. The herd moves as one group, driven by individual needs.

2. Identify some benefits of schooling to a:

   (a) Predatory fish species: _____

   (b) Prey fish species: _____

   _____

3. (a) How does flying in V formation help a bird during migration? _____

   _____

   _____

   (b) How would this help survival in the longer term? _____

   _____

   _____

   _____

4. Explain how herding in grazing mammals provides a survival advantage: _____

   _____

   _____

   _____

© 2014 **BIOZONE** International
ISBN: 978-1-927173-84-8

137 Migration

Key Idea: Migration is an energy expensive behavior. Grouping together reduces the energy used and improves survival of migrating individuals.

Migration is the long distance movement of individuals from one place to another. Migration usually occurs on a seasonal basis and for a specific purpose, e.g. feeding, breeding, or over-wintering.

▶ In order for migratory behavior to evolve, the advantages of migration must outweigh the disadvantages. Migration is an energy expensive behavior, and animals must spend a lot of time building energy stores that will fuel the effort. The destination provides enough food or shelter to enhance survival of individuals and their offspring.

▶ Although some animals migrate individually, many migrate in large groups (right).

Benefits of group migration

▶ Group migration helps navigation in what is called the "many wrongs principle" in which the combining of many inaccurate navigational compasses produces a more accurate single compass. Thus, if an animal navigates by itself with a slightly inaccurate internal compass, or inaccurately interprets environmental cues, it may arrive in the wrong location. In a group, each member can adjust its heading according to the movement of the others, thus an average direction is produced and each member is more likely to arrive in the correct place (right).

▶ Birds flying together increase aerodynamic efficiency to each other, saving energy. In schooling fish, individuals in the center of the school use less effort for movement. Flocking and schooling also provide feeding benefits and decreased risk of predation along the migration route.

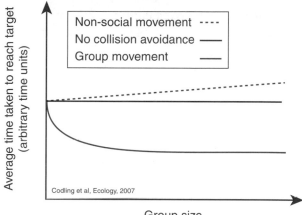

Codling et al, Ecology, 2007

Increasing group size decreases the time taken to reach a navigational target when the group is moving as a social unit. Non-social groups take longer with increasing size because of the need to avoid others in the group.

1. What is migration?_____

_____

2. Describe an advantage and a disadvantage of migration: _____

_____

_____

3. (a) How does grouping together increase navigational efficiency? _____

_____

_____

   (b) How does this enhance individual survival?_____

_____

_____

CONNECT   WEB

95 **137** **KNOW**

138 Social Organization

Key Idea: In social groups, members of the group interact regularly. Social species organize themselves in a way that divides resources and roles between group members.

Eusocial animals

▶ Eusocial animals are those in which a single female produces the offspring and non-reproductive individuals care for the young. They have the highest form of social organization. Individuals are divided into castes that carry out specific roles. In most cases there is a queen which produces the young and members of the group are normally directly related to the queen. Non-reproductive members of the group may be involved in care of the young or defense of the nest site. Examples include ants, honey bees, termites, and naked mole rats.

Honeybees   Termites

Presocial animals

▶ Presocial animals exhibit more than just sexual interactions with members of the same species, but do not have all of the characteristic of eusocial animals. They may live in large groups based around a single breeding pair. Offspring may be looked after by relations (e.g. aunts/older siblings). These groups often form hierarchies where the breeding pair are the most dominant. There may also be separate hierarchies for male and female members of the group. Examples include canine species that live in packs (e.g. wolves), many primates, and some bird species.

▶ The number of males in a social group varies between species. Wild horse herds have a single stallion, which controls a group of mares. Young males are driven away when they are old enough. Female elephants and their offspring form small groups lead by the eldest female (the matriarch). Adult male elephants only visit the group during the reproductive season.

1. Describe the organization of a eusocial animal group: _____

   _____

   _____

   _____

2. What is the difference between eusocial and presocial groups? _____

   _____

   _____

3. In eusocial animals, worker and soldier castes never breed but are normally all genetically related. How might their contribution to the group help pass their own genes to the next generation?

   _____

   _____

   _____

© 2014 **BIOZONE** International
ISBN: 978-1-927173-84-8
Photocopying Prohibited

139 How Social Behavior Improves Survival

Key Idea: By working together (directly or indirectly) members of a group increase each other's chances of survival. The level of help given depends on the level of relatedness.

▶ Living in a group can improve the survival of the members, e.g. improving foraging success or decreasing the chances of predation. Animals such as meerkats, ground squirrels, and prairie dogs decrease the chances of predation by using sentries, which produce alarms calls when a predator approaches.

Gunnison's prairie dogs

Gunnison's prairie dogs (right) live in large communities called towns in the grasslands of western North America. The towns are divided into territories which may have up to 20 individuals in them. During their foraging, above-ground individuals may produce alarm calls if a predator approaches, at which nearby prairie dogs will take cover. However, whether or not an alarm call is given depends on the relatedness of the individuals receiving the call to the individual giving it. Gunnison's prairie dogs put themselves at risk when giving an alarm call by attracting the attention of the predator.

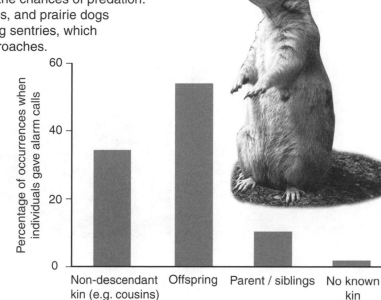

White fronted bee-eaters

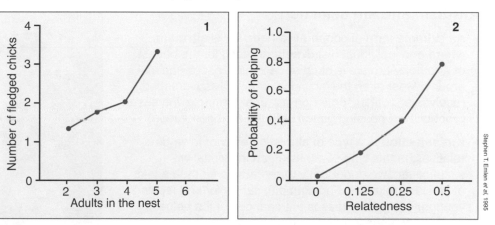

White fronted bee-eaters (left) live in family groups which include a breeding pair and non-breeding pairs. All adults help provide for the chicks. Graph 1 shows the relationship between the number of adults in the nest and the number of chicks fledged. Graph 2 shows how relatedness affects the amount of help the pairs give the chicks.

1. Use an example to explain how living in a group improves survival: _____

_____

_____

2. (a) The level of help between group members often depends on relatedness. Using the examples above, explain how relatedness to the helper affects the level of help given:

_____

_____

_____

(b) With respect to this, what is unexpected about the prairie dog data? _____

_____

CONNECT **183**   CONNECT **98**   WEB **139**   KNOW

140 Cooperative Behaviors

Key Idea: Cooperative behavior is where two or more individuals work together to achieve a common goal. It increases the probability of survival for all individuals involved.

▶ **Cooperative behavior** involves behavior in which two or more individuals work together to achieve a common goal such as defense, food acquisition, or rearing young. Examples include hunting as a coordinated team (e.g. wolf packs, chimpanzee hunts), responding to the actions of others doing much the same thing (e.g. predatory fish hunting), or acting to benefit others (e.g. oxpecker birds feeding on giraffe ticks). Cooperative behavior can occur between species or between members of the same species.

Coordinated behavior is used by many social animals for attack or defense. Kills are shared between the group.

Sometimes animals work together but not in a coordinated way, i.e. each is working to its own individual benefit.

Some cooperative behaviors are opportunistic. When bees gather nectar from flowers they incidentally pollinate the flower at the same time.

Altruism and kin selection

▶ An extreme form of cooperative behavior is **altruism**, where one individual disadvantages itself for the benefit of another. Altruism is often seen in highly social animal groups. Most often the individual who is disadvantaged receives benefit in some non-material form (e.g. increased probability of passing genes onto the next generation).

▶ **Kin selection** is a type of altruistic behavior towards relatives. In this type of behavior, an individual will sacrifice its own opportunity to reproduce for the benefit of its close relatives. For example, caring for and feeding younger siblings increases the chances of the helper's genes being passed on to the next generation.

1. Define altruism: _____

_____

2. (a) Why would altruism be uncommon between unrelated individuals? _____

_____

_____

(b) Why would altruism be common in family groups? _____

_____

_____

3. How do cooperative interactions enhance the survival of both individuals and the group they are part of?

_____

_____

© 2014 **BIOZONE** International
ISBN: 978-1-927173-84-8
Photocopying Prohibited

141 Cooperative Defense and Attack

Key Idea: Working together in defense and attack can help increase or maintain resources for the group, therefore increasing the chances of survival for individuals.

▶ Many animals engage in **group defense**, as a large number of defenders presents an intimidating obstacle to an attacker. Defense may be against a predator or defense of territories from members of the same species.

Bees and wasps form defensive swarms (right). If an attacker is stung, alarm pheromones released from the sting incite others in the swarm to attack.

Musk oxen form a defensive circle when confronted with predators such as wolves, protecting the young in the circle's interior.

Birds often mob possible predators (above). Here two crows chase a hawk which has ventured too close to their nest.

▶ **Group attack** is often used for hunting for food, but may be used by some species for raiding nests or territories to gain other resources.

Photo: April Noble from www.AntWeb.org

Some ant species, known as slavemaker ants, raid other ant nests, killing workers and capturing grubs. The grubs are carried back to the home nest where they grow and tend the slavemaker ants' own young.

Chimpanzee families living in the Kibale National Park in Uganda have been observed attacking and killing members of neighboring families to expand their own territories and acquire more resources. Some attacks lasted up to an hour.

Lionesses hunt as a coordinated group. Several lionesses hide downwind of the prey, while others circle upwind and stampede the prey towards the lionesses in wait. Group cooperation reduces the risk of injury and increases the chance of a kill. Only 15% of hunts by a solitary lioness are successful. Those hunting in a group are successful 40% of the time.

1. Describe two benefits of cooperative defense: _____

_____

_____

2. (a) Suggest two reasons for cooperative attacks: _____

_____

(b) Suggest why cooperative attacks are more likely to be successful that individuals attacks: _____

_____

_____

142 Cooperative Food Gathering

Key Idea: Cooperative behavior in gathering food increases the chances of foraging success and improves efficiencies.

▶ Cooperating to gather food can be much more efficient that finding it alone. It increases the chances of finding food or capturing prey.

▶ Cooperative hunting will evolve in a species if the following circumstances apply:
  • If there is a sustained benefit to the hunting participants
  • If the benefit for a single hunter is less than that of the benefit of hunting in a group
  • Cooperation within the group is guaranteed

Worker castes in army ants

Orcas hunting

Wild beehive

Cooperative food gathering in ants often involves division of labor. Leaf-cutter ants harvest parts of leaves and use them to cultivate a fungus, which they eat. Workers that tend the fungus gardens have smaller heads than the foragers, which cut and transport the leaves. Similarly, **army ants** have several distinct worker castes. The smaller castes collect small prey, and larger porter ants collect larger prey. The largest workers defend the nest.

Dolphins herd fish into shallow water and trap them against the shore where they can be easily caught. When a pod of orcas (killer whales) spot a seal on an iceberg they swim towards the flow at high speed before ducking under the ice. This causes a large wave to wash over the iceberg, knocking the seal into the sea where it can be captured. If the seal fails to fall into the water, one of the whales will land itself onto the iceberg tipping the seal into the water.

Sometimes different species work together to gather food. The greater honeyguide, a bird that is found in the sub-Sahara, is notable for its behavior of guiding humans (either deliberately or not) to wild beehives. When the humans retrieve the honey, they leave behind some of the wax comb which the honeyguide eats. It is not known why this behavior evolved, because the honeyguide can enter a beehive without human help.

Cooperative hunting in chimpanzees

Hunting in chimpanzees benefits from cooperation between members of the group. As the number of hunters increase, the number of successful hunts also increases, as does the amount of meat acquired per hunt and the energy gained per individuals to a peak at four hunters. Importantly members of the hunt share their meat other who did not hunt. On average, hunters share 45% of their meat with others, helping to build bonds between members of the group.

Number of hunters	Number of hunts	Hunting success (%)	Meat per hunt (kg)	Net benefit per hunter(kJ)
1	30	13	1.23	4015
2	34	29	0.82	1250
3	39	49	3.12	3804
4	25	72	5.47	5166
5	12	75	4.65	3471
6	12	42	3.17	1851
>6	10	90	9.27	5020

Christophe Boesch 1994

1. What are the advantages of cooperative food gathering? _____

2. What conditions favor cooperative food gathering? _____

3. Why do chimpanzees share almost half their meat from a hunt? _____

© 2014 **BIOZONE** International
ISBN: 978-1-927173-84-8
Photocopying Prohibited

143 Chapter Review

Summarize what you know about this topic so far under the headings provided. You can draw diagrams or mind maps, or write short notes to organize your thoughts. Use the images and hints and guidelines included to help you:

Group behavior and social organization

HINT: How does group behavior enhance survival?

Migration, flocking, schooling and herding

HINT: Include definitions for each, and describe why these behaviors occur.

REVISE

144 KEY TERMS: Did You Get It?

1. Test your vocabulary by matching each term to its definition, as identified by its preceding letter code.

altruism

cooperative behavior

flocking

herding

kin selection

migration

schooling

social group

A A behavior in which individuals work together in order to reach a common goal, e.g. finding food.

B The long distance movement of individuals from one place to another.

C A behavior seen in hoofed mammals for protection against predators.

D Behavior that favors the reproductive success of an organism's relatives, even at a cost to their own survival and/or reproduction.

E Behavior in which an animal sacrifices its own well-being for the benefit of another animal.

F The grouping together of a large number of fish.

G The grouping together of a large number of birds.

H Examples of this may be loosely associated groups, or complex groups with clear social structures.

2. Describe the difference between solitary, non-social group, and social group behavior in animals:

_____

_____

_____

3. (a) Identify the structure formed by the fish in the photo right:

_____

(b) What benefits are there to the fish forming this structure?

_____

_____

_____

_____

4. Draw line to match up the first half and second half of the sentences below:

Cooperative behavior in a group...

Cooperative behavior evolves when there is a...

By displaying altruistic behavior to family members, an individual can indirectly increase...

The energetic costs involved in migration are outweighed...

Cooperation has been taken to the extreme by eusocial animals in which the majority of members of the group...

...sacrifice their individual reproductive chances to ensure the collective's genes (and therefore their own genes) are passed to the next generation.

...by the reproductive benefits.

...the likelihood of their own genes being passed on to the next generation.

....benefits all members of that group.

...sustained benefit to the members of the group.

© 2014 **BIOZONE** International
ISBN: 978-1-927173-84-8
Photocopying Prohibited

VOCAB

Heredity: Inheritance and Variation of Traits

Concepts and Connections
Use arrows to make your own connections between related concepts in this section of this workbook

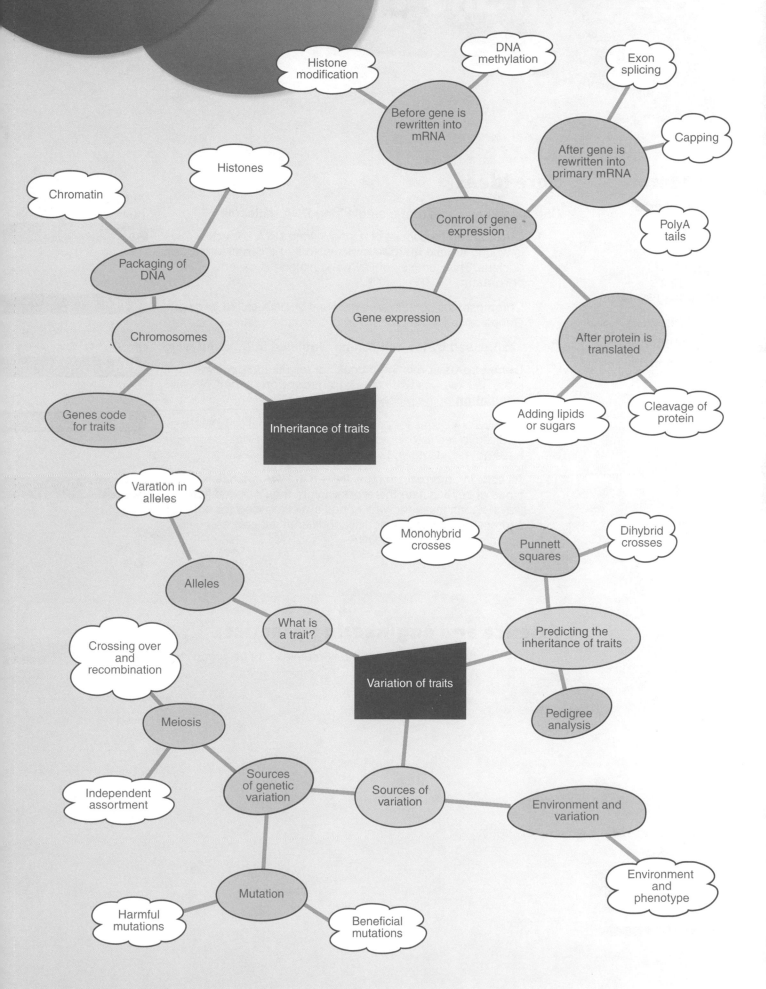

Inheritance of Traits

Key terms

chromosome

chromatin

DNA

gene

gene expression

histone

transcription

translation

Core ideas

Activity number

Chromosomes consist of a single long DNA molecule

☐ 1. **Chromosomes** consist of a single, long **DNA** molecule. DNA is wound around **histone** proteins so that it can fit into a cell's nucleus. This complex of DNA and histone is called **chromatin**.

`145`

☐ 2. Chromosomes contain long sections of DNA called **genes**. Genes code for the production of specific proteins.

`145`

Genes expressed by the cell can be regulated in different ways

☐ 3. **Gene expression** is the process of rewriting a gene into a protein. It involves two stages: **transcription** of the DNA and **translation** of the mRNA into protein.

`146`

☐ 4. Not all DNA codes for proteins. Some segments of DNA are involved in regulatory or structural functions. Some segments have no as-yet known function.

`146`

☐ 5. All cells in an organism have the same DNA but different types of cell regulate the expression of the DNA in different ways. Variations in the way genes are expressed (or not expressed) can cause significant differences between cells or organisms with identical DNA.

`146  147`

Science and engineering practices

☐ 1. Use models to explain how many proteins can be produced from just one gene.

`145  146`

145 Chromosomes

Key Idea: Chromosomes are a single long molecule of DNA coiled around histone proteins. Chromosomes contain protein-coding regions called genes.

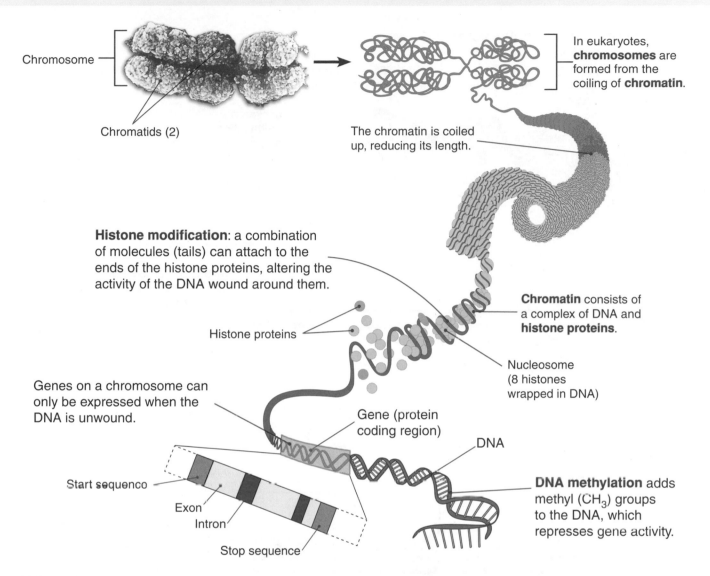

Chromosome

Chromatids (2)

In eukaryotes, **chromosomes** are formed from the coiling of **chromatin**.

The chromatin is coiled up, reducing its length.

Histone modification: a combination of molecules (tails) can attach to the ends of the histone proteins, altering the activity of the DNA wound around them.

Histone proteins

Chromatin consists of a complex of DNA and **histone proteins**.

Nucleosome (8 histones wrapped in DNA)

Genes on a chromosome can only be expressed when the DNA is unwound.

Gene (protein coding region)

DNA

Start sequence

Exon

Intron

Stop sequence

DNA methylation adds methyl (CH_3) groups to the DNA, which represses gene activity.

Eukaryotic genes are typically preceded by segments of DNA called promoters and enhancers which are involved in starting the process of transcribing the gene into mRNA. The part of the gene that is transcribed includes areas called introns and exons and is enclosed between a start and stop sequence. The start and stop sequences indicate where transcription starts and stops. The introns are removed from the mRNA after transcription so that only the exons appear in the final mRNA.

1. Suggest a purpose for DNA coiling: _____

   _____

2. What is the purpose of DNA methylation and histone modification? _____

   _____

   _____

3. Identify three parts of the gene that do not appear in the final mRNA:

   _____

4. What has to happen before a gene can be expressed (made into a protein)?: _____

   _____

 © 2014 **BIOZONE** International
ISBN: 978-1-927173-84-8
Photocopying Prohibited

146 Gene Expression

Key Idea: Different factors determine whether or not a gene is expressed. Much of the information in a gene may be removed or modified before a protein is produced.

Gene expression in eukaryotes

▶ Gene expression in eukaryotes is complex. A gene may be modified after transcription and after translation, so that many different gene products (e.g. proteins) can be produced from just one gene. In addition, not all mRNAs are translated into a protein. Some have roles in regulating gene expression.

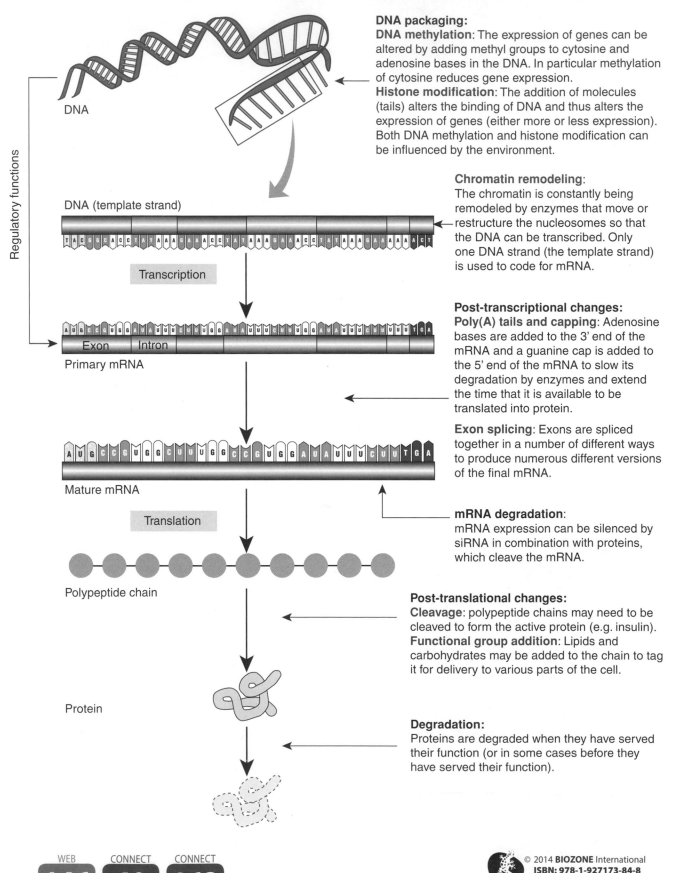

DNA packaging:
DNA methylation: The expression of genes can be altered by adding methyl groups to cytosine and adenosine bases in the DNA. In particular methylation of cytosine reduces gene expression.
Histone modification: The addition of molecules (tails) alters the binding of DNA and thus alters the expression of genes (either more or less expression). Both DNA methylation and histone modification can be influenced by the environment.

Chromatin remodeling: The chromatin is constantly being remodeled by enzymes that move or restructure the nucleosomes so that the DNA can be transcribed. Only one DNA strand (the template strand) is used to code for mRNA.

Post-transcriptional changes:
Poly(A) tails and capping: Adenosine bases are added to the 3' end of the mRNA and a guanine cap is added to the 5' end of the mRNA to slow its degradation by enzymes and extend the time that it is available to be translated into protein.

Exon splicing: Exons are spliced together in a number of different ways to produce numerous different versions of the final mRNA.

mRNA degradation: mRNA expression can be silenced by siRNA in combination with proteins, which cleave the mRNA.

Post-translational changes:
Cleavage: polypeptide chains may need to be cleaved to form the active protein (e.g. insulin).
Functional group addition: Lipids and carbohydrates may be added to the chain to tag it for delivery to various parts of the cell.

Degradation:
Proteins are degraded when they have served their function (or in some cases before they have served their function).

KNOW

WEB
146

CONNECT
40

CONNECT
163

1. Identify three ways in which gene expression can be supressed: _____

   _____

2. Identify a way in which gene expression can be increased: _____

   _____

3. Identical twins have identical DNA. However, as they grow and age, differences in personality and physical appearance can occur (people who are familiar with twins can easily tell them apart). How does DNA methylation and histone modification account for some of these differences?

   _____

   _____

   _____

   _____

   _____

   _____

   _____

4. What is the purpose of the poly(A) tails and guanine cap on mRNA? _____

   _____

   _____

5. How does exon splicing lead to a variety of possible polypeptide chains? _____

   _____

   _____

   _____

6. How do post-translational changes to the polypeptide chain affect its functioning? _____

   _____

   _____

   _____

   _____

7. Human DNA contains about 25,000 genes, but it is estimated that there are more than 250,000 proteins in the human body. In *Drosophila*, the Dscam gene can produce 38,016 different mRNA molecules. Explain how these vast numbers of proteins are produced from such a few genes.

   _____

   _____

   _____

   _____

   _____

   _____

   _____

   _____

 © 2014 **BIOZONE** International
ISBN: 978-1-927173-84-8
Photocopying Prohibited

147 Controlling Gene Expression

Key Idea: Variations in the way genes are expressed can cause significant differences between cells or organisms, even if their DNA is identical.

Same genes, different result

All the cells in an multicellular organism have identical DNA. As an organism develops from the zygote, differences in the way the DNA is expressed in developing cells cause them to differentiate into different types (right). Cells of the same type can express different proteins or amounts of protein depending on the environment they are exposed to during development or due to random gene inactivations (below).

Examples of plant cells

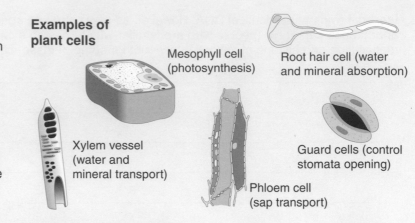

Mesophyll cell (photosynthesis)

Root hair cell (water and mineral absorption)

Xylem vessel (water and mineral transport)

Guard cells (control stomata opening)

Phloem cell (sap transport)

Females have two X chromosomes, one of which is inactivated in every cell. However, which X chromosome is inactivated is random. In cats, the gene for coat color is found on the X chromosome. Cats with two different alleles for coat color will only have one active allele per pigment cell, so each cell will produce one or the other color, giving a patchwork (calico) coat.

All worker bees and the queen bee (circled above) in a hive have the same genome, yet the queen looks and behaves very differently from the workers. Bee larvae fed a substance called royal jelly develop into queens. Research shows that royal jelly contains factors that silence the activation of a gene called Dnmt3, which itself silences many other genes.

Twins have the same genome but over the years of their life become different in both appearance and behavior. Studies have found that 35% of twins have significant differences in gene expression. More importantly, the older the twins, the more difference there is in the gene expression.

1. How do cells with identical DNA differentiate into different types? _____

_____

2. Why are patchwork (calico) cats always female? _____

_____

_____

3. Studies on bee development focused on the Dnmt3 gene. One study switched off the Dnmt3 gene in 100 bee larvae. All the larvae developed into queens. Leaving the gene switched on in larvae causes them to develop into workers. Compare these results to feeding larvae royal jelly. Which results mimic feeding larvae royal jelly?

_____

_____

_____

 © 2014 **BIOZONE** International
ISBN: 978-1-927173-84-8
Photocopying Prohibited

148 Chapter Review

Summarize what you know about this topic so far under the headings provided. You can draw diagrams or mind maps, or write short notes to organize your thoughts. Use the images and hints and guidelines included to help you:

The structure of chromosomes

HINT: Include how the chromosome is packaged into the nucleus.

The regulation of gene expression

HINT: Include post transcriptional and post translation changes.

The effects of gene expression

HINT: Include the effects of switching genes on and off.

REVISE

149 KEY TERMS: Did You Get It?

1. Use the wordlist to label the diagram below: *chromatid, chromatin, chromosome, DNA, exon, gene, histone, intron, nucleosome, start sequence, stop sequence*:

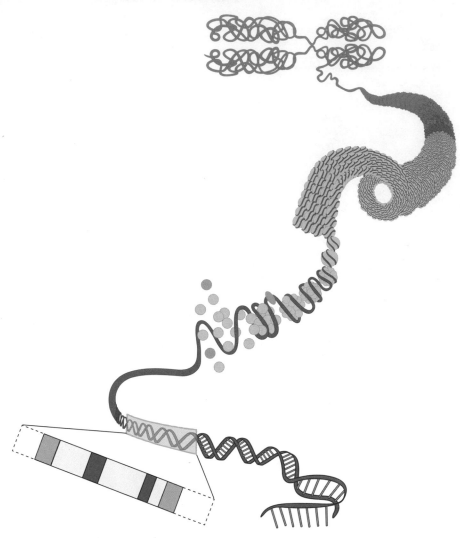

2. Test your vocabulary by matching each term to its definition, as identified by its preceding letter code.

chromosome

chromatin

DNA

gene

gene expression

histone

histone tail

methylation

A Deoxyribonucleic acid. Macromolecule consisting of many millions of units, each containing a phosphate group, sugar and a base (A,T, C or G). Stores the genetic information of the cell.

B The basic unit of inheritance.

C The process of transferring the information encoded in a gene into a gene product.

D Single piece of DNA that contains many genes and associated regulatory elements and proteins.

E The addition of a CH_3 group to a DNA base, altering gene activity.

F A structure added to a histone protein that alters the way DNA binds to it.

G A protein found in the nuclei of eukaryotic cells that packages and orders the DNA into structural units called nucleosomes.

H A complex of DNA and histone proteins.

TEST

© 2014 **BIOZONE** International
ISBN: 978-1-927173-84-8
Photocopying Prohibited

LS3.B

Variation of Traits

Key terms

allele

crossing over

dihybrid

heterozygous

chromosome

gene

independent assortment

homozygous

meiosis

monohybrid

mutation

phenotype

recombination

trait

Core ideas

Variation can result from genetic processes

	Activity number
☐ 1. **Traits** are phenotypic variants controlled by **genes**, which can be inherited.	**150 151**
☐ 2. **Variation** in individuals is caused by both genetic and environmental factors.	**152-154**
☐ 3. **Homologous chromosomes** can be **heterozygous** or **homozygous**.	**151**
☐ 4. The process of **meiosis** gives the opportunity to produce genetic variation. **Chromosomes** may exchange genetic material in a process called **crossing over**, which results in **recombination** of **alleles**. **Independent assortment** allows random pairing of alleles.	**155 156**
☐ 5. **Mutations** are the ultimate source of all new genetic variation (new alleles). Most mutations are harmful, but some may be beneficial.	**157-161**

Variation can result from environmental influences

☐ 6. A **phenotype** is the expression of both genetic and environmental influences.	**150**
☐ 7. The environmental influences experienced by one generation may affect subsequent generations.	**162 163**
☐ 8. Punnett squares can be used to predict the probable outcomes of **monohybrid** and **dihybrid** crosses.	**164-167**

Science and engineering practices

☐ 1. Use a model based on evidence to explain that heritable genetic variation arises through meiosis and mutation.	**156-158 160**
☐ 2. Argue from evidence that the environment can affect the phenotype of subsequent generations.	**159 163**
☐ 3. Apply concepts of statistics and probability to explain the variation and expression of traits in a population.	**164-167**
☐ 4. Use evidence to build and defend the claim that variation in a population is essential for its survival and evolution.	**152 158 159**

BIOZONE APP
Student Review Series
Variation of Traits

150 What is a Trait?

Key Idea: Traits are controlled by genes, which can be inherited and passed from one generation to the next.

Traits are inherited

▶ **Traits** are particular variants of phenotypic (observed physical) characters, e.g. blue eye color. Traits may be controlled by one gene or many genes and can show continuous variation, e.g. height in humans, or discontinuous variation, e.g. flower color in pea plants.

▶ **Gregor Mendel**, an Austrian monk (1822-1884), used pea plants to study inheritance. Using several phenotypic characteristics he was able to show that their traits were inherited in predictable ways.

Mendel's experiments

Mendel studied seven phenotypic characters of the pea plant:
- Flower color (violet or white)
- Pod color (green or yellow)
- Height (tall or short)
- Position of the flowers on the stem (axial or terminal)
- Pod shape (inflated or constricted)
- Seed shape (round of wrinkled)
- Seed color (yellow or green)

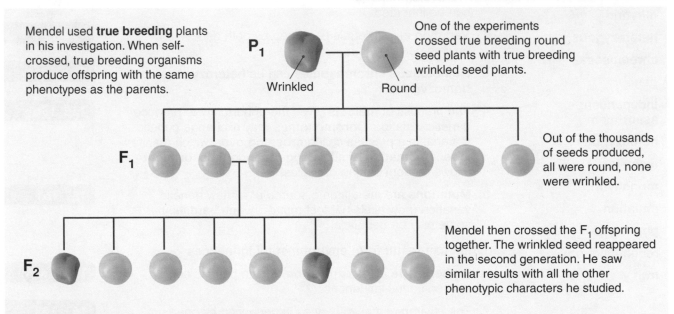

Mendel used **true breeding** plants in his investigation. When self-crossed, true breeding organisms produce offspring with the same phenotypes as the parents.

P_1 — Wrinkled — Round

One of the experiments crossed true breeding round seed plants with true breeding wrinkled seed plants.

F_1

Out of the thousands of seeds produced, all were round, none were wrinkled.

F_2

Mendel then crossed the F_1 offspring together. The wrinkled seed reappeared in the second generation. He saw similar results with all the other phenotypic characters he studied.

Three conclusions could be made from these results:
- Traits are determined by a unit that passes unchanged from parent to offspring (these units are now called genes).
- Each individual inherits one unit (gene) for each trait from each parent (each individual has two units).
- Traits may not physically appear in an individual, but the units (genes) for them can still be passed to its offspring.

1. Define a trait: _____

_____

2. Define true breeding: _____

_____

3. (a) What was the ratio of smooth seeds to wrinkled seeds in the F_2 generation? _____

_____

   (b) Suggest why the wrinkled seed trait did not appear in the F_1 generation: _____

_____

_____

© 2014 **BIOZONE** International
ISBN: 978-1-927173-84-8
Photocopying Prohibited

151 Different Alleles For Different Traits

Key Idea: Eukaryotes generally have paired chromosomes. Each chromosome contains many genes, and each gene may have a number of versions, called alleles.

Homologous chromosomes

In sexually reproducing organisms, chromosomes are generally found in pairs. Each parent contributes one chromosome to the pair. The pairs are called **homologues** or **homologous pairs**. Each homologue carries an identical assortment of genes, but the version of the gene (the **allele**) from each parent may differ. This diagram shows the position of three different genes on the same chromosome that control three different traits (A, B and C).

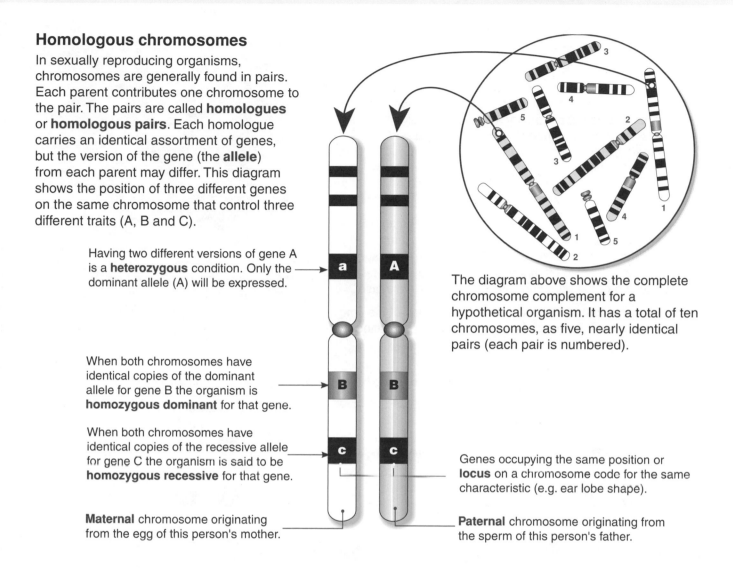

Having two different versions of gene A is a **heterozygous** condition. Only the dominant allele (A) will be expressed.

The diagram above shows the complete chromosome complement for a hypothetical organism. It has a total of ten chromosomes, as five, nearly identical pairs (each pair is numbered).

When both chromosomes have identical copies of the dominant allele for gene B the organism is **homozygous dominant** for that gene.

When both chromosomes have identical copies of the recessive allele for gene C the organism is said to be **homozygous recessive** for that gene.

Genes occupying the same position or **locus** on a chromosome code for the same characteristic (e.g. ear lobe shape).

Maternal chromosome originating from the egg of this person's mother.

Paternal chromosome originating from the sperm of this person's father.

1. Define the following terms describing the allele combinations of a gene in a sexually reproducing organism:

   (a) Heterozygous: _____

   _____

   (b) Homozygous dominant: _____

   _____

   (c) Homozygous recessive: _____

   _____

2. For a gene given the symbol '**A**', write down the alleles present in an organism that is:

   (a) Heterozygous:_____   (b) Homozygous dominant: _____   (c) Homozygous recessive:_____

3. What is a **homologous pair** of chromosomes?_____

   _____

   _____

 © 2014 **BIOZONE** International
ISBN: 978-1-927173-84-8
Photocopying Prohibited

152 Why is Variation Important?

Key Idea: Variation in a population or species is important in a changing environment. Both sexually and asexually reproducing species have strategies to increase variation.

▸ **Variation** refers to the diversity of phenotypes or genotypes within a population or species. Variation helps organisms survive in a changing environment.

▸ Sexual reproduction produces variability, which provides ability to adapt to a changing physical environment. However, environments can change very slowly, it may take millions of years for a mountain range to rise from the seabed. This is more than enough time for even asexually reproducing species to acquire the variability needed to adapt. However changes in the biotic environment, such as the appearance of new strains of disease, require a fast response.

▸ Variation is important for defending against disease. Species that evolve to survive a disease flourish. Those that do not, die out. It is thought that sexual reproduction is an adaptation to increase variability in offspring and so provide a greater chance that any one of the offspring will survive a given disease.

▸ Even species that reproduce asexually for much of the time can show a large amount of variation within a population.

Aphids can reproduce sexually and asexually. Females hatch in spring and give birth to clones. Many generations are produced asexually. Just before fall, the aphids reproduce sexually. The males and females mate and the females produce eggs which hatch the following spring. This increases variability in the next generation.

Diagrams to show how three beneficial mutations could be combined through sexual or asexual reproduction.

Variation by sexual reproduction

Variation from recombining alleles

Mutation A

Mutation B

AB   AB   AB

AB   AB   AB

Mutation C

ABC   ABC   ABC

During meiosis, alleles are recombined in new combinations. Some combinations of alleles may be better suited to a particular environment than others. This variability is produced without the need for mutation. Beneficial mutations in separate lineages can be quickly combined through sexual reproduction.

Variation by asexual reproduction

Clones   ◀   D   ◀ Mutation D

D   D   D

Mutation E

DE   D   D

Mutation F

DEF   DEF   DEF   DEF

Most asexually reproducing species are able to exchange genes occasionally. Bacteria exchange genes with other bacteria during a process called conjugation (thicker blue line). This allows mutations that arise in one lineage to be passed to another.

1. Why is variation important in populations or species? _____

_____

_____

© 2014 **BIOZONE** International
ISBN: 978-1-927173-84-8
Photocopying Prohibited

153 Sources of Variation

Key Idea: Variation may come from changes to the genetic material (mutation), through sexual reproduction, and as a result of the effects of the environment.

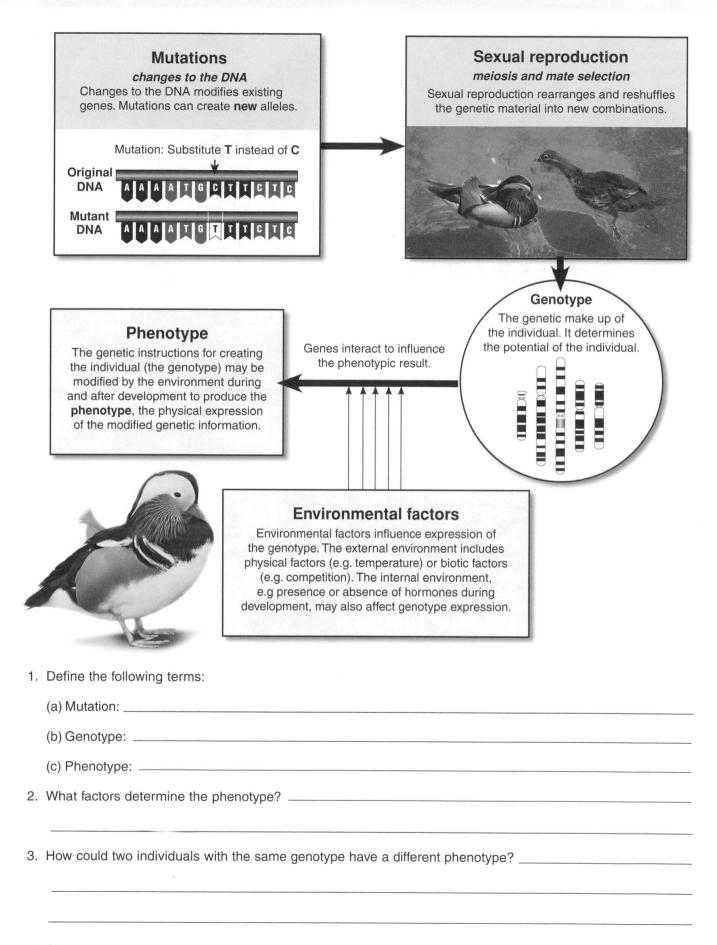

Mutations
changes to the DNA
Changes to the DNA modifies existing genes. Mutations can create **new** alleles.

Mutation: Substitute **T** instead of **C**

Original DNA: A A A A T G C T T C T C

Mutant DNA: A A A A T G T T T C T C

Sexual reproduction
meiosis and mate selection
Sexual reproduction rearranges and reshuffles the genetic material into new combinations.

Genotype
The genetic make up of the individual. It determines the potential of the individual.

Phenotype
The genetic instructions for creating the individual (the genotype) may be modified by the environment during and after development to produce the **phenotype**, the physical expression of the modified genetic information.

Genes interact to influence the phenotypic result.

Environmental factors
Environmental factors influence expression of the genotype. The external environment includes physical factors (e.g. temperature) or biotic factors (e.g. competition). The internal environment, e.g presence or absence of hormones during development, may also affect genotype expression.

1. Define the following terms:

   (a) Mutation: _____

   (b) Genotype: _____

   (c) Phenotype: _____

2. What factors determine the phenotype? _____

   _____

3. How could two individuals with the same genotype have a different phenotype? _____

   _____

CONNECT CONNECT
182 **157** KNOW

154 Examples of Genetic Variation

Key Idea: Genetic variation may be continuous as a result of quantitative traits (e.g. leaf length) or it may be discontinuous as a result of qualitative traits (e.g. gender).

Individuals show particular variants of phenotypic characters called traits, e.g. blue eye color. Traits that show continuous variation are called quantitative traits. Traits that show discontinuous variation are called qualitative traits.

Quantitative traits

Quantitative traits are determined by a large number of genes. For example, skin color has a continuous number of variants from very pale to very dark. Individuals fall somewhere on a normal distribution curve of the phenotypic range. Other examples include height in humans for any given age group, length of leaves in plants, grain yield in corn, growth in pigs, and milk production in cattle. Most quantitative traits are also influenced by environmental factors.

Leaf length in ivy

Leaf length in ivy is determined by a number of factors. Here the lengths approximate a normal distribution.

Graph: x-axis "Length of leaf (mm)" from 15 to 105; y-axis "Number of leaves" from 0 to 16

Grain yield in corn

Pig growth

Qualitative traits

Qualitative traits are determined by a single gene with a very limited number of variants present in the population. For example, blood type (ABO) in humans has four discontinuous traits A, B, AB or O. Individuals fall into discrete categories. Comb shape in poultry (right) is a qualitative trait and birds have one of four phenotypes depending on which combination of four alleles they inherit. The dash (missing allele) indicates that the allele may be recessive or dominant. Albinism is the result of the inheritance of recessive alleles for melanin production. Those with the albino phenotype lack melanin pigment in the eyes, skin, and hair. Flower color in snapdragons (right) is also a qualitative trait determined by two alleles (red and white).

Single comb	Walnut comb	Pea comb	Rose comb
rrpp	**R_P_**	**rrP_**	**R_pp**

Both the comb shape in chickens and the flower color in snapdragons are qualitative traits. They are either one or the other, but not in-between.

C^R C^R
C^W C^W

Snapdragons

1. What is the difference between **continuous** and **discontinuous** variation? _____

_____

_____

2. Identify each of the following phenotypic traits as continuous (quantitative) or discontinuous (qualitative):

(a) Wool production in sheep: _____   (d) Albinism in mammals: _____

(b) Hand span in humans: _____   (e) Body weight in mice: _____

(c) Blood groups in humans: _____   (f) Flower color in snapdragons: _____

© 2014 **BIOZONE** International
ISBN: 978-1-927173-84-8
Photocopying Prohibited

KNOW

155 Meiosis

Key Idea: Meiosis is a special type of cell division for the purposes of sexual reproduction.

Meiosis

▶ **Meiosis** is a special type of cell division necessary for the production of sex cells (gametes) for the purpose of sexual reproduction. Meiosis involves a single chromosomal duplication followed by two successive nuclear divisions, and it halves the diploid chromosome number. An overview of meiosis is shown on the right. Meiosis occurs in the sex organs of plants and animals.

Meiosis produces variation

▶ Meiosis is an important way of introducing genetic variation. The assortment of chromosomes into the gametes (the proportion from the parental father or mother) is random and can produce a huge number of possible chromosome combinations.

▶ During meiosis, a process called **crossing over** may occur when homologous chromosomes may exchange genes. This further adds to the variation in the gametes.

▶ DNA replication precedes meiosis. If genetic mistakes (mutations) occur here, they will be passed on (inherited) by the offspring.

The process of meiosis

2n cell (diploid. n is the number of haploid chromosomes). Each chromosome is replicated.

Crossing over has occurred.

Homologous chromosomes pair up. The pairs are separated.

An intermediate cell forms with a replicated copy of each chromosome.

Chromatid

The chromosomes line up again. This time the chromatids are separated.

Gametes (n. haploid) form with one copy of each chromosome

1. (a) What is the purpose of meiosis? _____

   _____

   _____

   (b) Where does meiosis take place? _____

   _____

2. Describe how variation can arise during meiosis: _____

   _____

   _____

   _____

 © 2014 **BIOZONE** International
ISBN: 978-1-927173-84-8
Photocopying Prohibited

CONNECT   WEB

171 **155** **KNOW**

156 Meiosis and Variation

Key Idea: Meiosis produces variation. Independent assortment and crossing over are two important ways of introducing variation into the gametes formed during meiosis.

▶ Independent assortment and crossing over (leading to recombination of alleles) are mechanisms that occur during meiosis. They increase the genetic variation in the gametes, and therefore the offspring.

Independent assortment

Independent assortment is an important mechanism for producing variation in gametes. The law of independent assortment states that allele pairs separate independently during meiosis. This results in the production of 2^x different possible combinations (where x is the number of chromosome pairs). For the example right, there are two chromosome pairs. The number of possible allele combinations in the gametes is therefore $2^2 = 4$ (only two possible combinations are shown).

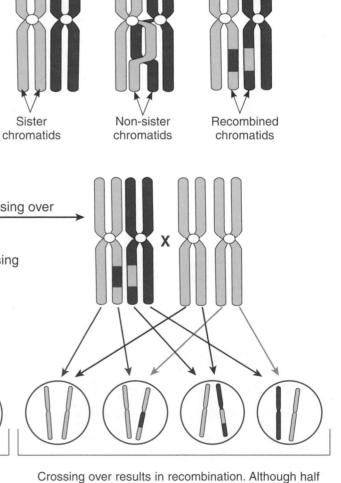

Crossing over and recombination

While they are paired during the first stage of meiosis, the non-sister chromatids of homologous chromosomes may become tangled and segments may be exchanged in a process called **crossing over**.

Crossing over results in the **recombination** of alleles, producing greater variation in the offspring than would otherwise occur. Alleles that are linked (on the same chromosome) may be exchanged and so become unlinked.

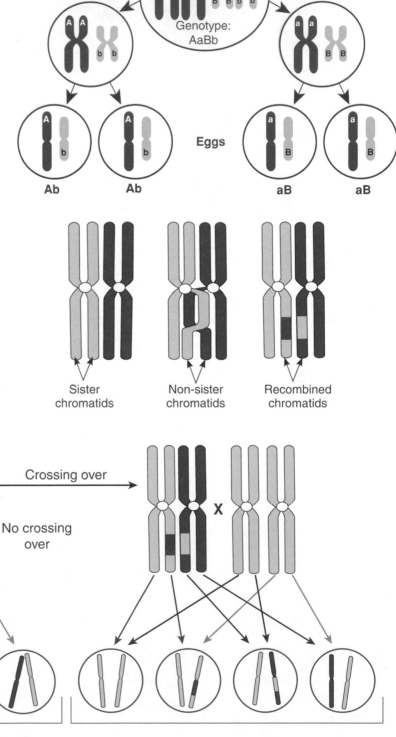

No crossing over (so no recombination) in this cross results in all the offspring having the same genotypes as the parents.

Crossing over results in recombination. Although half of the offspring are the same as the parents, half have a new genetic combination.

© 2014 **BIOZONE** International
ISBN: 978-1-927173-84-8
Photocopying Prohibited

1. (a) Using the diagram on independent assortment (previous page) draw the other two gamete combinations not shown in the diagram:

Gamete 3

Gamete 4

(b) For each of the following chromosome numbers, calculate the number of possible gamete combinations:

i. 8 chromosomes: _____

ii. 24 chromosomes: _____

iii 64 chromosomes: _____

2. What are sister and non-sister chromatids? _____

_____

_____

3. (a) What is crossing over? _____

_____

_____

(b) How does crossing over increase the variation in the gametes (and hence the offspring)?

_____

_____

_____

4. Crossing over occurs at a single point between the chromosomes below.

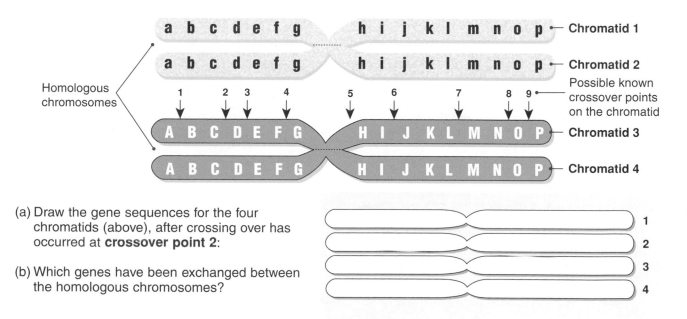

a b c d e f g   h i j k l m n o p ← Chromatid 1

a b c d e f g   h i j k l m n o p ← Chromatid 2

Homologous chromosomes

1   2 3   4   5   6   7   8 9 ← Possible known crossover points on the chromatid

A B C D E F G   H I J K L M N O P ← Chromatid 3

A B C D E F G   H I J K L M N O P ← Chromatid 4

(a) Draw the gene sequences for the four chromatids (above), after crossing over has occurred at **crossover point 2**:

1

2

3

4

(b) Which genes have been exchanged between the homologous chromosomes?

_____

157 Mutations

Key Idea: Changes to the DNA sequence are called mutations. Mutations are the ultimate source of new genetic variation (i.e. new alleles).

▶ **Mutations** are changes to the DNA sequence and occurs through errors in DNA copying. Changes to the DNA modifies existing genes and can create variation in the form of new alleles. Ultimately, mutations are the source of all new genetic variation.

▶ There are several types of mutation. Some change only one nucleotide base, while others change large parts of chromosomes. Bases may be inserted into, substituted, or deleted from the DNA. Most mutations are harmful (e.g. those that cause cancer) but very occasionally a mutation can be beneficial.

▶ An example of a mutation producing a new allele is described below. This mutation causes the most common form of genetic hearing loss (called NSRD) in children. The mutation occurs in the gene coding for a protein called connexin 26.

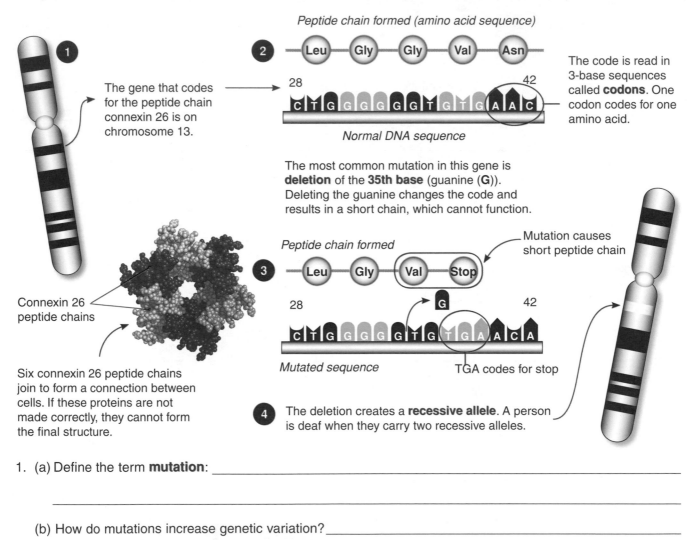

Peptide chain formed (amino acid sequence)

2 Leu — Gly — Gly — Val — Asn

The gene that codes for the peptide chain connexin 26 is on chromosome 13.

28 ... CTGGGGGGTGTGAAC ... 42

Normal DNA sequence

The code is read in 3-base sequences called **codons**. One codon codes for one amino acid.

The most common mutation in this gene is **deletion** of the **35th base** (guanine (**G**)). Deleting the guanine changes the code and results in a short chain, which cannot function.

Peptide chain formed

3 Leu — Gly — Val — Stop

Mutation causes short peptide chain

28 ... CTGGGGGTGTGAACA ... 42

Mutated sequence   TGA codes for stop

Connexin 26 peptide chains

Six connexin 26 peptide chains join to form a connection between cells. If these proteins are not made correctly, they cannot form the final structure.

4 The deletion creates a **recessive allele**. A person is deaf when they carry two recessive alleles.

1. (a) Define the term **mutation**: _____

    _____

   (b) How do mutations increase genetic variation? _____

    _____

    _____

2. Identify the type of mutation that has occurred in the DNA sequences shown below:

    Normal DNA sequence: CCT GAT GCG AAG TTA TCA GTA CCA

    DNA sequence 1:   CCT GAT GCG TTA TCA GTA CCA   _____

    DNA sequence 2:   CCT GAT GCG AAG CCC TTA TCA GTA CCA _____

    DNA sequence 3:   CCT GAT GCG AAG TTA TGA GTA CCA   _____

© 2014 **BIOZONE** International
ISBN: 978-1-927173-84-8
Photocopying Prohibited

158 The Effects of Mutations

Key Idea: Most mutations produce harmful effects, but some can produce beneficial effects, and others have no effect at all.

▶ Most mutations have a harmful effect on the organism. This is because changes to the DNA sequence of a gene can potentially change the amino acid chain encoded by the gene. Proteins need to fold into a precise shape to function properly. A mutation may change the way the protein folds and prevent it from carrying out its usual biological function.

▶ However, sometimes a mutation can be beneficial. The mutation may result in a more efficient protein, or produce an entirely different protein that can improve the survival of the organism.

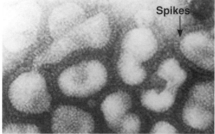

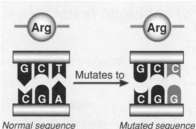

Beneficial mutations

Some mutations aid survival. In viruses (e.g. *Influenzavirus* above) genes coding for the glycoprotein spikes (arrowed) are constantly mutating, producing new strains that avoid detection by the host's immune system.

Silent mutations

Silent mutations do not change the amino acid sequence nor the final protein. In the genetic code, several codons may code for the same amino acid. Silent mutations may be neutral if they do not alter an organism's fitness.

Harmful mutations

Most mutations cause harmful effects, usually because they stop or alter the production of a protein (often an enzyme). Albinism (above) is one of the more common mutations in nature, and leaves an animal with no pigmentation.

Beneficial mutations in *E. coli*

An experiment known as the *E.coli* long term evolution experiment has incubated 12 lines of *E. coli* bacteria for more than 20 years.

After 31,000 generations a mutation in one of the *E. coli* populations enabled it to feed off citrate, a component of the medium they were grown in (*E. coli* are usually unable to do this). This ability gave these *E. coli* an advantage because they could use another food source.

The mutation was noticed when the optical density (cloudiness) of the flask containing the *E. coli* increased (right), indicating an increase in bacteria numbers.

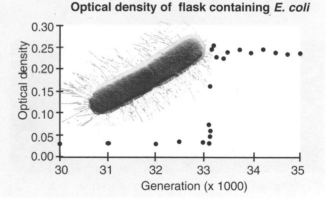

1. How might a mutation cause a beneficial effect on protein function? _____

2. How might a mutation have a harmful effect on protein function? _____

3. Why does a silent mutation have no apparent affect on an organism? _____

4. How would the citrate mutation in *E. coli* give it an advantage over other *E. coli* populations?

© 2014 **BIOZONE** International
ISBN: 978-1-927173-84-8
Photocopying Prohibited

159 Evolution of Antibiotic Resistance

Key Idea: Resistance to antibiotics can arise in bacteria by mutation. Antibiotic resistant bacteria can pass this resistance on to the next generation and to other populations.

▶ Antibiotic resistance arises when a genetic change allows bacteria to tolerate levels of an antibiotic that might normally kill it or stop its growth. This resistance may arise spontaneously through mutation or by transfer of genetic material between microbes.

▶ Genomic analyses from 30,000 year old permafrost sediments show that antibiotic resistant genes are not new. They have long been present in the bacterial genome, pre-dating the modern selective pressure of antibiotic use. In the current selective environment, these genes have proliferated and antibiotic resistance has spread.

The evolution of antibiotic resistance in bacteria

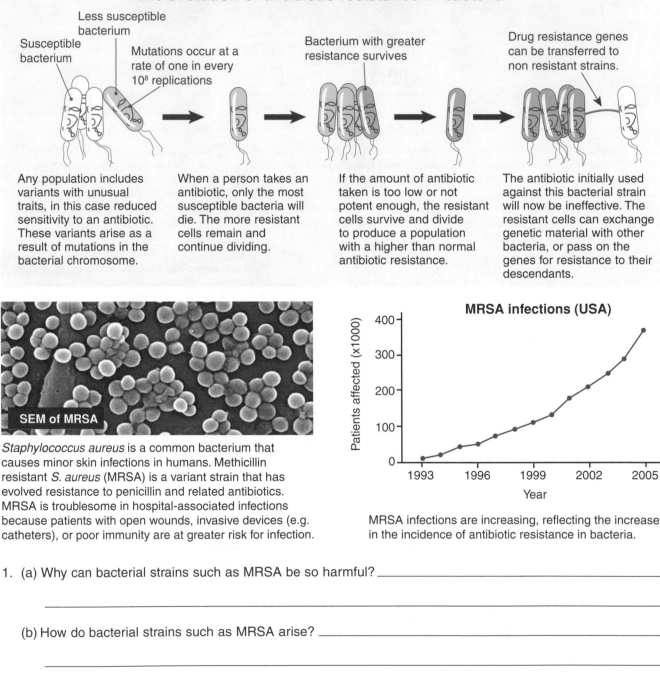

Less susceptible bacterium

Susceptible bacterium

Mutations occur at a rate of one in every 10^8 replications

Bacterium with greater resistance survives

Drug resistance genes can be transferred to non resistant strains.

Any population includes variants with unusual traits, in this case reduced sensitivity to an antibiotic. These variants arise as a result of mutations in the bacterial chromosome.

When a person takes an antibiotic, only the most susceptible bacteria will die. The more resistant cells remain and continue dividing.

If the amount of antibiotic taken is too low or not potent enough, the resistant cells survive and divide to produce a population with a higher than normal antibiotic resistance.

The antibiotic initially used against this bacterial strain will now be ineffective. The resistant cells can exchange genetic material with other bacteria, or pass on the genes for resistance to their descendants.

SEM of MRSA

Staphylococcus aureus is a common bacterium that causes minor skin infections in humans. Methicillin resistant *S. aureus* (MRSA) is a variant strain that has evolved resistance to penicillin and related antibiotics. MRSA is troublesome in hospital-associated infections because patients with open wounds, invasive devices (e.g. catheters), or poor immunity are at greater risk for infection.

MRSA infections (USA)

Patients affected (x1000) — 400, 300, 200, 100, 0
Year — 1993, 1996, 1999, 2002, 2005

MRSA infections are increasing, reflecting the increase in the incidence of antibiotic resistance in bacteria.

1. (a) Why can bacterial strains such as MRSA be so harmful? _____

_____

(b) How do bacterial strains such as MRSA arise? _____

_____

_____

2. How can the resistance become widespread? _____

_____

© 2014 **BIOZONE** International
ISBN: 978-1-927173-84-8
Photocopying Prohibited

160 Beneficial Mutations in Humans

Key Idea: Beneficial mutations increase the fitness of the organisms that possess them. Beneficial mutations are relatively rare.

▶ **Beneficial mutations** are mutations that increase the fitness of the organisms that possess them. Although beneficial mutations are rare compared to those that are harmful, there are a number of well documented beneficial mutations in humans.

▶ Some of these mutations are not very common in the human population. This is because the mutations have been in existence for a relatively short time, so the mutations have not had time to become widespread in the human population.

▶ Scientists often study mutations that cause disease. By understanding the genetic origin of various diseases it may be possible to develop targeted medical drugs and therapies against them.

The village of Limone, Italy

Apolipoprotein A1-Milano is a well documented mutation to apolipoprotein A1 that helps transport cholesterol through the blood. The mutation causes a change to one amino acid and increases the protein's effectiveness by ten times, dramatically reducing incidence of heart disease. The mutation can be traced back to its origin in Limone, Italy, in 1644. Another mutation to a gene called PCSK9 has a similar effect, lowering the risk of heart disease by 88%.

Lactose is a sugar found in milk. All infant mammals produce an enzyme called lactase that breaks the lactose into the smaller sugars glucose and galactose. As mammals become older, their production of lactase declines and they lose the ability to digest lactose. As adults, they become lactose intolerant and feel bloated after drinking milk. About 10,000 years ago a mutation appeared in humans that maintained lactase production into adulthood. This mutation is now carried in people of mainly European, African and Indian descent.

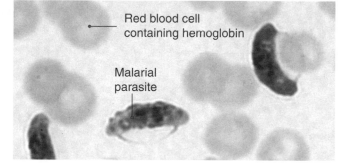
Red blood cell containing hemoglobin

Malarial parasite

Malaria resistance results from a mutation to the hemoglobin gene (Hb^S) that also causes sickle cell disease. This mutation in beneficial in regions where malaria is common. A less well known mutation (Hb^C) to the same gene, discovered in populations in Burkina Faso, Africa, results in a 29% reduction in the likelihood of contracting malaria if the person has one copy of the mutated gene and a 93% reduction if the person has two copies. In addition, the anemia that person suffers as a result of the mutation is much less pronounced than in the Hb^S mutation.

1. Why is it that many of the recent beneficial mutations recorded in humans have not spread throughout the entire human population?

_____

_____

2. What selection pressure could act on Apolipoprotein A1-Milano to help it spread through a population?

_____

_____

3. Why would it be beneficial to be able to digest milk in adulthood?_____

_____

_____

 © 2014 **BIOZONE** International
ISBN: 978-1-927173-84-8
Photocopying Prohibited

CONNECT

184 KNOW

161 Harmful Effects of Mutations in Humans

Key Idea: Many mutations are harmful. Changes to the DNA coding for the protein may prevent the protein from functioning correctly.

Cystic fibrosis

▶ Cystic fibrosis (CF) is an inherited disorder caused by a mutation of the CFTR gene. It is one of the most common lethal autosomal recessive conditions affecting white skinned people of European descent.

▶ The CFTR gene's protein product is a membrane-based protein that regulates chloride transport in cells. The δF508 mutation produces an abnormal CFTR protein, which cannot take its position in the plasma membrane (below, right) or perform its transport function. The δF508 mutation is the most common mutation causing CF.

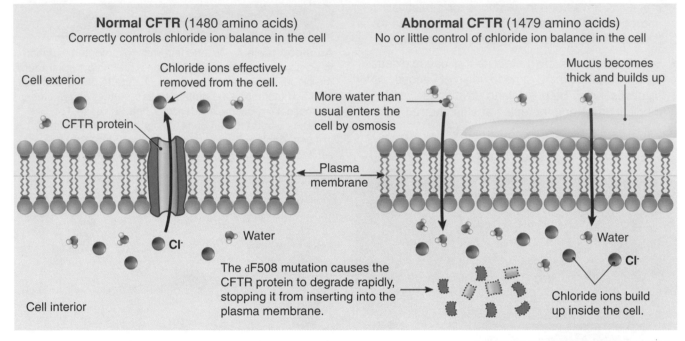

Normal CFTR (1480 amino acids)
Correctly controls chloride ion balance in the cell

Abnormal CFTR (1479 amino acids)
No or little control of chloride ion balance in the cell

Cell exterior

Chloride ions effectively removed from the cell.

CFTR protein

More water than usual enters the cell by osmosis

Mucus becomes thick and builds up

Plasma membrane

Water

Cl⁻

The dF508 mutation causes the CFTR protein to degrade rapidly, stopping it from inserting into the plasma membrane.

Chloride ions build up inside the cell.

Water

Cl⁻

Cell interior

Huntington's disease

Al Aumuller, NY World Telegram and the Sun, Public Domain

▶ Huntington's disease is a progressive genetic disorder in which nerve cells in certain parts of the brain waste away, or degenerate. Symptoms include shaky hands and an awkward gait.

▶ Huntington's disease is caused by a mutation of the HTT gene on chromosome 4. The HTT gene has a repeating base sequence CAG. Normally this section repeats between 10 and 28 times, but in people with Huntington's disease this sequence repeated between 36 to 120 times. The greater the number of repeats, the greater the effects appear to be and the earlier the onset of the disease.

▶ Woody Guthrie (right) was an influential folk singer-songwriter who died in 1967 due to complications related to Huntington's disease.

1. In what way is the amino acid sequence for the CFTR protein affected in the δF508 mutation? _____

_____

2. (a) What causes Huntington's disease? _____

_____

(b) How is the effect of the disease affected by the extent of the mutation: _____

_____

_____

WEB
161

CONNECT
30

CONNECT
184

KNOW

© 2014 **BIOZONE** International
ISBN: 978-1-927173-84-8
Photocopying Prohibited

162 Environment and Variation

Key Idea: The environment can have a large effect on an organism's phenotype without affecting its genotype.

▶ Environmental factors can modify the phenotype encoded by genes without changing the genotype. This can occur both during development and later in life. Environmental factors that affect the phenotype of plants and animals include nutrients or diet, temperature, altitude or latitude, and the presence of other organisms.

The effect of temperature

▶ The sex of some animals is determined by the incubation temperature during their embryonic development. Examples include turtles, crocodiles, and the American alligator. In some species, high incubation temperatures produce males and low temperatures produce females. In other species, the opposite is true. Temperature regulated sex determination may provide an advantage by preventing inbreeding (since all siblings will tend to be of the same sex).

▶ Color-pointing is a result of a temperature sensitive mutation to one of the melanin-producing enzymes. The dark pigment is only produced in the cooler areas of the body (face, ears, feet, and tail), while the rest of the body is a pale color, or white. Color-pointing is seen in some breeds of cats and rabbits (e.g. Siamese cats and Himalayan rabbits).

The effect of other organisms

▶ The presence of other individuals of the same species may control sex determination for some animals. Some fish species, including Sandager's wrasse (right), show this characteristic. The fish live in groups consisting of a single male with attendant females and juveniles. In the presence of a male, all juvenile fish of this species grow into females. When the male dies, the dominant female will undergo physiological changes to become a male. The male and female look very different.

▶ Some organisms respond to the presence of other, potentially harmful, organisms by changing their body shape. Invertebrates such as *Daphnia* will grow a large helmet when a predatory midge larva is present. The helmet makes *Daphnia* more difficult to attack and handle. Such changes are usually in response to chemicals produced by the predator (or competitor) and are common in plants as well as animals.

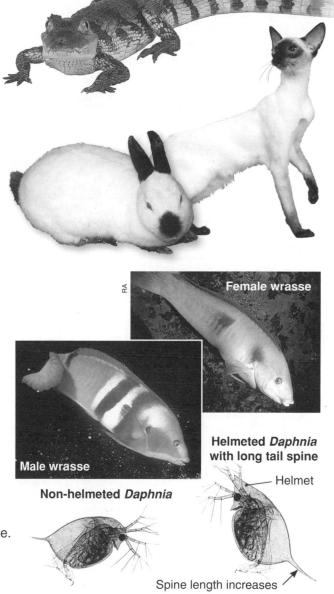

RA

Female wrasse

Male wrasse

Non-helmeted *Daphnia*

Helmeted *Daphnia* **with long tail spine**

Helmet

Spine length increases

1. (a) Give two examples of how temperature affects a phenotypic characteristic in an organism:

_____

(b) Why are the darker patches of fur in color-pointed cats and rabbits found only on the face, paws, and tail?

_____

2. How is helmet development in *Daphnia* an adaptive response to environment? _____

_____

_____

© 2014 **BIOZONE** International
ISBN: 978-1-927173-84-8
Photocopying Prohibited

The effect of altitude

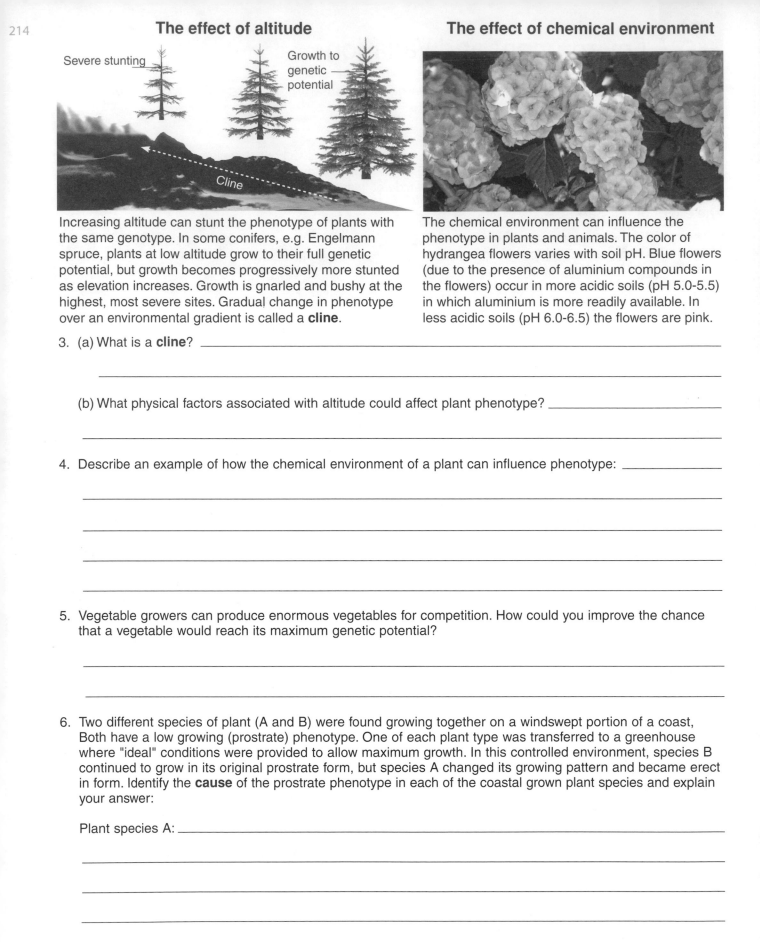

Severe stunting

Growth to genetic potential

Cline

Increasing altitude can stunt the phenotype of plants with the same genotype. In some conifers, e.g. Engelmann spruce, plants at low altitude grow to their full genetic potential, but growth becomes progressively more stunted as elevation increases. Growth is gnarled and bushy at the highest, most severe sites. Gradual change in phenotype over an environmental gradient is called a **cline**.

The effect of chemical environment

The chemical environment can influence the phenotype in plants and animals. The color of hydrangea flowers varies with soil pH. Blue flowers (due to the presence of aluminium compounds in the flowers) occur in more acidic soils (pH 5.0-5.5) in which aluminium is more readily available. In less acidic soils (pH 6.0-6.5) the flowers are pink.

3. (a) What is a **cline**? _____

    _____

    (b) What physical factors associated with altitude could affect plant phenotype? _____

    _____

4. Describe an example of how the chemical environment of a plant can influence phenotype: _____

    _____

    _____

    _____

    _____

5. Vegetable growers can produce enormous vegetables for competition. How could you improve the chance that a vegetable would reach its maximum genetic potential?

    _____

    _____

6. Two different species of plant (A and B) were found growing together on a windswept portion of a coast, Both have a low growing (prostrate) phenotype. One of each plant type was transferred to a greenhouse where "ideal" conditions were provided to allow maximum growth. In this controlled environment, species B continued to grow in its original prostrate form, but species A changed its growing pattern and became erect in form. Identify the **cause** of the prostrate phenotype in each of the coastal grown plant species and explain your answer:

    Plant species A: _____

    _____

    _____

    _____

    Plant species B: _____

    _____

    _____

    _____

 © 2014 **BIOZONE** International
ISBN: 978-1-927173-84-8
Photocopying Prohibited

163 Genes and Environment Interact

Key Idea: The environment or experiences of an individual can affect the development of following generations.

▶ Studies of heredity have found that the environment or lifestyle of an ancestor can have an effect on future generations. Certain environments or diets can affect the methylation and packaging of the DNA (rather than the DNA itself) determining which genes are switched on or off and so affecting the development of the individual. These effects can be passed on to offspring, and even on to future generations. It is thought that these inherited effects may provide a rapid way to adapt to particular environmental situations.

▶ The destruction of New York's Twin Towers on September 11, 2001, traumatized thousands of people. In those thousands were 1700 pregnant women. Some of them suffered (often severe) post-traumatic stress disorder, others did not. Studies on the mothers who developed PTSD found very low levels of the stress-related hormone cortisol in their saliva. The children of these mothers also had much lower levels of cortisol than those whose mothers had not suffered PTSD. The environment of the mother had affected the offspring.

Mice and environmental effects

The effect of the environment and diet of mothers on later generations exposed to a breast cancer trigger was investigated in mice fed a high fat diet or a diet high in estrogen. The length of time taken for breast cancer to develop in later generations after the trigger for breast cancer was given was recorded and compared. The data is presented below. F_1= daughters, F_2= granddaughters, F_3 = great granddaughters.

	Cumulative percent mice with breast cancer (high fat diet (HFD))					
Weeks since trigger	F_1%		F_2%		F_3%	
	HFD	Control	HFD	Control	HFD	Control
6	0	0	5	0	3	0
8	15	0	20	5	3	20
10	22	8	30	5	10	25
12	22	18	50	20	20	30
14	22	18	50	30	25	40
16	29	18	60	30	25	40
18	29	18	60	40	40	42
20	40	18	65	40	50	60
22	80	60	79	50	50	60

Data source: Science News April 6, 2013

	Cumulative percent mice with breast cancer (high estrogen dosage (HED))					
Weeks since trigger	F_1%		F_2%		F_3%	
	HED	Control	HED	Control	HED	Control
6	5	0	10	0	0	0
8	10	0	10	0	15	10
10	30	15	15	20	30	20
12	38	19	30	30	40	20
14	50	22	30	40	50	20
16	50	22	30	40	50	30
18	60	35	40	40	75	40
20	60	42	50	50	80	45
22	80	55	50	50	80	60

1. Use the data on the previous page to complete the graphs below. The first graph is done for you:

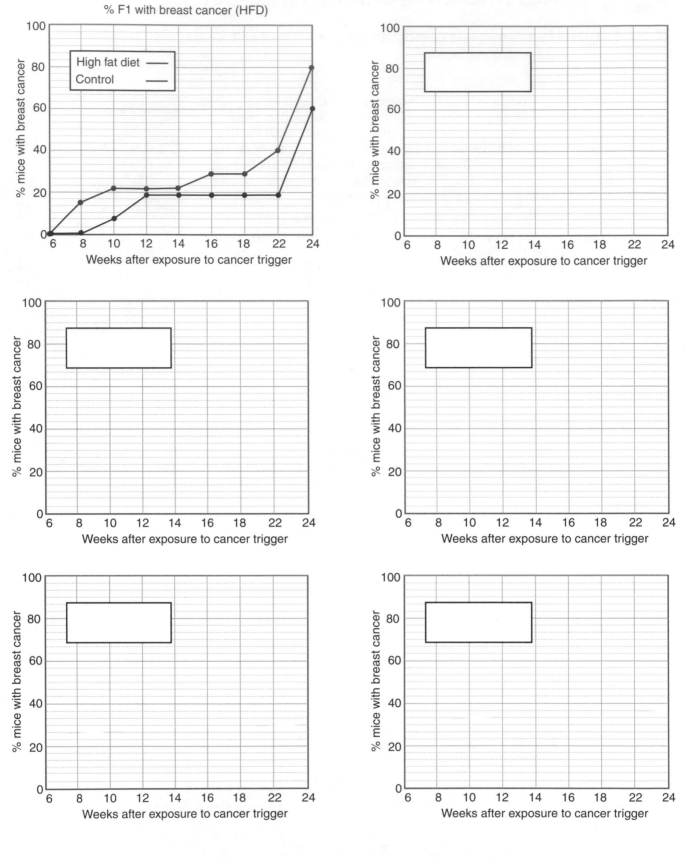

% F1 with breast cancer (HFD)

2. (a) Which generations are affected by the original mother eating a high fat diet? _____

   (b) Which generations are affected by the mother eating a high estrogen diet? _____

   (c) Which diet had the longest lasting effect? _____

3. What do these experiments show with respect to diet and generational effects? _____

_____

164 Predicting Traits

Key Idea: The outcome of a cross depends on the parental genotypes. A true breeding parent is homozygous for the gene or genes involved.

Monohybrid cross

A cross between parents differing in one phenotypic characteristic.

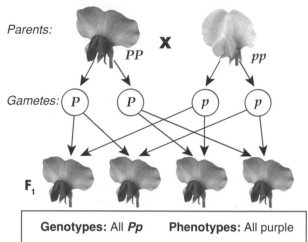

Homozygous purple   **Homozygous white**

Parents: *PP*   **X**   *pp*

Gametes: P   P   p   p

F₁

| Genotypes: All *Pp* | Phenotypes: All purple |

The F₁ (first filial generation) offspring of a cross between two **true breeding** pea plants all have purple flowers (*Pp*). A cross between these offspring (*Pp x Pp*) would yield a 3:1 ratio in the F₂ of purple (*PP, Pp, Pp*) to white (*pp*).

Test cross

A test cross reveals an unknown indivdual's genotype.

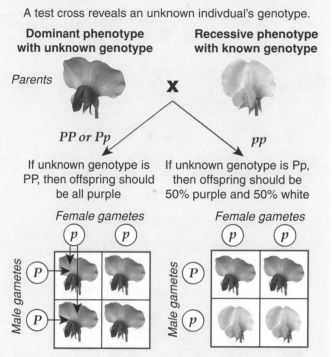

Dominant phenotype with unknown genotype   **Recessive phenotype with known genotype**

Parents   **X**

PP or Pp   *pp*

If unknown genotype is PP, then offspring should be all purple

If unknown genotype is Pp, then offspring should be 50% purple and 50% white

Female gametes

Male gametes   P | p |
                 P | |

Female gametes

Male gametes   P | |
                 p | |

Using Punnett squares makes determining the outcome of the cross a simple case of pairing up the gametes.

Dihybrid cross

A cross between parent pea plants differing in two phenotypic characteristics.

In pea seeds, yellow colour (*Y*) is dominant to green (*y*) and round shape (*R*) is dominant to wrinkled (r). Each parent is **true breeding** for both alleles (*YYRR* or *yyrr*). F₁ offspring will all have the same genotype and phenotype (yellow-round: *YyRr*).

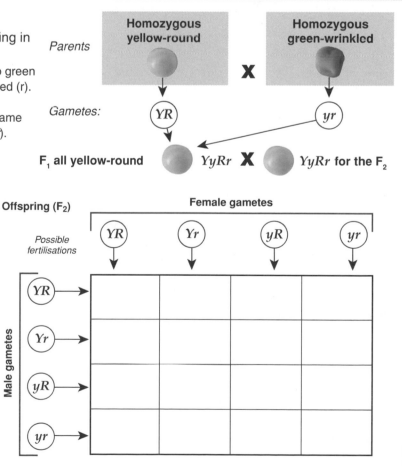

Parents   **Homozygous yellow-round**   **X**   **Homozygous green-wrinkled**

Gametes: YR   yr

F₁ all yellow-round   *YyRr* **X** *YyRr* **for the F₂**

1. Fill in the Punnett square (below right) to show the genotypes of the F₂ generation.

2. In the boxes below, use fractions to indicate the numbers of each phenotype produced from this cross.

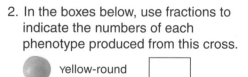
   Yellow-round ☐

   Green-round ☐

   Yellow-wrinkled ☐

   Green-wrinkled ☐

3. Express these numbers as a ratio:
_____

Offspring (F₂)

Possible fertilisations

Female gametes

YR   Yr   yR   yr

Male gametes   YR →
                 Yr →
                 yR →
                 yr →

© 2014 **BIOZONE** International
ISBN: 978-1-927173-84-8
Photocopying Prohibited

KNOW

165 Monohybrid Cross

Key Idea: A monohybrid cross studies the inheritance pattern of one gene. The offspring of these crosses occur in predictable ratios.

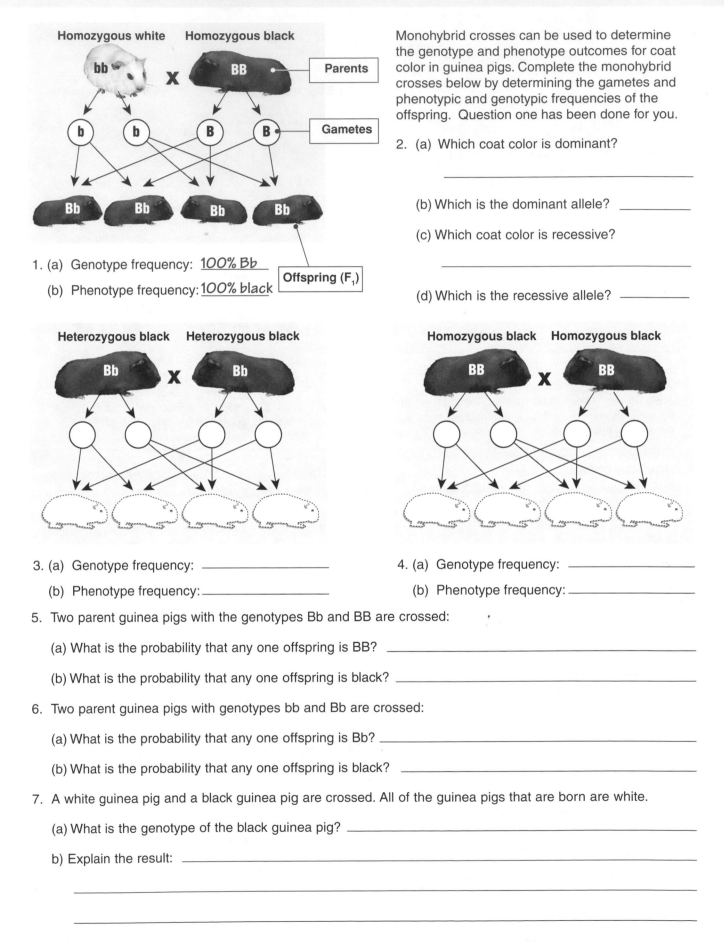

Monohybrid crosses can be used to determine the genotype and phenotype outcomes for coat color in guinea pigs. Complete the monohybrid crosses below by determining the gametes and phenotypic and genotypic frequencies of the offspring. Question one has been done for you.

Homozygous white bb X **Homozygous black** BB **Parents**

b b B B **Gametes**

Bb Bb Bb Bb **Offspring (F₁)**

1. (a) Genotype frequency: _100% Bb_
   (b) Phenotype frequency: _100% black_

2. (a) Which coat color is dominant?

   _____

   (b) Which is the dominant allele? _____

   (c) Which coat color is recessive?

   _____

   (d) Which is the recessive allele? _____

Heterozygous black Bb X **Heterozygous black** Bb

Homozygous black BB X **Homozygous black** BB

3. (a) Genotype frequency: _____
   (b) Phenotype frequency: _____

4. (a) Genotype frequency: _____
   (b) Phenotype frequency: _____

5. Two parent guinea pigs with the genotypes Bb and BB are crossed:

   (a) What is the probability that any one offspring is BB? _____

   (b) What is the probability that any one offspring is black? _____

6. Two parent guinea pigs with genotypes bb and Bb are crossed:

   (a) What is the probability that any one offspring is Bb? _____

   (b) What is the probability that any one offspring is black? _____

7. A white guinea pig and a black guinea pig are crossed. All of the guinea pigs that are born are white.

   (a) What is the genotype of the black guinea pig? _____

   b) Explain the result: _____

   _____

© 2014 **BIOZONE** International
ISBN: 978-1-927173-84-8
Photocopying Prohibited

166 Dihybrid Cross

Key Idea: A dihybrid cross studies the inheritance pattern of two genes. The offspring of these crosses occur in predictable ratios.

1. In guinea pigs, rough coat **R** is dominant over smooth coat **r** and black coat **B** is dominant over white **b**. The genes are not linked. A homozygous rough black animal was crossed with a homozygous smooth white:

   (a) State the genotype of the F_1: _____

   (b) State the phenotype of the F_1: _____

   (c) Use the Punnett square (top right) to show the outcome of a cross between the F_1 (the F_2):

   (d) Using ratios, state the phenotypes of the F_2 generation: _____

   _____

   (e) Use the Punnett square (right) to show the outcome of a back cross of the F_1 to the rough, black parent:

   (f) Using ratios, state the phenotype of the offspring of this back cross:

   _____

   (g) A rough black guinea pig was crossed with a rough white guinea pig produced the following offspring: 3 rough black, 2 rough white, and 1 smooth white. What are genotypes of the parents?

   _____

   _____

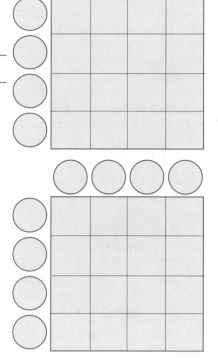

2. In humans, two genes affecting the appearance of the hands are the gene for thumb hyperextension (curving) and the gene for mid-digit hair. The allele for curved thumb, **H**, is dominant to the allele for straight thumb, **h**. The allele for mid digit hair, **M**, is dominant to that for an absence of hair, **m**.

   (a) Give all the genotypes of individuals who are able to curve their thumbs, but have no mid-digit hair:

   _____

   (b) Complete the Punnett square to show the possible genotypes from a cross between two individuals heterozygous for both alleles:

   (c) State the phenotype ratios of the F_1 progeny: _____

   _____

   _____

   (d) What is the probability one of the offspring would have mid-digit hair?

   _____

3. In rabbits, spotted coat **S** is dominant to solid color **s**, while for coat color: black **B** is dominant to brown **b** (the genes are not linked). A SSbb rabbit is mated with a SsBb.

   (a) What is the probability of any one offspring being spotted and black? _____

   (b) What is the probability of any one of the offspring carrying a b allele? _____

© 2014 **BIOZONE** International
ISBN: 978-1-927173-84-8
Photocopying Prohibited

WEB

166 KNOW

167 Pedigree Analysis

Key Idea: Pedigree charts illustrate inheritance patterns over a number of generations. They allow a genetic disorder to be traced back to its origin.

Sample Pedigree Chart

Pedigree charts are a way of graphically illustrating inheritance patterns over a number of generations. They are used to study the inheritance of genetic disorders. The key (below the chart) should be consulted to make sense of the various symbols. Particular individuals are identified by their generation number and their order number in that generation. For example, **II-6** is the sixth person in the second row. The arrow indicates the **propositus**; the person through whom the pedigree was discovered (i.e. who reported the condition).

If the chart on the right were illustrating a human family tree, it would represent three generations: grandparents (I-1 and I-2) with three sons and one daughter. Two of the sons (II-3 and II-4) are identical twins, but did not marry or have any children. The other son (II-1) married and had a daughter and another child (sex unknown). The daughter (II-5) married and had two sons and two daughters (plus a child that died in infancy).

For the particular trait being studied, the grandfather was expressing the phenotype (showing the trait) and the grandmother was a carrier. One of their sons and one of their daughters also show the trait, together with one of their granddaughters.

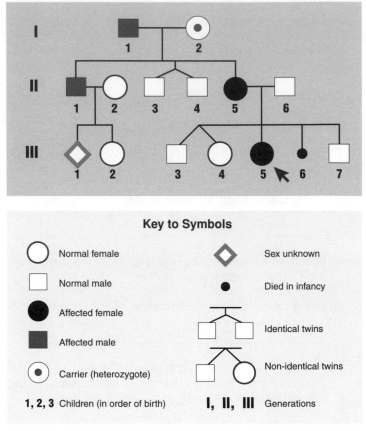

Key to Symbols

- ◯ Normal female
- ◻ Normal male
- ⬤ Affected female
- ◼ Affected male
- ◉ Carrier (heterozygote)
- **1, 2, 3** Children (in order of birth)
- ◇ Sex unknown
- ● Died in infancy
- Identical twins
- Non-identical twins
- **I, II, III** Generations

1. **Pedigree chart of your family**

   Using the symbols in the key above and the example illustrated as a guide, construct a pedigree chart of your own family (or one that you know of) starting with the parents of your mother and/or father on the first line. Your parents will appear on the second line (II) and you will appear on the third line (III). There may be a fourth generation line (IV) if one of your brothers or sisters has had a child. Use a ruler to draw up the chart carefully.

© 2014 **BIOZONE** International
ISBN: 978-1-927173-84-8
Photocopying Prohibited

2. The pedigree chart right shows the inheritance of the allele A in a flower that can be blue or white. Answer the questions below:

   (a) Which color is produced by the dominant allele (white or blue)?

   _____

   (b) Write on the chart the genotype for each of the generation I individuals.

   (c) III4 is crossed with a white flower. What is the probability that any one offspring also has a white flower?

   _____

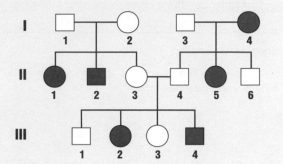

3. The pedigree chart right shows the inheritance of the allele B in a mammal that can have a coat color of black or white. Answer the questions below:

   (a) Which color is produced by the dominant allele (black or white)?

   _____

   (b) Explain how you know this: _____

   _____

   _____

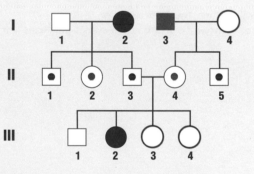

4. **Autosomal recessive traits**
   Albinos lack pigment in the hair, skin and eyes. This trait is inherited as an autosomal recessive allele (i.e. it is not carried on the sex chromosome).

   Albinism in humans

   (a) Write the genotype for each of the individuals on the chart using the following letter codes: **PP** normal skin color; **P-** normal, but unknown if homozygous; **Pp** carrier; **pp** albino.

   (b) Why must the parents (II-3) and (II-4) be **carriers** of a **recessive** allele:

   _____

   _____

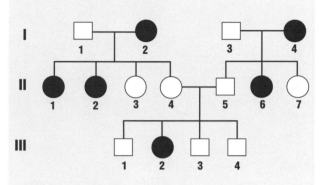

5. **Autosomal dominant traits**
   An unusual trait found in some humans is woolly hair (not to be confused with curly hair). Each affected individual will have at least one affected parent.

   Woolly hair in humans

   (a) Write the genotype for each of the individuals on the chart using the following letter codes: **WW** woolly hair, **Ww** woolly hair (heterozygous), **W-** woolly hair, but unknown if homozygous, **ww** normal hair.

   (b) Describe a feature of this inheritance pattern that suggests the trait is the result of a **dominant** allele:

   _____

   _____

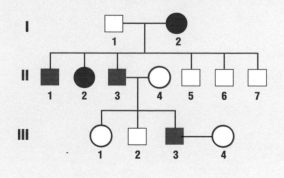

168 Chapter Review

Summarize what you know about this topic under the headings provided.
You can draw diagrams or mind maps, or write short notes to organize
your thoughts. Use the images, hints, and guidelines included to help you:

Sources of variation

HINT: Describe the sources of variation in sexually reproducing organisms.
Why is variation important?

Meiosis produces variation

Hint: Define the term meiosis.
Use diagrams to show how meiosis produces variation in the offspring.

REVISE

Mutations

HINT: What is a mutation? Name some causes of mutations.
Describe the importance of mutations on genetic variation.

Genetic crosses

HINT: Describe the difference between monohybrid and dihybrid crosses.

© 2014 **BIOZONE** International
ISBN: 978-1-927173-84-8
Photocopying Prohibited

169 KEY TERMS: Comprehension and Vocabulary

1. Test your vocabulary by matching each term to its definition, as identified by its preceding letter code.

allele ....

dominant ....

genotype ....

heterozygous ....

homologous
chromosomes ....

meiosis ....

mutation ....

phenotype ....

recessive ....

recombination ....

trait ....

A Observable characteristics in an organism.

B Allele that will only express its trait in the absence of the dominant allele.

C Phenotypic characteristic (e.g. red hair).

D Possessing two different alleles of a particular gene, one inherited from each parent.

E Sequences of DNA occupying the same gene locus (position) on different, but homologous, chromosomes.

F The process of double nuclear division (reduction division) to produce four nuclei, each containing half the original number of chromosomes (haploid).

G A change to the DNA sequence of an organism. This may be a deletion, insertion, duplication, inversion or translocation of DNA in a gene or chromosome.

H The exchange of alleles between homologous chromosomes as a result of crossing over.

I Allele that expresses its trait irrespective of the other allele.

J Chromosome pairs, one paternal and one maternal, of the same length, centromere position, and staining pattern with genes for the same characteristics at corresponding loci.

K The allele combination of an organism.

2. The allele for wrinkled seeds in pea plants is considered recessive because:

A It is not expressed in the F_2 generation

B It is not expressed in the heterozygote

C Individuals who have the allele are less likely to pass genes on to the next generation

D Round peas are smaller than wrinkled peas.

3. True breeding individuals:

A Only ever breed with others of the same kind

B Always have the same coat or flower color

C Are homozygous for a particular allele

D Always produce heterozygous offspring.

4. Use lines to match the statements in the table below to form complete sentences:

Mutations are the ultimate...

Alleles are variations...

A person carrying two of the same alleles (one on each homologous chromosome)...

If the person carries two different alleles...

Alleles may be...

A dominant allele...

A recessive allele only...

...of a gene.

...dominant or recessive

...for the gene, they are heterozygous.

...is said to be homozygous

...expresses its trait if it is homozygous.

...source of new alleles.

...always expresses its trait whether it is homozygous or heterozygous

 © 2014 **BIOZONE** International
ISBN: 978-1-927173-84-8

VOCAB

Biological Evolution: Unity and Diversity

Concepts and Connections
Use arrows to make your own connections between related concepts in this section of this workbook

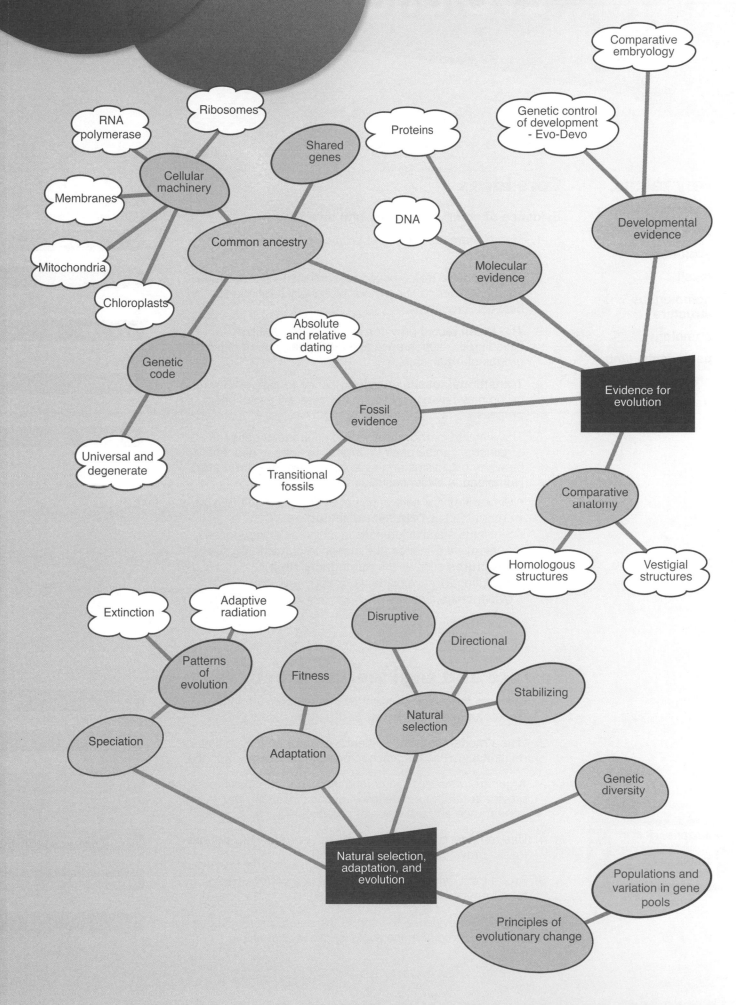

Evidence for Evolution

Key terms

common ancestor

DNA hybridization

evolution

fossil

homologous structure

homology

pentadactyl limb

phylogenetic tree

transitional fossil

Core ideas

Evidence of common ancestry and diversity.

Activity number

☐ 1. Many different branches of science provide evidence for **evolution**.

170

☐ 2. **Phylogenetic trees**, based on biological evidence, can be constructed to show the evolutionary relationships between organisms.

171

☐ 3. The **fossil** record provides evidence for evolution, including the appearance and disappearance of different groups of organisms.

172 173

☐ 4. **Transitional fossils** provide evidence to support how one group may have given rise to the other by evolutionary processes.

174 175

☐ 5. The similarities (**homology**) and differences between organisms can be used to determine lines of descent between organisms and establish a **common ancestor**. Homologous evidence includes:

176-179

- Similarities and differences of DNA sequences (e.g as determined by **DNA hybridization**).
- Similarities and differences in amino acid sequences.
- Anatomical evidence (the presence of **homologous structures** such as the **pentadactyl limb**).
- Evolutionary developmental biology (evo-devo) (the evidence from embryology and development).

Science and engineering practices

☐ 1. Evaluate the evidence from a variety of sources for the common ancestry of life on Earth.

171

☐ 2. Use a model based on evidence to show how evolutionary relationships can be constructed from biological evidence.

171

☐ 3. Argue from evidence to defend the use of the fossil record to provide a relative sequence of events, including the appearance and extinction of organisms.

173

☐ 4. Use a model based on evidence to show how transitional fossils provide evidence for evolution.

175

☐ 5. Argue from evidence to support the use of DNA homology to determine evolutionary relationships.

177

☐ 6. Argue from evidence to support the use of protein sequence similarities to determine evolutionary relationships.

178

170 Evidence for Evolution

Key Idea: Evolution describes the heritable changes in a population's gene pool over time. Evidence for the fact that populations evolve comes from many fields of science.

What is evolution?

Evolution is defined as the heritable genetic changes seen in a population over time. There are two important points to take from this definition. The first is that evolution refers to populations, not individuals. The second is that the changes must be passed on to the next generation (i.e. be inherited). The evidence for evolution comes from many branches of science (below) and includes evidence from living populations as well as from the past.

Comparative anatomy

Comparative anatomy examines the similarities and differences in the anatomy of different species. Similarities in anatomy (e.g. the bones forming the arms in humans and the wings in birds and bats) indicate descent from a common ancestor.

Geology

Geological strata (the layers of rock, soil, and other deposits such as volcanic ash) can be used to determine the relative order of past events and therefore the relative dates of fossils. Fossils in lower strata are older than fossils in higher (newer) strata, unless strata have been disturbed.

DNA comparisons

DNA can be used to determine how closely organisms are related to each other. The greater the similarities between the DNA sequences of species, the more closely related the species are.

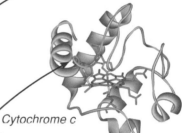
Cytochrome c

Protein evidence

Similarities (and differences) between proteins provides evidence for determining shared ancestry. Fewer differences in amino acid sequences reflects closer genetic relatedness.

EVOLUTION

Fossil record

Fossils, like this shark's tooth (left) are the remains of long-dead organisms. They provide a record of the appearance and extinction of organisms.

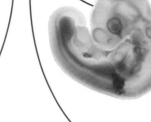

Developmental evidence

The study of developmental processes and the genes that control them gives insight into evolutionary processes. This field of study is called evolutionary developmental biology (evo-devo).

Biogeography

The geographical distribution of living and extinct organisms provides evidence of common ancestry and can be explained by speciation, extinction, and continental drift. The biogeography of islands, e.g the Galápagos Islands, provides evidence of how species evolve when separated from their ancestral population on the mainland.

Chronometric dating

Radiometric dating techniques (such as carbon dating) allow scientists to determine an absolute date for a fossil by dating it or the rocks around it. Absolute dating has been used to assign ages to strata, and construct the geological time scale.

CONNECT CONNECT CONNECT
186 **185** **184** REFER

171 The Common Ancestry of Life

Key Idea: Molecular studies have enabled scientists to clarify the earliest beginnings of the eukaryotes. Such studies provide powerful evidence of the common ancestry of life.

How do we know about the relatedness of organisms?

▶ Traditionally, the phylogeny (evolutionary history) of organisms was established using morphological comparisons. In recent decades, molecular techniques involving the analysis of DNA, RNA, and proteins have provided more information about how all life on Earth is related.

▶ These newer methods have enabled scientists to clarify the origin of the eukaryotes and to recognize two prokaryote domains. The universality of the genetic code and the similarities in the molecular machinery of all cells provide powerful evidence for a common ancestor to all life on Earth.

There is a universal genetic code

DNA encodes the genetic instructions of all life. The form of these genetic instructions, called the **genetic code**, is effectively universal, i.e. the same combination of three DNA bases code for the same amino acid in almost all organisms. The very few exceptions in which there are coding alternatives are restricted to some bacteria and to mitochondrial DNA.

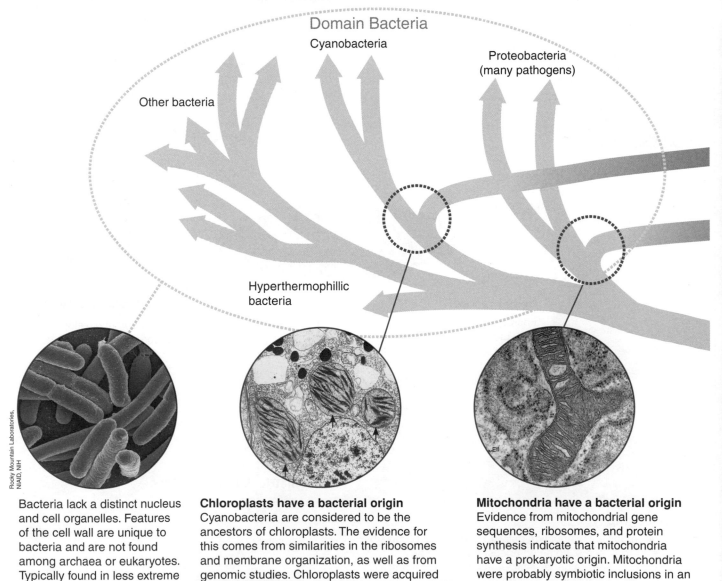

Domain Bacteria

Cyanobacteria

Proteobacteria (many pathogens)

Other bacteria

Hyperthermophillic bacteria

Rocky Mountain Laboratories, NIAID, NIH

Bacteria lack a distinct nucleus and cell organelles. Features of the cell wall are unique to bacteria and are not found among archaea or eukaryotes. Typically found in less extreme environments than archaea.

Chloroplasts have a bacterial origin
Cyanobacteria are considered to be the ancestors of chloroplasts. The evidence for this comes from similarities in the ribosomes and membrane organization, as well as from genomic studies. Chloroplasts were acquired independently of mitochondria, from a different bacterial lineage, but by a similar process.

Mitochondria have a bacterial origin
Evidence from mitochondrial gene sequences, ribosomes, and protein synthesis indicate that mitochondria have a prokaryotic origin. Mitochondria were probably symbiotic inclusions in an early eukaryotic ancestor.

WEB CONNECT CONNECT CONNECT

KNOW 171 38 177 178

© 2014 **BIOZONE** International
ISBN: 978-1-927173-84-8
Photocopying Prohibited

1. Identify three features of the metabolic machinery of cells that support a common ancestry of life:

   (a) _____

   (b) _____

   (c) _____

2. Suggest why scientists believe that mitochondria were acquired before chloroplasts: _____

   _____

   _____

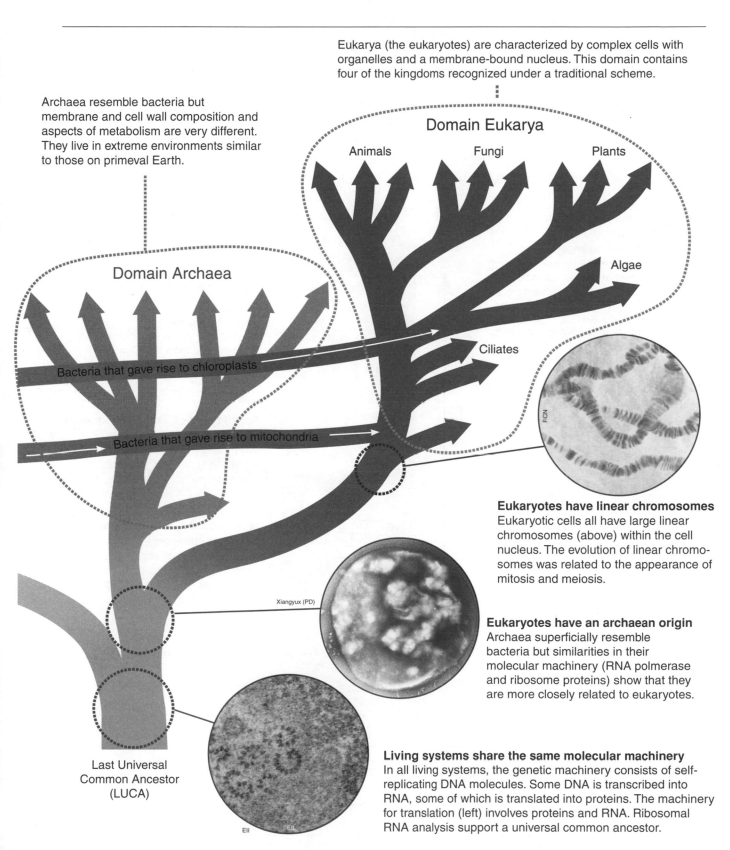

Eukarya (the eukaryotes) are characterized by complex cells with organelles and a membrane-bound nucleus. This domain contains four of the kingdoms recognized under a traditional scheme.

Archaea resemble bacteria but membrane and cell wall composition and aspects of metabolism are very different. They live in extreme environments similar to those on primeval Earth.

Domain Eukarya

Animals   Fungi   Plants

Algae

Ciliates

Domain Archaea

Bacteria that gave rise to chloroplasts

Bacteria that gave rise to mitochondria

FICN

Eukaryotes have linear chromosomes
Eukaryotic cells all have large linear chromosomes (above) within the cell nucleus. The evolution of linear chromosomes was related to the appearance of mitosis and meiosis.

Xiangyux (PD)

Eukaryotes have an archaean origin
Archaea superficially resemble bacteria but similarities in their molecular machinery (RNA polmerase and ribosome proteins) show that they are more closely related to eukaryotes.

Last Universal
Common Ancestor
(LUCA)

EII   EII

Living systems share the same molecular machinery
In all living systems, the genetic machinery consists of self-replicating DNA molecules. Some DNA is transcribed into RNA, some of which is translated into proteins. The machinery for translation (left) involves proteins and RNA. Ribosomal RNA analysis support a universal common ancestor.

172 The Fossil Record

Key Idea: Fossils provide a record of the appearance and extinction of organisms. The fossil record can be used to establish the relative order of past events.

The importance of the fossil record

▶ **Fossils** are the remains of long-dead plants and animals that have become preserved in the Earth's crust.

▶ Fossils provide a record of the appearance and extinction of organisms, from species to whole taxonomic groups.

▶ The fossil record can be calibrated against a time scale (using dating techniques), to build up a picture of the evolutionary changes that have taken place.

Fossilized fern frond

Gaps in the fossil record

The fossil record contains gaps and without a complete record, it can sometimes be difficult to determine an evolutionary sequence. Scientists use other information (e.g. associated fossils and changes in morphology) to produce a order of events that best fits all the evidence.

Gaps in the fossil can occur because:

▶ Fossils are destroyed.

▶ Some organisms do not fossilize well.

▶ Fossils have not yet been found.

Profile with sedimentary rocks containing fossils

Rock strata are layered through time

Rock strata are arranged in the order that they were deposited (unless they have been disturbed by geological events). The most recent layers are near the surface and the oldest are at the bottom. Fossils can be used to establish the sequential (relative) order of past events in a rock profile.

New fossil types mark changes in environment

In the strata at the end of one geological period, it is common to find many new fossils that become dominant in the next.

Each geological period had a different environment from the others. Their boundaries coincided with drastic environmental changes and the appearance of new niches. These produced new selection pressures, resulting in new adaptive features in the surviving species as they responded to the changes.

Ground surface

Youngest sediments

Oldest sediments

Recent fossils are found in more recent sediments
The more recent the layer of rock, the more resemblance there is between the fossils found in it and living organisms.

Extinct species
The number of extinct species is far greater than the number of species living today.

Fossil types differ in each stratum
Fossils found in a given layer of sedimentary rock are generally significantly different to fossils in other layers.

More primitive fossils are found in older sediments
Fossils in older layers tend to have quite generalized forms. In contrast, organisms alive today have specialized forms.

1. Discuss the importance of fossils as a record of evolutionary change over time: _____

_____

_____

_____

2. Why can gaps in the fossil record make it difficult to determine an evolutionary sequence? _____

_____

_____

© 2014 **BIOZONE** International
ISBN: 978-1-927173-84-8
Photocopying Prohibited

173 Interpreting the Fossil Record

Key Idea: Analyzing the fossils within rock strata allows scientists to order past events in a rock profile, from oldest to most recent.

The diagram below shows a hypothetical rock profile from two locations separated by a distance of 67 km. There are some differences between the rock layers at the two locations. Apart from layers D and L which are volcanic ash deposits, all other layers comprise sedimentary rock. Use the information on the diagram to answer the questions below.

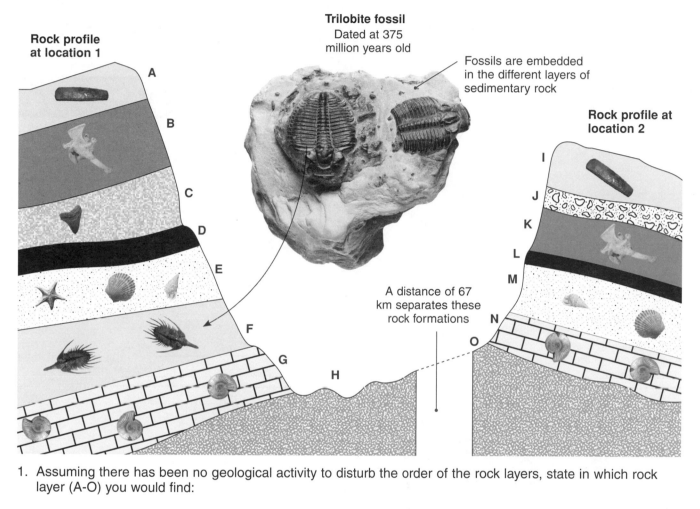

Trilobite fossil
Dated at 375 million years old

Fossils are embedded in the different layers of sedimentary rock

Rock profile at location 1

Rock profile at location 2

A distance of 67 km separates these rock formations

1. Assuming there has been no geological activity to disturb the order of the rock layers, state in which rock layer (A-O) you would find:

   (a) The youngest rocks at Location 1: _____   (c) The youngest rocks at Location 2: _____

   (b) The oldest rocks at Location 1: _____   (d) The oldest rocks at Location 2: _____

2. (a) State which layer at location 1 is of the same age as layer M at location 2: _____

   (b) Explain the reason for your answer in 2 (a): _____

   _____

   _____

3. (a) State which layers present at location 1 are missing at location 2: _____

   (b) State which layers present at location 2 are missing at location 1: _____

4. The rocks in layer H and O are sedimentary rocks. Why are there no visible fossils in these layers?

   _____

   _____

174 Transitional Fossils

Key Idea: Transitional fossils show intermediate states between two different, but related, groups. They provide important links in the fossil record.

Transitional fossils are fossils which have a mixture of features that are found in two different, but related, groups. Transitional fossils provide important links in the fossil record and provide evidence to support how one group may have given rise to the other by evolutionary processes.

Important examples of transitional fossils include horses, whales, and *Archaeopteryx* (below).

Archaeopteryx was a transitional form between birds and reptiles. It is regarded as the earliest known bird. *Archaeopteryx* was a crow-sized animal (50 cm length), which lived about 150 million year ago. It had many reptilian features, but also a number of birdlike (avian) features, including feathers.

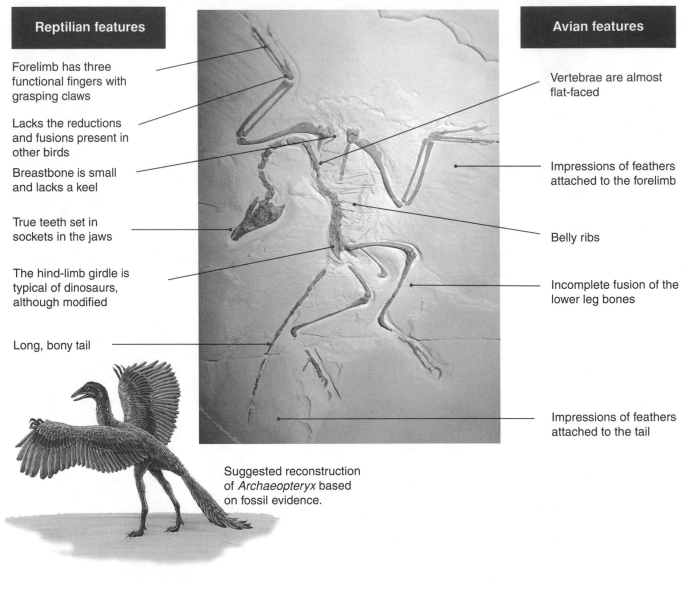

Reptilian features

Forelimb has three functional fingers with grasping claws

Lacks the reductions and fusions present in other birds

Breastbone is small and lacks a keel

True teeth set in sockets in the jaws

The hind-limb girdle is typical of dinosaurs, although modified

Long, bony tail

Avian features

Vertebrae are almost flat-faced

Impressions of feathers attached to the forelimb

Belly ribs

Incomplete fusion of the lower leg bones

Impressions of feathers attached to the tail

Suggested reconstruction of *Archaeopteryx* based on fossil evidence.

1. (a) What is a transitional fossil? _____

_____

(b) Why are transitional fossils important in understanding evolution? _____

_____

_____

_____

© 2014 **BIOZONE** International
ISBN: 978-1-927173-84-8
Photocopying Prohibited

175 Case Study: Whale Evolution

Key Idea: The evolution of whales is well documented in the fossil record, with many transitional forms recording the shift from a terrestrial to an aquatic life.

Whale evolution

The evolution of modern whales from an ancestral land mammal is well documented in the fossil record. The fossil record of whales includes many transitional forms, which has enabled scientists to develop an excellent model of whale evolution. The evolution of the whales (below) shows a gradual accumulation of adaptive features that have equipped them for life in the open ocean.

Modern whales are categorized into two groups.

▶ Toothed whales have full sets of teeth throughout their lives. (e.g. sperm whale and orca).

▶ Baleen whales. These are toothless whales and they use a comb-like structure (baleen) to filter food (e.g. humpback whale).

Humpback whale

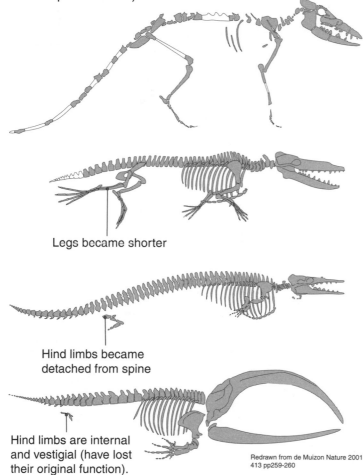

Legs became shorter

Hind limbs became detached from spine

Hind limbs are internal and vestigial (have lost their original function).

Redrawn from de Muizon Nature 2001 413 pp259-260

50 mya *Pakicetus*

Pakicetus was a transitional species between carnivorous land mammals and the earliest true whales. It was mainly terrestrial (land dwelling), but foraged for food in water. It had four, long limbs. Its eyes were near the top of the head and its nostrils were at the end of the snout. It had external ears, but they showed features of both terrestrial mammals and fully aquatic mammals.

45 mya *Rhodocetus*

Rhodocetus was mainly aquatic (water living). It had adaptations for swimming, including shorter legs and a shorter tail. Its eyes had moved to the side of the skull, and the nostrils were located further up the skull. The ear showed specializations for hearing in water.

40 mya *Dorudon*

Dorudon was fully aquatic. Its adaptations for swimming included a long, streamlined body, a broad powerful muscular tail, the development of flippers and webbing. It had very small hind limbs (not attached to the spine) which would no longer bear weight on land.

***Balaena* (recent whale ancestor)**

The hind limbs became fully internal and vestigial. Studies of modern whales show that limb development begins, but is arrested at the limb bud stage. The nostrils became modified as blowholes. This recent ancestor to modern whales diverged into two groups (toothed and baleen) about 36 million years ago. Baleen whales have teeth in their early fetal stage, but lose them before birth.

1. Why does the whale fossil record provide a good example of the evolutionary process? _____

_____

_____

2. Briefly describe the adaptations of whales for swimming that evolved over time: _____

_____

_____

176 Anatomical Evidence for Evolution

Key Idea: Homologous structures are anatomical similarities that are the result of common origin. The indicate the evolutionary relationship between groups of organisms.

Homologous structures

Homologous structures are structures found in different organisms that are the result of their inheritance form a common ancestor. Their presence indicates the evolutionary relationship between organisms. Homologous structures have a common origin, but they may have different functions.

For example, the forelimbs of birds and seals are homologous structures. They have the same basic skeletal structure, but have different functions. A bird's wings have been adapted for flight, and a seal's flippers are modified as paddles for swimming.

The pentadactyl limb

A **pentadactyl limb** is a limb with five fingers or toes (e.g. hands and feet), with the bones arranged in a specific pattern (below, left).

Early land vertebrates were amphibians and had pentadactyl limbs. All vertebrates that descended from these early amphibians have limbs that have evolved from this same basic pentadactyl pattern. They illustrate adaptive radiation, where the basic limb plan has been adapted to meet the requirements of organisms in different niches.

Generalized pentadactyl limb

The forelimbs and hind limbs have the same arrangement of bones. In many cases bones in different parts of the limb have been modified for a specialized locomotory function.

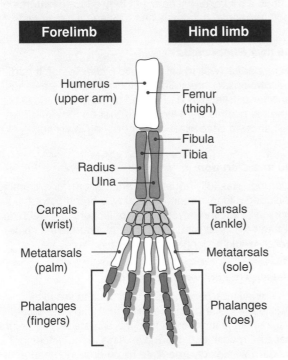

Forelimb	Hind limb

- Humerus (upper arm) — Femur (thigh)
- Fibula
- Tibia
- Radius
- Ulna
- Carpals (wrist) — Tarsals (ankle)
- Metatarsals (palm) — Metatarsals (sole)
- Phalanges (fingers) — Phalanges (toes)

Specializations of pentadactyl limbs

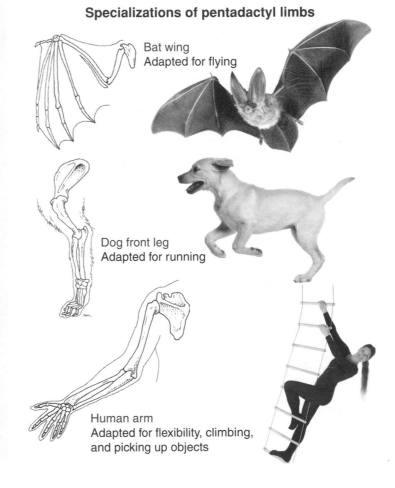

Bat wing
Adapted for flying

Dog front leg
Adapted for running

Human arm
Adapted for flexibility, climbing, and picking up objects

1. What is a pentadactyl limb? _____

_____

2. Explain how homology in the pentadactyl limb provides evidence for adaptive radiation: _____

_____

_____

_____

_____

© 2014 **BIOZONE** International
ISBN: 978-1-927173-84-8
Photocopying Prohibited

177 DNA Evidence for Evolution

Key Idea: DNA homology can be used to determine evolutionary relationships. An increased similarity in DNA sequences indicates closer genetic relatedness.

DNA can be used to measure relatedness

▶ Similarity between DNA sequences (**DNA homology**) can be used to determine evolutionary relatedness. The more closely two species are related, the fewer differences there is between their DNA sequences.

▶ Changes in the DNA sequence are a result of mutations. Organisms which are closely related have similar DNA sequences because there has been less time for mutations to occur and accumulate.

▶ **DNA hybridization** can be used to determine genetic relationships between species. In DNA hybridization, single strands of DNA from two species are mixed together, and allowed to bind. The degree of binding between the two strands indicates the level of DNA similarity.

The universal code

The structure of DNA is common to all life and, with very few exceptions (e.g. mitochondrial DNA) living organisms use the same genetic code. This universality supports a common ancestry for all life.

Using DNA hybridization

By comparing the degree of DNA hybridization between species, phylogenetic trees can be constructed. This one on the right shows the evolutionary relationship between several species of birds.

DNA hybridization shows these birds shared a common ancestor around 50 million years ago (A on the diagram). Further splits subsequently occurred, resulting in each of the species seen here.

Genetically closer organisms have a smaller difference in DNA than organisms which shared a common ancestor long ago. For example, the pelicans and the shoebills are more closely related (DNA difference of 7-8), than the pelican and the flamingo (DNA difference greater than 10).

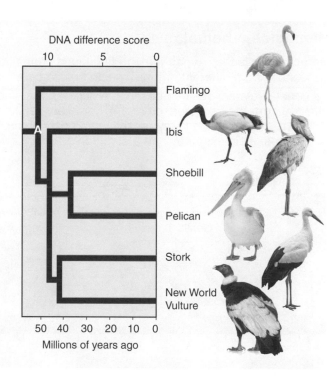

DNA difference score
10   5   0

A

Flamingo
Ibis
Shoebill
Pelican
Stork
New World Vulture

50 40 30 20 10 0
Millions of years ago

1. How does **DNA homology** measure genetic relatedness?

_____

_____

_____

2. The diagram (right) shows the results of DNA hybridization between human DNA and that of other primates.

   (a) Which is the most closely related primate to humans?

   _____

   (b) Which is the most distantly related primate to humans? _____

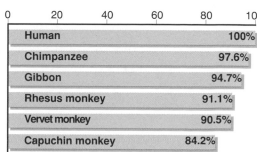

DNA similarity (%)

Primate species	0	20	40	60	80	100
Human						100%
Chimpanzee						97.6%
Gibbon						94.7%
Rhesus monkey						91.1%
Vervet monkey						90.5%
Capuchin monkey					84.2%	
Galago			58.0%			

178 Protein Evidence for Evolution

Key Idea: Protein homology can be used to determine evolutionary patterns. As genetic relatedness increases, the number of amino acid differences decreases.

Protein homology

The amino acid sequence of proteins can be used to establish homologies (similarities) between organisms. Any change in the amino acid sequence reflects changes in the DNA sequence.

Some proteins are common to many different species. These proteins are often highly conserved, meaning they change (mutate) very little over time. This is because they have critical roles (e.g. in cellular respiration) and mutations are likely to be detrimental to their function.

Evidence indicates that these highly conserved proteins are homologous and have been derived from a common ancestor. Because they are highly conserved, changes in the amino acid sequence are likely to represent major divergences between groups during the course of evolution.

The Pax-6 protein provides evidence for evolution

▶ The Pax-6 protein regulates eye formation during embryonic development.

▶ The Pax-6 gene is so highly conserved that the gene from one species can be inserted into another species, and still produce a normally functioning eye.

▶ This suggests the Pax-6 proteins are homologous, and the gene has been inherited from a common ancestor.

An experiment inserted mouse Pax6 gene into fly DNA and turned it on in a fly's legs. The fly developed fly eyes on its legs!

Hemoglobin homology

Hemoglobin is the oxygen-transporting blood protein found in most vertebrates. The beta chain hemoglobin sequences from different organisms can be compared to determine evolutionary relationships.

As genetic relatedness decreases, the number of amino acid differences between the hemoglobin chains of different vertebrates increases (right). For example, there are no amino acid differences between humans and chimpanzees, indicating they recently shared a common ancestor. Humans and frogs have 67 amino acid differences, indicating they had a common ancestor a very long time ago.

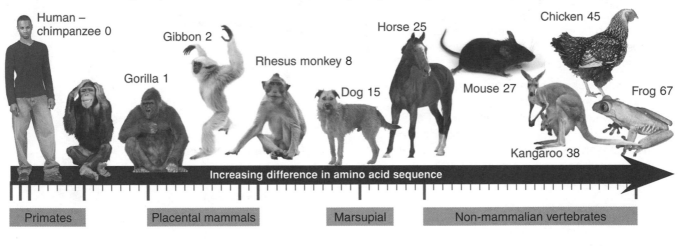

Human – chimpanzee 0 | Gorilla 1 | Gibbon 2 | Rhesus monkey 8 | Dog 15 | Horse 25 | Mouse 27 | Kangaroo 38 | Chicken 45 | Frog 67

Increasing difference in amino acid sequence

Primates | Placental mammals | Marsupial | Non-mammalian vertebrates

1. (a) What is a highly conserved protein? _____

_____

(b) Why are highly conserved proteins good for constructing phylogenies? _____

_____

_____

2. Compare the differences in the hemoglobin sequence of humans, rhesus monkeys, and horses. What do these tell you about the relative relatedness of these organisms?

_____

_____

179 Developmental Evidence for Evolution

Key Idea: Similarities in the development of embryos, including the genetic control of development, provides strong evidence for evolution.

Developmental biology

Developmental biology studies the process by which organisms grow and develop. In the past, it was restricted to the appearance (morphology) of a growing fetus. Today, developmental biology focuses on the genetic control of development and its role in producing the large differences we see in the adult appearance of different species.

During development, vertebrate embryos pass through the same stages, in the same sequence, regardless of the total time period of development. This similarity is strong evidence of their shared ancestry. The stage of embryonic development is identified using a standardized system based on the development of structures, not by size or the number of days of development. The Carnegie stages (right) cover the first 60 days of development.

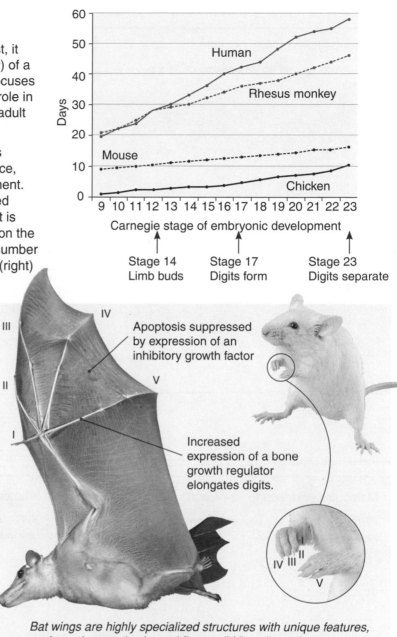

Carnegie stage of embryonic development

Stage 14
Limb buds

Stage 17
Digits form

Stage 23
Digits separate

Limb homology and the control of development

As we have seen, homology (e.g. in limb structure) is evidence of shared ancestry. How do these homologous structures become so different in appearance? The answer lies in the way the same genes are regulated during development.

All vertebrate limbs form as buds at the same stage of development. At first, the limbs resemble paddles, but **apoptosis** (programmed cell death) of the tissue between the developing bones separates the digits to form fingers and toes.

Like humans, mice have digits that become fully separated by interdigital apoptosis during development. In bat forelimbs, this controlled destruction of the tissue between the forelimb digits is inhibited. The developmental program is the result of different patterns of expression of the same genes in the two types of embryos.

Apoptosis suppressed by expression of an inhibitory growth factor

Increased expression of a bone growth regulator elongates digits.

Bat wings are highly specialized structures with unique features, such as elongated wrist and fingers (I-V) and membranous wing surfaces. The forelimb structures of bats and mice are homologous, but how the limb looks and works is quite different.

1. Describe a feature of vertebrate embryonic development that supports evolution from a common ancestor:

   _____

   _____

2. Explain how different specialized limb structures can arise from a basic pentadactyl structure:

   _____

   _____

   _____

CONNECT 176 CONNECT 171 CONNECT 147 CONNECT 65 WEB 179 **KNOW**

180 Chapter Review

Summarize what you know about this topic under the headings provided. You can draw diagrams or mind maps, or write short notes to organize your thoughts. Use the image (right) and the hints and guidelines included to help you.

Wing bone   Sand and tar matrix

Evidence for common ancestry

HINT: What do the major domains of life have in common?

Fossil evidence for evolution

HINT: Include the role of transitional fossils in determining evolutionary relationships.

Molecular evidence for evolution

HINT: Include DNA sequencing and amino acid sequencing of proteins.

Anatomical evidence for evolution

HINT: Include definitions and examples, e.g. the pentadactyl limb.

Developmental evidence for evolution

HINT: Include reference to features expressed in early embryos and the importance of gene control.

© 2014 **BIOZONE** International
ISBN: 978-1-927173-84-8

REVISE

181 KEY TERMS: Crossword

Complete the crossword below to test your understanding of key terms in this chapter and their meanings

Clues Across

1. Evidence of evolution using the distribution of organisms over geographical areas is called _ _ _ _ _ _ _ _ _ _ _ _ _ _ evidence.

6. A transitional fossil between birds and reptiles.

8. Layers of rock or soil with distinguishing characteristics. (2 words: 4, 6)

12. A technique used to show the homology between DNA sequences from different organisms. (2 words: 3, 13)

14. A limb with five digits is called this.

Clues Down

2. Name given to a combination of two biological fields: evolutionary biology and developmental biology. (Acronym 3,4)

3. Features or structures of organisms that are the result of common ancestry are this.

4. A fossil that illustrates an evolutionary transition between two groups is called this.

5. The remains of a long-dead organism that has escaped decay.

7. Dating method that gives the time a material was laid down relative to the formation of other materials. Usually done by matching an object's location in strata to strata of known age and/or fossils or objects. (2 words: 8, 6)

9. The movement of the land masses (continents) is called this. (2 words: 11, 5)

10. The individual from which all organisms in a taxon are directly descended. (2 words: 6, 8)

11. The heritable changes seen in a population's gene pool over time.

13. The evolutionary history of a related group of organisms.

VOCAB

LS**4.B**
LS**4.C**
LS**4.D**
ETS**1.B**

Natural Selection, Adaptation, and Evolution

Key terms

adaptation

biodiversity

evolution

extinction

fitness

genotype

natural selection

phenotype

trait

variation

Core ideas

Activity number

Natural selection: differential survival of favorable phenotypes

☐ 1. **Natural selection** is the process by which favorable **phenotypes** (and therefore **genotypes**) become relatively more or less common in a population. It is a consequence of **variation** in the genes and how they are expressed in individuals, which leads to differences in individual survival and reproductive success in the environment (i.e. **fitness**).

`182 184-186`

☐ 2. **Traits** that increase fitness are more likely to be passed on to the next generation and become more common over time.

`182 183`

Natural selection leads to adaptation

☐ 3. **Evolution** is a result of four factors: (1) the potential for a species to increase in number (2) the genetic variation between members of the species (3) competition for the limited resources, and (4) survival and reproduction of individuals which are better able to survive and reproduce.

`182 184`

☐ 4. Natural selection leads to **adaptation**, i.e. to a population dominated by individuals with the characteristics most suited to survival and reproduction within their environment.

`183-186`

☐ 5. Adaptation allows for a shift in the distribution of traits in a population when conditions change.

`184 185 186`

☐ 6. Changes in the environment may lead species to expand or contract their range. New species can emerge when populations diverge in different environments. In this way, evolution has produced the diversity of present and past life on Earth (**biodiversity**).

`189-192`

☐ 7. Species unable to adapt to a changing environment may become extinct. **Extinction** is natural process.

`193 194`

Humans depend on biodiversity

☐ 8. The activity of humans may adversely affect the biodiversity. The enhancement of biodiversity has positive benefits to humans, who depend on it.

`194 195`

☐ 9. Solutions to problems created by human use of resources must consider practical constraints, such as costs, and social, cultural, and environmental impacts.

Science and engineering practices

☐ 1. Argue from evidence for the role of natural selection in bringing about phenotypic change in populations.

`184 185 186`

☐ 2. Use a model to show the process of natural selection.

`187 188`

☐ 3. Use a model to explain different patterns of evolution.

`191`

☐ 4. Analyze data to show the effect of human activity on the extinction of a named animal species.

`194`

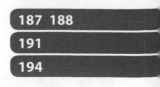
BIOZONE APP
Student Review Series
Natural Selection, Adaptation & Evolution

182 How Evolution Occurs

Key Idea: Evolution by natural selection describes how organisms that are better adapted to their environment survive to produce a greater number of offspring.

Evolution

Evolution is the change in inherited characteristics in a population over generations. Evolution is the consequence of interaction between four factors: (1) The potential for populations to increase in numbers, (2) Genetic variation as a result of mutation and sexual reproduction, (3) competition for resources, and (4) proliferation of individuals with better survival and reproduction.

Natural selection is the term for the mechanism by which better adapted organisms survive to produce a greater number of viable offspring. This has the effect of increasing their proportion in the population so that they become more common. This is the basis of Darwin's theory of evolution by natural selection.

We can demonstrate the basic principles of evolution using the analogy of a 'population' of M&M's candy.

#1

In a bag of M&M's, there are many colors, which represents the variation in a population. As you and a friend eat through the bag of candy, you both leave the blue ones, which you both dislike, and return them to bag.

#2

The blue candy becomes more common...

#3

Eventually, you are left with a bag of blue M&M's. Your selective preference for the other colors changed the make-up of the M&M's population. This is the basic principle of selection that drives evolution in natural populations.

Darwin's theory of evolution by natural selection

Darwin's theory of evolution by natural selection is outlined below. It is widely accepted by the scientific community today and is one of founding principles of modern science.

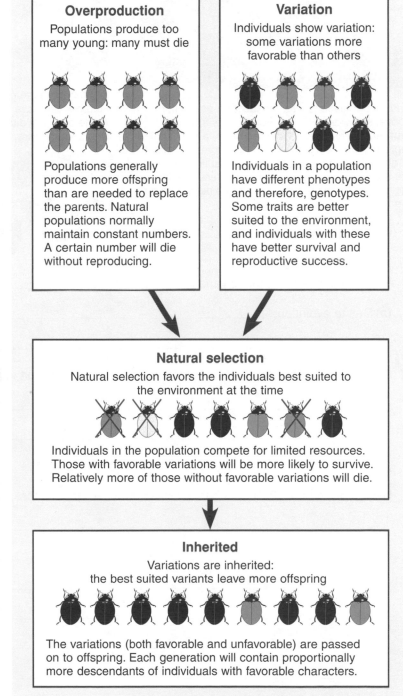

Overproduction
Populations produce too many young: many must die

Populations generally produce more offspring than are needed to replace the parents. Natural populations normally maintain constant numbers. A certain number will die without reproducing.

Variation
Individuals show variation: some variations more favorable than others

Individuals in a population have different phenotypes and therefore, genotypes. Some traits are better suited to the environment, and individuals with these have better survival and reproductive success.

Natural selection
Natural selection favors the individuals best suited to the environment at the time

Individuals in the population compete for limited resources. Those with favorable variations will be more likely to survive. Relatively more of those without favorable variations will die.

Inherited
Variations are inherited: the best suited variants leave more offspring

The variations (both favorable and unfavorable) are passed on to offspring. Each generation will contain proportionally more descendants of individuals with favorable characters.

CONNECT 162   CONNECT 157   CONNECT 153   WEB 182

KNOW

Variation, selection, and population change

1. Variation through mutation and sexual reproduction:
In a population of brown beetles, mutations independently produce red coloration and 2 spot marking on the wings. The individuals in the population compete for limited resources.

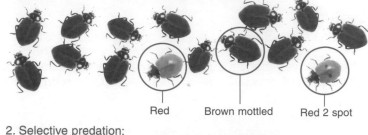

Red   Brown mottled   Red 2 spot

2. Selective predation:
Brown mottled beetles are eaten by birds but red ones are avoided.

Natural populations, like the ladybug population above, show genetic variation. This is a result of **mutation** (which creates new alleles) and sexual reproduction (which produces new combinations of alleles). Some variants are more suited to the environment of the time than others. These variants will leave more offspring, as described for the hypothetical population (right).

3. Change in the genetics of the population:
Red beetles have better survival and fitness and become more numerous with each generation. Brown beetles have poor fitness and become rare.

1. What produces the genetic variation in populations? _____

2. (a) Define evolution: _____

   _____

   (b) Identify the four factors that interact to bring about evolution in populations: _____

   _____

   _____

   _____

3. Using your answer 2(b) as a basis, explain how the genetic make-up of a population can change over time:

   _____

   _____

   _____

   _____

   _____

   _____

   _____

© 2014 **BIOZONE** International
ISBN: 978-1-927173-91-6
Photocopying Prohibited

183 Adaptation

Key Idea: An adaptation is any heritable trait that suits an organism to its natural functional role in the environment (its niche).

Adaptation and fitness

▶ An **adaptation**, is any heritable trait that equips an organism for its niche, enhancing its exploitation of the environment and contributing to its survival and successful reproduction.

▶ Adaptations are a product of natural selection and can be structural (morphological), physiological, or behavioral traits. Traits that are not helpful to survival and reproduction will not be favoured and will be lost.

▶ Adaptation is important in an evolutionary sense because adaptive features promote fitness. **Fitness** is a measure of an organism's ability to maximize the numbers of offspring surviving to reproductive age.

Adaptive features of the North American beaver

North American beavers (*Castor canadensis*) are semi-aquatic and are able remain submerged for up to 15 minutes. Their adaptations enable them to exploit both aquatic and terrestrial environments.

Beavers are strict herbivores and eat leaves, bark, twigs, roots, and aquatic plants. They do not hibernate. They live in domelike homes called lodges, which they build from mud and branches. Lodges are usually built in the middle of a pond or lake, with an underwater entrance, making it difficult for predators to capture them.

Ears and nostrils
Valves in the ears and nose close when underwater.
These keep water out.

Lips
Their lips can close behind their front teeth.
This lets them carry objects and gnaw underwater, but keeps water out and stops them drowning.

Front feet
Front paws are good at manipulating objects.
The paws are used in dam and lodge construction to pack mud and manipulate branches.

Teeth
Large, strong chisel-shaped front teeth (incisors) grow constantly.
These let them fell trees and branches for food and to make lodges with.

Waterproof coat
A double-coat of fur (coarse outer hairs and short, fine inner hairs). An oil is secreted from glands and spread through the fur.
The underfur traps air against the skin for insulation and the oil acts as a waterproofing agent and keeps the skin dry in the water.

Eyes
A clear eyelid (nictitating membrane).
This protects the eye and allows the beaver to still see while swimming.

Oxygen conservation
During dives, beavers slow their heartbeat and reduce blood flow to the extremities to conserve oxygen and energy.
This enables them to stay submerged for 15 minutes even though they are not particularly good at storing oxygen in the tissues.

Thick insulating fat
Thick fat layer under the skin.
Insulates the beaver from the cold water and helps to keep it warm.

Large, flat paddle-like tail
Assists swimming and acts like a rudder. Tail is also used to slap the water in communication with other beavers, to store fat for the winter, and as a means of temperature regulation in hot weather because heat can be lost over the large unfurred surface area.

Large, webbed, hind feet
The webbing between the toes acts like a diver's swimming fins, and helps to propel the beaver through the water.

CONNECT **184**   CONNECT **175**   WEB **183**   KNOW

Adaptations for diving in air-breathing animals

Air breathing animals that dive must cope with lack of oxygen (which limits the length of the dive) and pressure (which limits the depth of the dive). Many different animal phyla have diving representatives, which have evolved from terrestrial ancestors and become adapted for an aquatic life. Diving air-breathers must maintain a supply of oxygen to the tissues during dives and can only stay underwater for as long as their oxygen supplies last. Their adaptations enable them to conserve oxygen and so prolong their dive time.

Species for which there is a comprehensive fossil record, e.g. whales (right), show that adaptations for a diving lifestyle accumulated slowly during the course of the group's evolution.

Humpback whale

Penguin

Green turtle

Diving birds
Penguins show many of the adaptations typical of diving birds. During dives, a bird's heart rate slows, and blood is diverted to the head, heart, and eyes.

Diving reptiles
Sea turtles have low metabolic rates and their issues are tolerant of low oxygen. These adaptations allow them to remain submerged for long periods and they surface only occasionally.

Diving mammals
Dolphins, whales, and seals are among the most well adapted divers. The exhale before diving, so that the lungs are compressed at depth and nitrogen cannot enter the blood. This prevents them getting the bends when they surface.

During dives, oxygen is conserved by reducing heart rate dramatically, and redistributing blood to supply only critical organs. Diving mammals have high levels of muscle myoglobin, which stores oxygen, but their muscles also function efficiently using anaerobic metabolism.

1. (a) What is an adaptation? _____

_____

_____

   (b) How can an adaptation increase an organism's fitness? _____

_____

_____

2. The following list identifies some adaptations in a beaver which allow it to survive in its environment. Identify each adaptation as structural, physiological, or behavioral and describe its survival advantage:

   (a) Large front teeth: _____

_____

   (b) Lodge built in middle of pond: _____

_____

   (c) Oil glands in coat: _____

_____

3. (a) What restricts the amount of time diving animals can spend underwater? _____

   (b) How does reducing heart rate during a dive enable animals to stay underwater longer?

_____

 © 2014 **BIOZONE** International
ISBN: 978-1-927173-91-6

184 Natural Selection

Key Idea: The effect of natural selection on a population can be verified by making quantitative measurements of phenotypic traits.

▶ **Natural selection** acts on the phenotypes of a population. Individuals with phenotypes that increase their fitness produce more offspring, increasing the proportion of the genes corresponding to that phenotype in the next generation.

▶ Numerous population studies have shown natural selection can cause phenotypic changes in a population relatively quickly.

▶ The finches on the Galápagos island (Darwin's finches) are famous in that they are commonly used as examples of how evolution produces new species. In this activity you will analyze data from the measurement of beaks depths of the medium ground finch (*Geospiza fortis*) on the island of Daphne Major near the center of the Galápagos Islands. The measurements were taken in 1976 before a major drought hit the island and in 1978 after the drought (survivors and survivors' offspring).

Beak depth (mm)	No. 1976 birds	No. 1978 survivors	Beak depth of offspring (mm)	Number of birds
7.30-7.79	1	0	7.30-7.79	2
7.80-8.29	12	1	7.80-8.29	2
8.30-8.79	30	3	8.30-8.79	5
8.80-9.29	47	3	8.80-9.29	21
9.30-9.79	45	6	9.30-9.79	34
9.80-10.29	40	9	9.80-10.29	37
10.30-10.79	25	10	10.30-10.79	19
10.80-11.29	3	1	10-80-11.29	15
11.30+	0	0	11.30+	2

1. Use the data above to draw two separate sets of histograms:

   (a) On the left hand grid draw side-by-side histograms for the number of 1976 birds per beak depth and the number of 1978 survivors per beak depth.

   (b) On the right hand grid draw a histogram of the beak depths of the offspring of the 1978 survivors.

2. (a) Mark on the graphs of the 1976 beak depths and the 1978 offspring the approximate mean beak depth.

   (b) How much has the average moved from 1976 to 1978? _____

   (c) Is beak depth heritable? What does this mean for the process of natural selection in the finches?

   _____

   _____

CONNECT   CONNECT   WEB

182 153 184 **DATA**

185 Natural Selection in Pocket Mice

Key Idea: The need to blend into their surroundings to avoid predation is an important selection pressure acting on the coat color of rock pocket mice.

Rock pocket mice are found in the deserts of southwestern United States and northern Mexico. They are nocturnal, foraging at night for seeds, while avoiding owls (their main predator). During the day they shelter from the desert heat in their burrows. The coat color of the mice varies from light brown to very dark brown. Throughout the desert environment in which the mice live there are outcrops of dark volcanic rock. The presence of these outcrops and the mice that live on them present an excellent study in natural selection.

▶ The coat color of the Arizona rock pocket mice is controlled by the Mc1r gene (a gene that in mammals is commonly associated with the production of the pigment melanin). Homozygous dominant (DD) and heterozygous mice (Dd) have dark coats, while homozygous recessive mice (dd) have light coats. Coat color of mice in New Mexico is not related to the Mc1r gene.

▶ 107 rock pocket mice from 14 sites were collected and their coat color and the rock color they were found on were recorded by measuring the percentage of light reflected from their coat (low percentage reflectance equals a dark coat). The data is presented right:

Site	Rock type (V volcanic)	Percent reflectance (%) Mice coat	Percent reflectance (%) Rock
KNZ	V	4	10.5
ARM	V	4	9
CAR	V	4	10
MEX	V	5	10.5
TUM	V	5	27
PIN	V	5.5	11
AFT		6	30
AVR		6.5	26
WHT		8	42
BLK	V	8.5	15
FRA		9	39
TIN		9	39
TUL		9.5	25
POR		12	34.5

1. (a) What is the genotype(s) of the dark colored mice? _____

   (b) What is the genotype of the light colored mice? _____

2. Using the data in the table above and the grids below and on the facing page, draw column graphs of the percent reflectance of the mice coats and the rocks at each of the 14 collection sites.

© 2014 **BIOZONE** International
ISBN: 978-1-927173-91-6
Photocopying Prohibited

3. (a) What do you notice about the reflectance of the rock pocket mice coat color and the reflectance of the rocks they were found on?

_____

_____

_____

(b) Suggest a cause for the pattern in 3(a). How do the phenotypes of the mice affect where the mice live?

_____

_____

_____

(c) What are two exceptions to the pattern you have noticed in 3(a)? _____

_____

_____

_____

(d) How might these exceptions have occurred? _____

_____

_____

4. The rock pocket mice populations in Arizona use a different genetic mechanism to control coat color than the New Mexico populations. What does this tell you about the evolution of the genetic mechanism for coat color?

_____

_____

_____

186 Insecticide Resistance

Key Idea: The application of insecticide provides a strong selection pressure on insects so that resistance to insecticides is constantly evolving in insect populations.

Insecticides are pesticides used to control insects considered harmful to humans, their livelihood, or environment. Insecticide use has increased since the advent of synthetic insecticides in the 1940s.

▶ The widespread but often inefficient use of insecticides can lead to resistance to the insecticide in insects (see right). Mutations may also produce traits that further assist with resistance.

▶ Ineffectual application may include application at the wrong time, e.g. before the majority of the population has established or close to rain, and applying contact sprays that may be avoided by hiding under leaves.

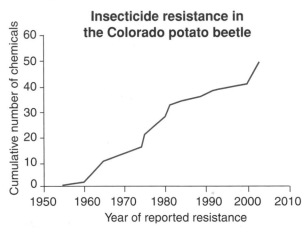

Insecticide resistance in the Colorado potato beetle

(Graph: x-axis "Year of reported resistance" from 1950 to 2010; y-axis "Cumulative number of chemicals" from 0 to 60.)

The Colorado potato beetle (*Leptinotarsa decemlineata*) is a major potato pest that was originally found living on buffalo-bur (*Solanum rostratum*) in the Rocky mountains. It has an extraordinary ability to develop resistance to synthetic pesticides. Since the 1940s, when these pesticides were first developed, it has become resistant to more than 50 different insecticides.

The evolution of resistance

The application of an insecticide can act as a potent selection pressure for resistance in pest insects. Insects with a low natural resistance die from an insecticide application, but those with a higher natural resistance may survive if the insecticide is not effectively applied. These will found a new generation which will on average have a higher resistance to the insecticide

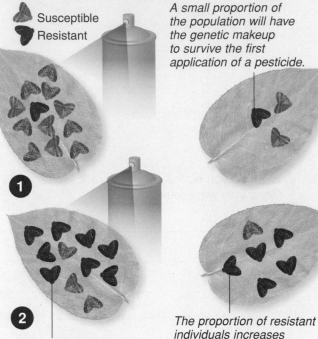

- Susceptible
- Resistant

A small proportion of the population will have the genetic makeup to survive the first application of a pesticide.

1

2

The genetic information for pesticide resistance is passed to the next generation.

The proportion of resistant individuals increases following subsequent applications of insecticide. Eventually, almost all of the population is resistant.

Mechanisms of resistance in insect pests

▶ Insecticide resistance in insects can arise through a combination of mechanisms.
  (1) Increased sensitivity to an insecticide will cause the pest to avoid a treated area.
  (2) Certain genes confer stronger physical barriers, decreasing the rate at which the chemical penetrates the insect's cuticle.
  (3) Detoxification by enzymes within the insect's body can render the insecticide harmless.
  (4) Structural changes to the target enzymes make the insecticide ineffective.
      No single mechanism provides total immunity, but together they transform the effect from potentially lethal to insignificant.

1. Why must farmers be sure that insecticides are applied correctly: _____

_____

_____

2. Describe two mechanisms that increase insecticide resistance in insects: _____

_____

_____

© 2014 **BIOZONE** International
ISBN: 978-1-927173-91-6
Photocopying Prohibited

187 Gene Pool Exercise

The set of all the versions of all the genes in a population (it genetic make-up) is called the **gene pool**. Cut out the squares below and use them to model the events described in *Modeling Natural Selection*.

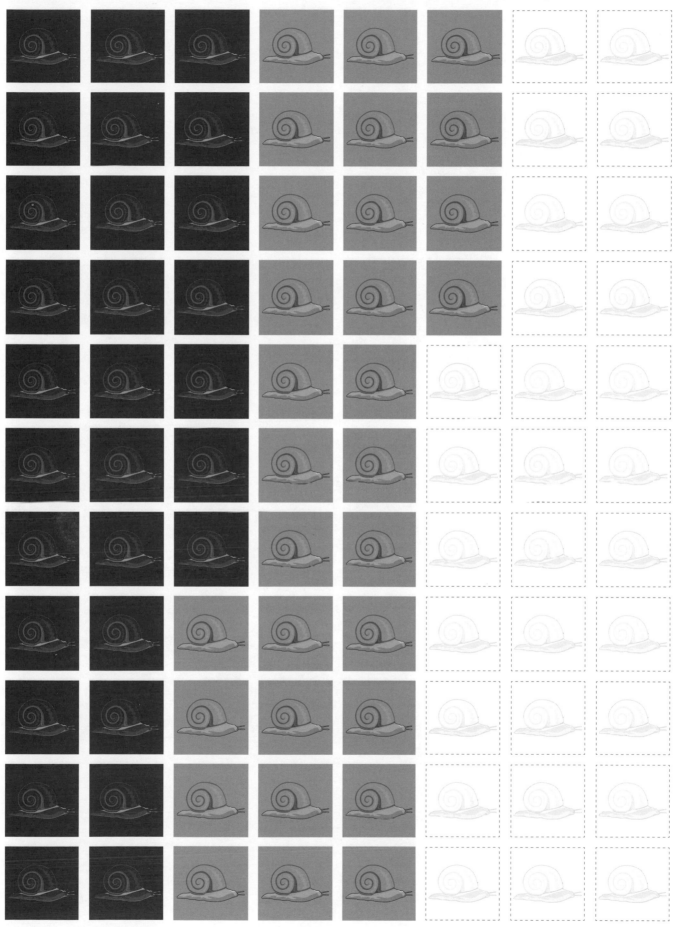

PRAC

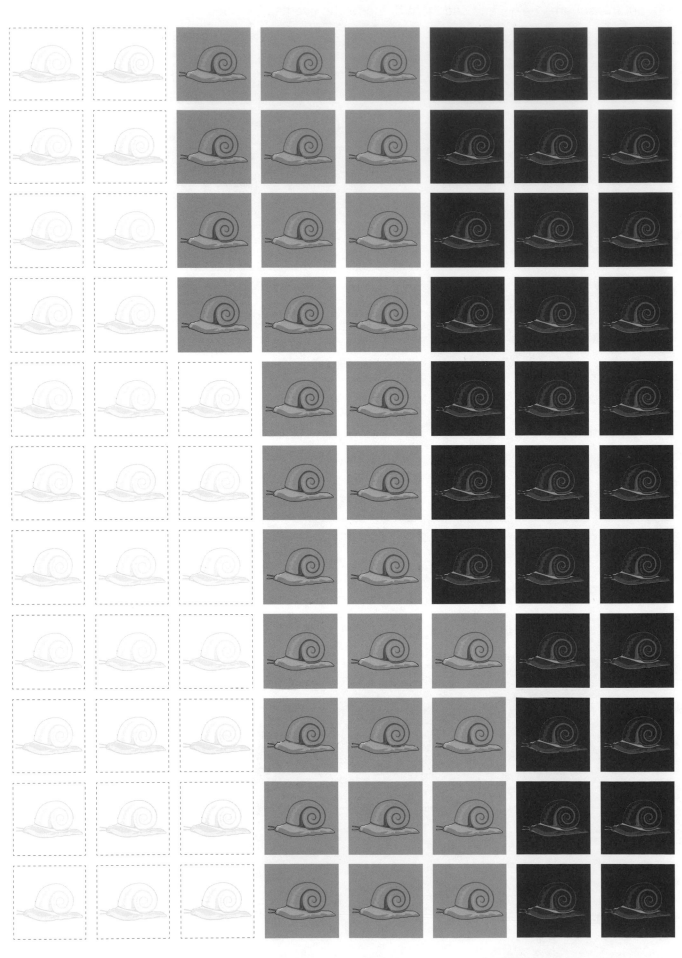

188 Modeling Natural Selection

Key Idea: Natural selection acts on phenotypes. Those individuals better suited to an environment will have a greater chance of reproductive success.

Natural selection can be modeled in a simple activity based on predation. You can carry out the following activity by yourself, or work with a partner to increase the size of the population. The black, gray, and white squares on the preceding pages represent phenotypes of a population. Cut them out and follow the instructions below to model natural selection. You will also need a sheet of white paper and a sheet of black paper.

1. Cut out the squares on the preceding pages and record the number of black, gray, and white squares. Work out the proportion of each phenotype in the population (e.g. 0.33 black 0.34 gray, 0.33 white) and place these values in the table below. This represents your starting population (you can combine populations with a partner to increase the population size for more reliable results).

2. For the first half of the activity you will also need a black sheet of paper or material that will act as the environment (A3 is a good size). For the second half of the activity you will need a white sheet of paper.

3. Place 14 each of the black, gray, and white squares in a bag and shake them up to mix them. Keep the others for making up population proportions later.

4. Now take the squares out of the bag and randomly distribute them over the sheet of black paper (this works best if your partner does this while you aren't looking).

5. For 20 seconds, pick up the squares that stand out (are obvious) on the black paper. These squares represent animals in the population that have been preyed upon and killed (you are acting the part of a predator on the snails). Place them to one side and pick up the rest of the squares. These represent the population that survived to reproduce.

6. Count the remaining phenotype colors and calculate the proportions of each phenotype. Record them in the table below in the proportions row of generation 2. Use the formula: Proportion = number of colored squares /total number of squares remaining. For example: for one student doing this activity: proportion of white after predation = 10/30 = 0.33.

7. Before the next round of selection, the population must be rebuilt to its original total number using the newly calculated proportions of colors and the second half of the squares from step 3. Use the following formula to calculate the number of each color: number of colored squares required = proportion x number of squares in original population (42 if you are by yourself, 84 with a partner). For example: for one student doing this activity: 0.33 x 42 = 13.9 = 14 (you can't have half a phenotype). Therefore in generation 2 there should be 14 white squares. Do this for all phenotypes using the spare colors to make up the numbers if needed. Record the numbers in the numbers row of generation 2. Place generation 2 into the bag.

8. Repeat steps 4 to 7 for generation 2, and 3 more generations (5 generations in total or more if you wish).

9. On separate graph paper, draw a line graph of the proportions of each color over the five generations. Which colors have increased, which have decreased?

10. Now repeat the whole activity using a white sheet background instead of the black sheet. What do you notice about the proportions this time?

Generation		Black	Gray	White
1	Number			
	Proportion			
2	Number			
	Proportion			
3	Number			
	Proportion			
4	Number			
	Proportion			
5	Number			
	Proportion			

CONNECT   WEB
185 188   KNOW

189 What is a Species?

Key Idea: A species is a very specific unit of taxonomy given to a group of organisms with similar characteristics in order to categorize them.

The **species** is the basic unit of taxonomy. A **biological species** is defined as a group of organisms capable of interbreeding to produce fertile offspring.

▶ However, there are some difficulties in applying the biological species concept (BSC) in practice as some closely related species are able to interbreed to produce fertile hybrids (e.g. *Canis* (dog) species).

▶ The BSC is also more successfully applied to animals than to plants. Plants hybridize easily and can reproduce vegetatively. For some, e.g. cotton and rice, first generation hybrids are fertile but second generation hybrids are not.

▶ In addition, the BSC cannot be applied to asexually reproducing or extinct organisms. Increasingly, biologists are using DNA analyses to clarify relationships between the related populations that we regard as one species.

These breeds of dogs may look and behave very differently, but they can all interbreed to produce viable offspring.

Another way of defining a species is by using the **phylogenetic species concept** (PSC). Phylogenetic species are defined on the basis of their shared evolutionary ancestry, which is determined on the basis of **shared derived characteristics**. These are characteristics that evolved in an ancestor and are present in all its descendants.

▶ The PSC defines a species as the smallest group that all share a derived character state. It is useful in paleontology because biologists can compare both living and extinct organisms.

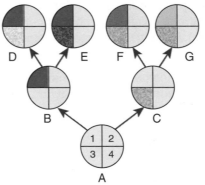

Species B and C are related to species A as they share three of four characteristics with it. However they only share two characteristics with each other. D and E share characteristics with B, while F and G share characteristics with C

1. Define a biological species: _____

_____

_____

2. Why is it difficult to fully define a species in practice?_____

_____

_____

3. How is the biological species different to the phylogenetic species?_____

_____

_____

4. There often appear to be greater differences between different breeds of dog than there are between different species of the *Canis* genus. Why then are dogs all considered one species?

_____

_____

_____

 © 2014 **BIOZONE** International
ISBN: 978-1-927173-91-6
Photocopying Prohibited

190 How Species Form

Key Idea: Gene flow is reduced when populations are separated. Continual reduction in gene flow by isolating mechanisms may eventually lead to the formation of new species.

Species formation

▶ Species evolve in response to selection pressures of the environment. These may be naturally occurring or caused by humans. The diagram below represents a possible sequence for the evolution of two hypothetical species of butterfly from an ancestral population. As time progresses (from top to bottom of the diagram below) the amount of genetic difference between the populations increases, with each group becoming increasingly isolated from the other.

▶ The isolation of two gene pools from one another may begin with **geographical barriers**. This may be followed by isolating mechanisms that occur before the production of a zygote (e.g. behavioral changes), and isolating mechanisms that occur after a zygote is formed (e.g. hybrid sterility). As the two gene pools become increasingly isolated and different from each other, they are progressively labelled: population, race, and subspecies. Finally they attain the status of separate species.

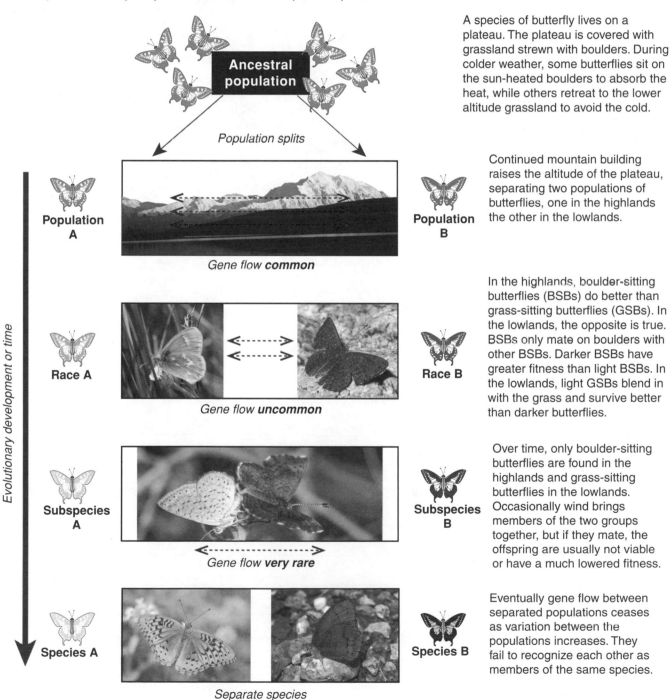

A species of butterfly lives on a plateau. The plateau is covered with grassland strewn with boulders. During colder weather, some butterflies sit on the sun-heated boulders to absorb the heat, while others retreat to the lower altitude grassland to avoid the cold.

Continued mountain building raises the altitude of the plateau, separating two populations of butterflies, one in the highlands the other in the lowlands.

In the highlands, boulder-sitting butterflies (BSBs) do better than grass-sitting butterflies (GSBs). In the lowlands, the opposite is true. BSBs only mate on boulders with other BSBs. Darker BSBs have greater fitness than light BSBs. In the lowlands, light GSBs blend in with the grass and survive better than darker butterflies.

Over time, only boulder-sitting butterflies are found in the highlands and grass-sitting butterflies in the lowlands. Occasionally wind brings members of the two groups together, but if they mate, the offspring are usually not viable or have a much lowered fitness.

Eventually gene flow between separated populations ceases as variation between the populations increases. They fail to recognize each other as members of the same species.

Ancestral population

Population splits

Evolutionary development or time

Population A — Population B — *Gene flow common*

Race A — Race B — *Gene flow uncommon*

Subspecies A — Subspecies B — *Gene flow very rare*

Species A — Species B — *Separate species*

CONNECT | WEB
182 **190** KNOW

Geographic isolation

- Geographical isolation describes the isolation of a species population (gene pool) by some kind of physical barrier, e.g. mountain range, water body, desert, or ice sheet. Geographical isolation is a frequent first step in the subsequent reproductive isolation of a species.

- An example of geographic isolation leading to speciation is the large variety of cichlid fish in the rift lakes of East Africa (right). Geologic changes to the lake basins have been important in the increase of cichlid fish species.

L. Victoria
L. Tanganyika
L. Malawi —

Reproductive isolating mechanisms

- Reproductive isolating mechanisms (RIMs) are reproductive barriers that are part of a species' biology (and therefore do not include geographical isolation). They prevent interbreeding (therefore gene flow) between species. Single barriers may not completely stop gene flow, so most species commonly have more than one type of barrier. Most RIMs operate before fertilization (prezygotic RIMs). They include mechanisms such as temporal isolation (e.g. differences in breeding season), behavioral isolation (e.g. differences in mating behaviors), and mechanical isolation (e.g. differences in copulatory structures). Postzyotic RIMs operate after fertilization and are important in preventing offspring between closely related species. They involve a mismatch of chromosomes in the zygote, and include hybrid sterility (right), hybrid inviability, and hybrid breakdown.

The white-tailed antelope squirrel (left) and the Harris' antelope squirrel (right) in the southwestern United States and northern Mexico, are separated by the Grand Canyon (center) and have evolved to occupy different habitats.

Mules are a cross between a male donkey and a female horse. The donkey contributes 31 chromosomes while the horse contributes 32, making 63 chromosomes in the mule. This produces sterility in the mule as meiosis cannot produce gametes with an even number of chromosomes.

1. Identify some geographical barriers that could separate populations: _____

_____

_____

2. Why is a geographical barrier not considered a reproductive isolating mechanism?

_____

_____

3. Identify the two categories of reproductive isolating mechanisms and explain the difference between them:

_____

_____

_____

_____

4. Why is more than one reproductive isolation barrier needed to completely isolate a species?

_____

_____

© 2014 **BIOZONE** International
ISBN: 978-1-927173-91-6

191 Patterns of Evolution

Key Idea: Populations moving into a new environment may diverge from their common ancestor and form new species.

▶ The diversification of an ancestral group into two or more species in different habitats is called **divergent evolution**. This is shown right, where two species diverge from a **common ancestor**. Note that another species arose, but became extinct.

▶ When divergent evolution involves the formation of a large number of species to occupy different niches, it is called an **adaptive radiation**.

▶ The evolution of species may not necessarily involve branching. A species may accumulate genetic changes that, over time, result in a new species. This is known as **sequential evolution**.

▶ Evolution of species does not always happen at the same pace. Two models describe the pace of evolution.

Phyletic gradualism proposes that populations diverge slowly by accumulating adaptive features in response to different selective pressures.

Punctuated equilibrium proposes that most of a species' existence is spent in stasis and evolutionary change is rapid. The stimulus for evolution is a change in some important aspect of the environment.

It is likely that both mechanisms operate at different times for different taxonomic groups.

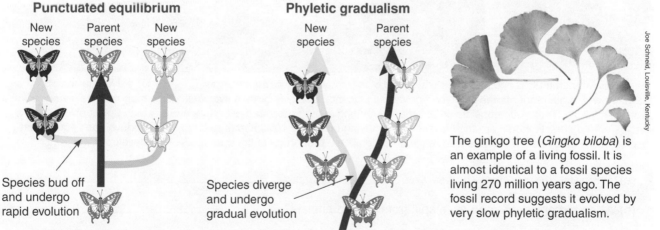

The ginkgo tree (*Gingko biloba*) is an example of a living fossil. It is almost identical to a fossil species living 270 million years ago. The fossil record suggests it evolved by very slow phyletic gradualism.

1. In the hypothetical example of divergent evolution illustrated at the top of the page:

   (a) Identify the type of evolution that produced species B from species D: _____

   (b) Identify the type of evolution that produced species P and H from species B: _____

   (c) Name all species that evolved from: **Common ancestor D**: _____ **Common ancestor B**: _____

2. When do you think punctuated evolution is most likely to occur and why? _____

_____

_____

CONNECT
174
WEB
191
KNOW

192 Evolution and Biodiversity

Key Idea: Mammals underwent an adaptive radiation around 65 million years ago and became a very diverse group.

▶ The mammals underwent a spectacular adaptive radiation following the sudden extinction of the non-avian dinosaurs, which left many niches vacant. Most of the modern groups of mammals appeared very quickly. Placental mammals have become dominant over other groups of mammals throughout the world, except in Australia where marsupials remained dominant after being isolated following the breakup of Gondwana. Placental mammals (bats) reached Australia 15 million years ago followed by rodents 10 million years ago.

▶ The diagram below shows a simple evolutionary tree for the mammals. The width of the bars indicates the number of species in each of the three mammalian groups.

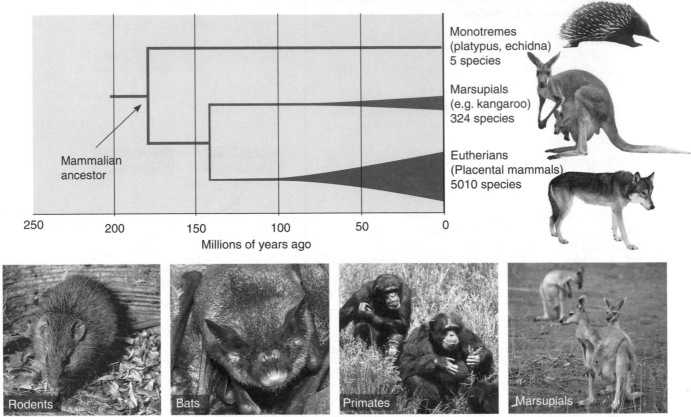

Monotremes (platypus, echidna) 5 species

Marsupials (e.g. kangaroo) 324 species

Eutherians (Placental mammals) 5010 species

Mammalian ancestor

250   200   150   100   50   0

Millions of years ago

Rodents | Bats | Primates | Marsupials

The placental mammals make up the largest group of mammals. They carry embryos internally, which get nutrients the from mother via the placenta, and give birth to live, well developed young. The group consists of 18 orders, although some orders contain many more species than others. Rodents are the largest order, making up nearly 40% of mammalian species, followed by bats. Primates also make up a significant proportion of the mammals.

The five species of monotremes lay eggs while marsupials give birth to very undeveloped young that develop in the pouch.

1. (a) When did the mammals first evolve? _____

    (b) When did placental mammals split from the marsupials? _____

    (c) Suggest why the mammals underwent such a major adaptive radiation after the dinosaurs died out?

    _____

    _____

    (d) Suggest a reason why marsupials remained the dominant mammalian group in Australia after the split between marsupial and placental mammals:

    _____

    _____

    _____

 © 2014 **BIOZONE** International
ISBN: 978-1-927173-91-6
Photocopying Prohibited

KNOW

193 Extinction

Key Idea: Extinctions are a natural process and may happen gradually or very rapidly. Human activity has played a large part in many recent extinctions.

Extinction is the death of an entire species, no individuals are left alive. Extinction is an important (and natural) process in evolution and describes the loss of a species forever. Extinction provides opportunities, in the form of vacant niches, for the evolution of new species. More than 98% of species that have ever lived are now extinct.

▶ Extinction is the result of a species being unable to adapt to an environmental change. Either it evolves into a new species (and the ancestral species becomes effectively extinct) or it and its lineage becomes extinct.

▶ A **mass extinction** describes the widespread and rapid (in geologic terms) decrease in life on Earth and involves not only the loss of species, but the loss of entire families (which comprise many genera and species). Such events are linked to major climate shifts or catastrophic events. The diagram below shows how the diversity of life has varied over the history of life on Earth. The five past mass extinctions are described. A sixth event, caused by the activities of humans, began in the current Holocene epoch and continues today.

Major Mass Extinctions

Ordovician extinction (450 mya) Second largest extinction of marine life: >60% of marine invertebrates died. One of the coldest periods in Earth's history.

Permian extinction (250 mya) Nearly all life on Earth perished. More than 90% of marine species and many terrestrial species die out. Many families and genera lost.

Cretaceous extinction (65 mya) Marked by the extinction of nearly all dinosaur species (their descendants the birds survive). Generally accepted as being caused by a massive asteroid that collided with Earth.

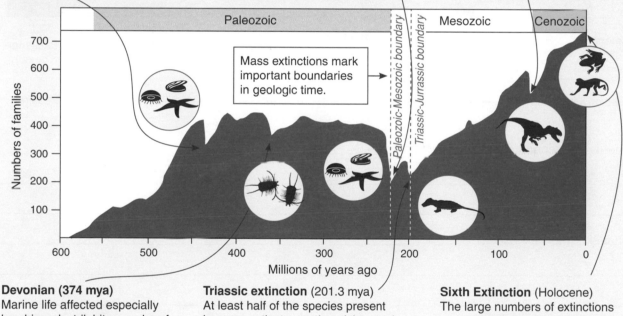

Mass extinctions mark important boundaries in geologic time.

Devonian (374 mya) Marine life affected especially brachiopods, trilobites, and reef building organisms.

Triassic extinction (201.3 mya) At least half of the species present became extinct, vacating niches and ushering in the age of the dinosaurs.

Sixth Extinction (Holocene) The large numbers of extinctions across many families is blamed on human activities.

Since the 1500s, at least 412 vertebrate or plant species have become extinct. The normal background extinction rate for these groups combined is estimated at 0.37 species a year (one species extinction every 3 years). The observed rate is about 0.82, more than twice that expected. The actual rate is probably higher because species numbers are underestimated.

The golden toad (left) from Costa Rica was officially declared extinct in 1989.

Organism*	Total number of species (approx)*	Known extinctions (since ~1500)*
Mammals	5487	87
Birds	9975	150
Reptiles	10,000	22
Amphibians	6700	39
Plants	300,000	114

* These numbers vastly underestimate the true numbers because so many species are undescribed.

CONNECT WEB
191 **193** KNOW

▶ Human activity is the cause of many recent extinctions. Famous extinctions include the extermination of the dodo within 70 years of its first sighting in 1598 and the passenger pigeon that went from an estimated 3 billion individuals before the arrival of Europeans to North America to being extinct by 1914. Recent extinctions include the Yangtze River dolphin and the Pinta Island tortoise (a giant tortoise).

▶ Human activities such as hunting and destruction of habitat (by logging, pollution, and land clearance) have caused many recent extinctions. The possible effects of climate change may also drive many other vulnerable species to extinction.

Logging

Climate change

Pollution

ICU Red list

Examples of threatened and endangered animal species

American Bison | Polar bear | Great white shark | Blue whale | Tiger | Cuban crocodile | Iberian lynx | Newfoundland wolf | Golden toad | Tasmanian tiger

| Near threatened | Vulnerable | Endangered or critically endangered | Extinct or extinct in the wild |

American Alligator | African elephant | Indian rhinoceros | Sea otter | Przewalski's gazelle | Walia ibex | Sea mink | Passenger pigeon

Population increasing ◄─────────────────────► Population decreasing

1. What is the general cause of extinction? _____

_____

_____

2. What is a mass extinction? _____

_____

_____

3. What are some reasons for recent extinctions? _____

_____

_____

4. How might extinction of a species or higher taxon act as a catalyst for the rapid evolution of other species?

_____

_____

5. Carry out some research to identify two species, other than those mentioned earlier, that have recently become extinct. Identify the date of their extinction, and possible reasons for their extinction:

_____

_____

© 2014 BIOZONE International
ISBN: 978-1-927173-91-6
Photocopying Prohibited

194 A Close Brush With Extinction

Key Idea: Many species have been brought back from the brink of extinction through conservation efforts. Losing one species can greatly affect an ecosystem.

▶ Although human activity has caused (either directly or indirectly) the extinction of numerous species, conservation efforts by humans have also managed to save many species from extinction.

▶ There are several examples, where species on the edge of extinction have recovered. The Chatham Island black robin, had just five individuals left in 1980, but has now recovered to more than 200 birds. In 1941, only 21 whooping cranes existed, now there are more than 400. The Aleutian Canada goose went from 790 birds in 1975 to more than 60,000 in 2005.

Chatham Island black robin

Whooping crane

Sea Otters

▶ Sometimes the likely effects of species loss can be indicated when a species declines rapidly to the point of near extinction. The sea otter example described below illustrates this.

▶ Sea otters live along the Northern Pacific coast of North America and have been hunted for hundreds of years for their fur. Commercial hunting didn't fully begin until about the mid 1700s when large numbers were killed and their fur sold to overseas markets.

▶ The drop in sea otter numbers had a significant effect on the local marine environment. Sea otters feed on shellfish, particularly sea urchins. Sea urchins eat kelp, which provides habitat for many marine creatures. Without the sea otters to control the sea urchin population, sea urchin numbers increased and the kelp forests were severely reduced.

Sea otter, *Enhydra lutris*

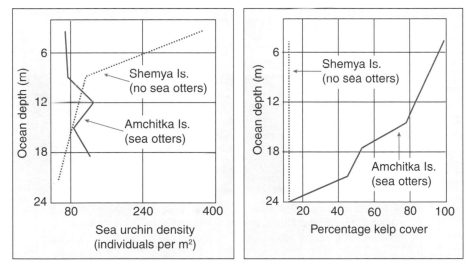

Sea otters are critical to ecosystem function. When their numbers were significantly reduced by the fur trade, sea urchin populations exploded and the kelp forests, on which many species depend, were destroyed.

The effect can be seen on Shemya and Amchitka Islands. Where sea otters are absent, large numbers of sea urchins are found, and kelp are almost absent.

1. Carry out some research and identify a species in your area which has successfully been brought back from the brink of extinction. Briefly described how this was achieved.

_____

_____

2. (a) What effect do sea otters have on sea urchin numbers? _____

   (b) What effect do sea urchins have on kelp cover? _____

   (c) Why would losing the sea otters destroy the kelp forests? _____

   _____

   _____

© 2014 **BIOZONE** International
ISBN: 978-1-927173-91-6
Photocopying Prohibited

195 Conservation and Genetic Diversity

Key Idea: Maintaining genetic diversity in a population is important in avoiding inbreeding depression, which is a cause of reduced fitness in individuals.

Conservation genetics involves many branches of science and uses genetic methods to restore genetic diversity in a declining species.

One of the biggest problems occurring when a species' population declines is loss of genetic diversity. This increases the relatedness between individuals because there are fewer individuals to breed with. Decreased genetic diversity and increased relatedness can result in inbreeding depression (the reduced fitness of individuals as a result of inbreeding), and can dramatically reduce population viability.

▶ The Florida panther (right) is an example of how conservation genetics has been used to restore genetic diversity in an endangered population.

▶ In the late 1970s, the Florida panther population had become critically low and occupied just 5% of its historical range. Population models showed it would be extinct within a few decades. Individuals often had several abnormalities including kinked tails, heart defects, and sperm defects. It was determined that these were due to inbreeding depression.

▶ In 1995, eight female panthers were translocated from Texas to increase genetic diversity. Over the last two decades there has been an increase in the population growth rate and an improvement in the survival and health of individuals (graph, right).

Population Florida panther

Conservation genetics is an important concept for conservation and breeding programmes. By keeping detailed breeding records of which individuals are bred together, and monitoring the genetic relatedness of populations, conservation scientists can make sure endangered species avoid inbreeding depression and maintain their genetic diversity.

Conservation genetics were used to rebuild the Illinois prairie chicken population.

Critically endangered birds, such as the New Zealand kakapo, have lost much of their genetic diversity.

1. What is conservation genetics? _____

_____

2. (a) Why were panthers from Texas translocated to Florida? _____

_____

(b) Why is it important to maintain genetic diversity in populations? _____

_____

_____

_____

KNOW

 © 2014 **BIOZONE** International
ISBN: 978-1-927173-91-6
Photocopying Prohibited

196 Chapter Review

Summarize what you know about this topic under the headings provided. You can draw diagrams or mind maps, or write short notes to organize your thoughts. Use the hints and guidelines included to help you.

Adaptation

HINT: How is adaptation a consequence of natural selection? How do adaptive features enhance fitness?

Evolution and natural selection

HINT: How does natural selection lead to evolution? What does natural selection act on?

REVISE

Species formation

HINT: How do species form and how do they remain distinct?.

Extinction

HINT: What causes extinction? Account for mass extinction.
How does genetic diversity affect a species' ability to avoid extinction?

197 Key Terms: Did You Get It?

1. Test your vocabulary by matching each term to its definition, as identified by its preceding letter code.

adaptation ....

biodiversity ....

evolution ....

extinction ....

fitness ....

genotype ....

natural selection ....

phenotype ....

variation ....

A The complete dying out of a species so that there are no representatives of the species remaining anywhere.

B The observable characteristics in an organism.

C A heritable characteristic of a species that equips it for survival and reproductive success in its environment.

D The allele combination of an organism e.g. AA.

E The differences between individuals in a population as a result of genes and environment.

F The process by which favorable heritable traits become more common in successive generations.

G Change in the genetic makeup of a population over time.

H Biological diversity, e.g. of a region or of the Earth

I A measure of an individual's relative genetic contribution to the next generation as a result of its combination of traits.

2. Complete the diagram below showing the four interacting factors involved in the evolution of the population:

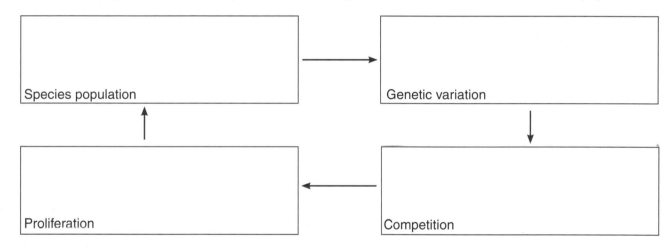

Species population → Genetic variation

Proliferation ← Competition

3.

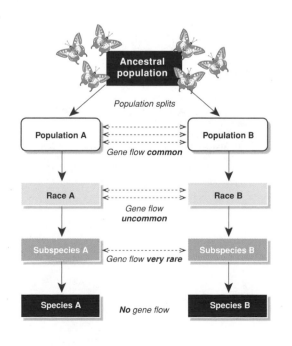

Fill in the boxes below to describe the mechanisms that lead to the formation of new species:

Geographic isolation:

Isolating mechanisms:

Species isolation:

© 2014 **BIOZONE** International
ISBN: 978-1-927173-91-6

TEST

Appendix

Questioning terms in biology

The following terms are often used when asking questions in examinations and assessments.

Analyze: Interpret data to reach stated conclusions.

Annotate: Add brief notes to a diagram, drawing or graph.

Apply: Use an idea, equation, principle, theory, or law in a new situation.

Calculate: Find an answer using mathematical methods. Show the working unless instructed not to.

Compare: Give an account of similarities between two or more items, referring to both (or all) of them throughout.

Construct: Represent or develop in graphical form.

Contrast: Show differences. Set in opposition.

Define: Give the precise meaning of a word or phrase as concisely as possible.

Derive: Manipulate a mathematical equation to give a new equation or result.

Describe: Define, name, draw annotated diagrams, give characteristics of, or an account of.

Design: Produce a plan, object, simulation or model.

Determine: Find the only possible answer.

Discuss: Show understanding by linking ideas. Where necessary, justify, relate, evaluate, compare and contrast, or analyze.

Distinguish: Give the difference(s) between two or more items.

Draw: Represent by means of pencil lines. Add labels unless told not to do so.

Estimate: Find an approximate value for an unknown quantity, based on the information provided and application of scientific knowledge.

Evaluate: Assess the implications and limitations.

Explain: Provide a reason as to how or why something occurs.

Identify: Find an answer from a number of possibilities.

Illustrate: Give concrete examples. Explain clearly by using comparisons or examples.

Interpret: Comment upon, give examples, describe relationships. Describe, then evaluate.

List: Give a sequence of answers with no elaboration.

Measure: Find a value for a quantity.

Outline: Give a brief account or summary. Include essential information only.

Predict: Give an expected result.

Solve: Obtain an answer using numerical methods.

State: Give a specific name, value, or other answer. No supporting argument or calculation is necessary.

Suggest: Propose a hypothesis or other possible explanation.

Summarize: Give a brief, condensed account. Include conclusions and avoid unnecessary details.

Credits

We acknowledge the generosity of those who have provided photographs for this edition: • Louise Howard and Katherine Connolly, Dartmouth College Electronic Microscope Facility • Kent Pryor • Marc King for photographs of comb types in poultry • Stephen Moore for his photos of aquatic invertebrates • PASCO for photographs of probeware • D. Dibenski for the photo of the flocking auklets.

We also acknowledge the photographers that have made their images available through **Wikimedia Commons** under Creative Commons Licences 2.0, 2.5. or 3.0: • James Hedberg • Brocken Inaglory • Matthias Zepper • Denali National Park Service • Gilles San Martin • Temsabulta • Uwe Kils • Milnette Layne • Sharon Mollerus • Jerald E. Dewey USDA • Mdf • Dan Pancamo • Sffubs • Mike Baird • Steve Maslowski, US Fish and Wildlife Service • Cephas • Paul Whippey • Hernán De Angelis Campephilus • Wendy Kaveney • Temsabuita • Eric Erbe • Gina Mikel • uwe Kils • WavyGeek • April Nobile (www.AntWeb.org) • Alex Wild • Jeffmock • Alastair Rae • Rocky Mountain Laboratories, NIAID, NIH • Xiangyux • National park service • Putney Mark • Allan and Elaine Wilson • UtahCamera • Dario • Jpbarrass • Kristian Peters • Wendy Kaveney • MEI Trust • it:Utente:Cits • Al Aumuller. NY World telegram and the Sun, Public Domain • Joe Schneid, Louisville, Kentucky.

Contributors identified by coded credits are: **BF**: Brian Finerran (Uni. of Canterbury), **BH**: Brendan Hicks (Uni. of Waikato), **CDC**: Centers for Disease Control and Prevention, Atlanta, USA, **DoC**: Department of Conservation (NZ), **EII**: Education Interactive Imaging, **IF**: I. Flux (DoC), **MPI**: Max Planck Institute, **NASA**: National Aeronautics and Space Administration, **NIH**: National Institutes of Health **KP**: Kent Pryor, **NOAA**: National Oceanic and Atmospheric Administration www.photolib.noaa.gov, **NYSDEC**: New York State Dept of Environmental Conservation, **RCN**: Ralph Cocklin, **RA**: Richard Allan, **USGS**: United States Geological Survey, **WBS**: Warwick Silvester (Uni. of Waikato), **WMU**: Waikato Microscope Unit, **USDA**: United States Department of Agriculture, **USFW**: United States Fish and Wildlife Service, **USGS**: United States Geological Survey.

Royalty free images, purchased by Biozone International Ltd, are used throughout this workbook and have been obtained from the following sources: Corel Corporation from their Professional Photos CD-ROM collection; IMSI (Intl Microcomputer Software Inc.) images from IMSI's MasterClips® and MasterPhotos™ Collection, 1895 Francisco Blvd. East, San Rafael, CA 94901-5506, USA; ©1996 Digital Stock, Medicine and Health Care collection; © 2005 JupiterImages Corporation www.clipart.com; ©Hemera Technologies Inc, 1997-2001; ©Click Art, ©T/Maker Company; ©1994., ©Digital Vision; Gazelle Technologies Inc.; PhotoDisc®, Inc. USA, www.photodisc.com. • TechPool Studios, for their clipart collection of human anatomy: Copyright ©1994, TechPool Studios Corp. USA (some of these images were modified by Biozone) • Totem Graphics, for their clipart collection • Corel Corporation, for use of their clipart from the Corel MEGAGALLERY collection • 3D images created using Bryce, Vue 6, Poser, and Pymol.

Index

Index

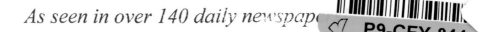

The Amish Cook Treasury

By Kevin Williams
with Lovina Eicher

Amish

Breakfasts

Breakfast is the hearty start to an Amish person's day. Even though the percentage of Amish that working as farmers has dropped significantly, the Amish have taken their legendary work ethic with them to other jobs like carpentry, factories, and tourism. This means their need for a filling start to the day has not diminished, so it's common to see huge breakfasts in Amish homes on an almost daily basis. No Pop-Tarts to be found. A typical Amish breakfast might consist of homemade cinnamon rolls, fresh eggs, cheese, cookies, toast, coffee, home-cured ham, and sometimes steak. A silent prayer will precede the meal, adding a moment of reflectivity before the rush of the day begins. Enjoy this culinary journey through the morning meals of an Amish family.

BREAKFAST CASSEROLE

1 pound sausage or ham
6 eggs
2 cups milk
3 slices toasted bread
1 cup grated cheese
1 teaspoon salt
1 teaspoon dry mustard

Brown sausage, crumble and drain. Beat eggs with milk, salt and mustard. Layer the sausage, bread, and cheese into a baking dish. Pour egg mixture over top. Refrigerate overnight. Bake at 350° for 45 minutes.

HAM & BACON OMELET

1/2 pound bacon, chopped
1/4 pound chopped smoked ham

4 small onions, sliced
4 eggs
1/4 cup milk or cream
Salt and pepper

Fry bacon in a skillet until crisp; remove from skillet. Fry the ham over low heat in the drippings until soft, stirring occasionally. Remove ham and fry onions in drippings until soft and lightly browned. Add bacon and ham to skillet and mix with the onion. Beat the eggs and add milk, salt and pepper. Pour into skillet and cook over medium heat for about 3 minutes, stirring occasionally. Serve immediately on hot toast.

PENNSYLVANIA EGGS

12 eggs, beaten
2-17 ounces cans creamed corn
2 cups cheddar cheese, shredded
2 cups Monterey Jack cheese, shredded
1 tablespoon grits
1 ½ teaspoons Worcestershire sauce
dash of pepper

Combine all ingredients and pour into greased baking dish. Bake at 325° for 1 hour or until firm and lightly browned on top.

HEARTY STEAK AND EGGS

6 strip steaks, 1/2-inch thick
1 large garlic clove, cut in half
1 teaspoon salt
1 teaspoon black pepper
1 teaspoon ground red chile pepper or paprika
6-1 inch slices sourdough or country bread
1/3 cup bacon fat or butter
12 eggs

Rub both sides of steak with garlic; sprinkle with half of salt, pepper and chile pepper. Cook steaks on hot griddle, turning once, until browned (2 to 3 minutes per side). Push steak to side of griddle to keep warm while cooking eggs.

Tear out the center of each bread slice, leaving 1/2 inch around edge. Reserve centers. Heat bacon fat on a griddle. Place bread rings on the griddle. Break 2 eggs into each ring. Sprinkle with remaining salt, pepper and chile pepper. Cook 2 to 3 minutes. Place bread centers on grill to toast.

Turn bread filled with eggs and toasted centers. Cook 1 to 3 minutes or until done. Serve eggs in toast along with breakfast steaks.

EGG SANDWICH

Even something as simple as an egg sandwich prepared with real butter in a heavy cast iron skillet has an almost wistful air of elegance, compared to the same zap-in-the-microwave concoction I might fix at my house – Kevin Williams, Editor

3 eggs
1 1/2 tablespoon vegetable butter
2 tablespoons chopped onions
1/3 cup stewed tomato

Put the butter and the onion into a small saucepan, sand let simmer together to soften the onion. Add the tomato, and bring to a boil. Add the beaten eggs all at once, and continue to stir rapidly until soft scrambled and evenly cooked. Salt to taste, spoon between slices of bread and serve.

WISCONSIN WAFFLES

2/3 cup all purpose flour
2/3 cup sifted cake flour
1 cup milk
1 3/4 teaspoons baking powder
1 1/3 whole eggs, well beaten
3 1/2 tablespoon butter or margarine
3/4 teaspoon vanilla extract

Topping:
1 cup water
7 tablespoons sugar
1 1/2 teaspoons white corn syrup

2 drops red food coloring
4 teaspoons cornstarch
2/3 whole (3 ounces) raspberry flavored gelatin
5 1/4 sol. ounces frozen blueberries, defrosted
5 1/4 sol. ounces frozen raspberries, defrosted
Vanilla ice cream, optional

Mix batter ingredients together in order just until smooth. Bake in waffle iron. For topping, combine water, sugar, corn syrup, food coloring and cornstarch in saucepan; cook over medium heat until thickened. Remove from heat; add gelatin, stirring until dissolved. Cook, add berries. Serve warm over waffles with a scoop of vanilla ice cream, if desired.

Pancakes sizzling on a hot griddle adds a yummy sight and sound to many Amish kitchens, although they might be considered a more leisurely dish would be reserved for the slower pace of Saturday and Sunday mornings.

AMISH CORNCAKE PANCAKES

1 1/3 cup buttermilk at room temperature
1 teaspoon baking powder
3/4 cup yellow cornmeal
1/4 flour
1 teaspoon salt
1 teaspoon sugar
1 egg
1 tablespoon corn oil

Whisk together buttermilk and baking soda let sit for 5 mins. Beat egg slightly, stir in oil, add dry ingredients. Add buttermilk mixture. The batter will be very thin. On hot skillet drop heaping Tablespoon of batter. 1 1/2 mins. on first side and 1 min. on bottom. Serve hot with syrup. They don't need any butter. The kids love them.

DUTCH PANCAKES

3 eggs, well beaten
1/2 cup flour
1/2 teaspoon salt
1/2 cup milk

Beat the eggs thoroughly. Add the flour and salt.   Continue to beat and add the milk while beating. Pour   into a well-buttered pan and bake at 375° for 25 minutes or until lightly browned.   Serve with powdered sugar and lemon juice.

STEAMED FRUIT MUSH

4 cups fruit*
2 cups sugar
1 teaspoon lemon juice
1 tablespoon butter
2 cups flour
4 teaspoons baking powder
1 teaspoon salt
3/4 cup milk

*For Blueberry Mush, use blueberries. For Raspberry Mush, use raspberries, etc. Blackberries, cherries or peaches also work well.
Sift flour, baking powder and salt together, and work in the butter. Add the milk and mix thoroughly. Combine sugar, fruit and lemon juice. Mix with first batter. Pour into a buttered mold, cover tightly and steam for 45 minutes. Serve with cream.

HOMEMADE GRAPE-NUTS

3 1/2 cups graham flour or whole wheat flour
1 cup brown sugar
2 cups buttermilk
1 teaspoon salt
1 teaspoon baking soda

Sift flour; measure and add soda and salt. Sift again. Add sugar and mix thoroughly. Add buttermilk to dry ingredients and beat until smooth.

Spread dough 1/4 inch thick on greased pans. Bake at 375° for 15 minutes or until crisp and golden brown. Let cool and crumble by hand or grind with a food chopper. Crisp in oven just before serving.

CRUMB CAKE

3 cups flour
1 tablespoon baking powder
1 teaspoon salt
1 1/2 cups sugar
2/3 cup butter or margarine
3 eggs
1 cup milk
2 teaspoons vanilla

Combine dry ingredients, then cut in butter until a coarse meal consistency. Reserve 1 1/2 cups of this mixture. Combine remaining ingredients and stir into dry mixture. Pour into 2 well buttered 9 inch cake pans. Sprinkle with reserved crumb mix. Bake at 350° for 25 to 30 minutes.

SWISS COFFEE CAKE

2 cup light brown sugar
2 cup flour
3/4 cup shortening
1 egg
2 teaspoon vanilla
1 cup hot coffee
1 teaspoon Soda

Mix sugar, flour and shortening until lumpy. Do not mix until creamy. Take out 1 cup for topping. Dissolve soda in hot coffee and add to the flour mixture. Also add egg and vanilla. Spread on sheet pan (9x12x2) and sprinkle on topping. This is a thin batter. Bake at 325-350° approximately 30 minutes. Sprinkle with powdered sugar after baked.

Amish

Soups

The Amish are legendary culinary creators when it comes to from-scratch soups. Gardens provide all the fresh vegetables to create a mouth-watering concert of carrots, tomatoes, beans, and broccoli. Homemade beef and chicken stock provide the most common bases for these soups. While many Amish may not be full-time farmers anymore, most families still maintain a small amount of livestock to feed their families. The smell of a homemade soup bubbling on the stove during a cold winter afternoon adds a cozy warmth to the house.

Chicken corn soup is a classic Pennsylvania Dutch dish, so much so that it has become one of their signature culinary creations. This dish pays homage to Amish self-sufficiency, because all of the ingredients are ones that would come directly from their farmstead. Variations have been added to the recipe over the years but these are three classic versions.

CHICKEN CORN SOUP

1/2 stewing hen or fowl
2 quarts chicken stock or broth
1/4 cup onion, coarsely chopped
1/2 cup carrots, coarsely chopped
1/2 cup celery, coarsely chopped
1 teaspoon saffron threads (optional)
3/4 cup fresh or frozen corn kernels
1 tablespoon fresh parsley, chopped
1 cup egg noodles, cooked

Combine stewing hen with chicken stock, coarsely chopped onions, carrots, 1/2 cup celery, and saffron threads. Bring the stock to a simmer. Simmer for about 1 hour, skimming the surface as necessary. Remove and reserve the stewing hen until cool enough to handle then pick the meat from the bones. Cut into little pieces. Strain the saffron broth through a fine sieve. (Note: The soup can be made through this step in advance. Simply refrigerate broth and diced chicken meat for 2 to 3 days, or freeze the broth and the chicken meat in separate convenient sized containers. Be sure to label and date them. To use, defrost, remove congealed fat, return the broth to a full boil, and add the diced meat. Continue with recipe.) Add the corn, parsley, and cooked noodles to the broth. Return the soup to a simmer and serve immediately

CHICKEN CORN NOODLE SOUP

6 chicken legs with thighs
¼ cup (1/2 stick) butter
1 onion, diced
1 can corn
1 bag noodles

Sugar

6 chicken legs with the thighs. (Legs have the most juice.) Boil in pot of 2 quarts of water with 1/2 stick of butter, diced onion and a dash of Poultry seasoning.
Cook until the chicken "falls" apart. Remove the chicken and debone it and put back into the pot. Now add noodles, corn and a dash of sugar.
Cook one hour.

GRANDMA'S CHICKEN NOODLE SOUP

3 pounds chicken
2 quarts water
2 teaspoons salt
1 1/2 cups chicken stock
2 cups celery, chopped
2 cups carrots, chopped
1 tart apple, chopped
1 cup onions, chopped
Dash of pepper
4 cups egg noodles

Place chicken in kettle with 2 quarts water. Cover and cook until tender (about 2 1/2 hours).

Remove chicken from kettle and strain broth. Debone chicken and return to kettle with strained broth. Add chicken stock, celery, carrots, apple, onions, and pepper and cook until vegetables are tender. Add noodles and cook for 8 to 10 minutes.

Rivels – sometimes spelled "rivvels" are a very common traditional Old Order Amish dish. It's a simple soup made up of rice-sized bits of dough called "rivels."

RIVELS

1/2 teaspoon salt
Pinch of ground white pepper
1 large egg, beaten
3/4 to 1 cup all-purpose flour, as needed

Stir salt and pepper into egg and add 2/3 cup flour and beat. Keep adding and beating in flour until mixture is crumbly but a bit sticky. Rub between hands or pinch off pea-size pieces and drop them into simmering soup. Cover loosely and let cook for about 15 minutes or until rivels solidify. To make the rivels ahead of time, cook them in lightly salted boiling water or some extra soup stock and then drain and reserve them to be reheated in the soup just before it is served.

CHICKEN CORN RIVEL SOUP

3-4 pound stewing chicken, free range if possible
2 tablespoons salt
1/4 teaspoon pepper, freshly ground
1 1/2 cups celery, chopped
1 medium onion, chopped
2 tablespoons flat leaf parsley, minced
1 quart corn kernels fresh, frozen or canned
rivels
Hard boiled eggs, diced or sieved or minced parsley for garnish.

Place chicken in a large soup pot, and add water to cover. Season with salt and pepper and simmer until meat is ready to fall from the bone.
Remove chicken from soup pot and let set until cool enough to handle. Remove skin and bones. Cut chicken into small pieces.
Reheat stock in pot to boiling and add remaining ingredients except chicken and rivels. Mix all together and serve.

CHICKEN CORN RIVEL SOUP (variation)

8 cups of chicken broth (you may use canned)
1 onion, diced
2 tablespoons dried parsley
2 cups flour
1 teaspoon salt
2 eggs, beaten
2 cans corn
2 cups cooked chicken, iced

Bring the broth to a boil.
In a bowl, mix flour, salt and eggs until you have a crummy mixture (not smooth, just until it makes "crumbs"). Rub mixture between your fingers over the broth dropping small amounts in, each about the size of a pea. These are called rivels. Add corn and cook about 10-15 minutes. Add the chicken just before you take it off the stove.

AMISH HAM BEAN SOUP

2 cups dried navy beans
1 ham bone
2 quarts water
salt and pepper
2 quarts milk
3 tablespoons butter

Soak beans overnight in enough water to cover. In the morning, drain beans. Add water and ham bone. Cook slowly until meat and beans are soft. Pick meat from bone. Chop and return to bean mixture. Add milk. Heat thoroughly. Season with salt and pepper to taste. Add butter and serve.

BEAN SOUP

1 pound navy beans
1 1/2 pounds butt end of ham
Cold water (about 8 cups)
2 cups diced celery
1 /2 cup chopped onion
1 /2 cup diced potatoes
1 1-lb can tomatoes, drained
2 teaspoons minced parsley
Salt
Pepper

Soak beans overnight in water to cover. Drain off the water. Add fresh water and cook until beans are soft. Cover ham with cold water, cover and cook until tender, about 2 1/ 2 hours. Skim fat from the broth and add the drained beans, celery, onion, and potatoes. Cook until vegetables are tender, about 15 minutes. Add the tomatoes and parsley and heat through. Season with salt and pepper to taste.

MICHIGAN CHILI SOUP

1/2 pound sausage
2 medium onions
3 cups tomato juice
1 can kidney beans
1 tablespoon chili powder
1/2 cup brown sugar
Salt to taste
3 tablespoons flour
4 ounces uncooked spaghetti

Brown sausage and onion in a deep pot. Add remaining ingredients, except spaghetti, stir well, and bring to a boil. Add some spaghetti to the soup. Cook until spaghetti is done, about 5 to 10 minutes over medium-high heat.

KENTUCKY AMISH SOUP

2 tablespoons butter or margarine
1 pound hamburger meat
1 onion, chopped
1 1/2 teaspoons salt
1 cup diced carrots
1/2 cup chopped celery
1 cup diced potatoes
2 cups tomato juice
2 cups milk
1/4 cup flour

Brown meat and onion in butter. Add remaining ingredients except milk and flour and cook until vegetables are tender. Combine milk and flour and stir until smooth, add to soup and stir until thickened.

ZUCCHINI & PORK SOUP

4 pork chops
1/2 cup all-purpose flour
2 teaspoons vegetable oil

1 onion, chopped
2 teaspoons chopped garlic
1 cup chopped red bell pepper
2 zucchinis, quartered and sliced
1/8 cup chopped sun-dried tomatoes
8 ounces fresh mushrooms, sliced
1-14.5 oz can diced tomatoes
2-14.5 ounces cans chicken broth
2 tablespoons oyster sauce
2 teaspoons dried basil
1 teaspoon dried oregano
Salt and pepper to taste
4 tablespoons grated Parmesan cheese

Place flour in a re-sealable plastic bag. Add pork cubes. Seal bag and shake to coat. In a large skillet, heat 1 teaspoon oil over medium high heat. Add pork and brown in oil for about 8 to 10 minutes. Set aside.
In a 5 to 8 quart soup pot, heat remaining teaspoon of oil over medium heat. Add the onions, garlic and bell pepper. Sauté for just a few minutes until tender, but not browned. Add reserved pork, zucchini, sun-dried tomato bits, mushrooms, diced tomatoes, chicken broth, oyster sauce, basil, oregano, salt and pepper. Heat until almost boiling, then reduce heat and simmer for 10 to 15 minutes. Sprinkle with cheese and serve.

GERMAN POTATO SOUP

2 1/2 quarts water & 1/3 cup of chicken base
OR
2 1/2 quarts of chicken stock   6-8 Diced potatoes
1 small diced onion   1 pound Amish rivels
(see recipe following)   1 cup cream or half &
half

Bring all ingredients except Amish rivels to a boil. Once boiling, add Amish Rivels. Cook 25 minutes or until tender. Add 1 cup of cream or half & half and salt pepper to taste.

Amish

Breads/Muffins

Homemade bread is considered a staple in an Amish kitchen. Each family and each Amish region have their own favorite special bread recipes, but a basic white loaf is most common. A golden brown loaf of soft-as-cotton billowing bread is stored in a clear plastic bag and served by the slice. It's not uncommon for an Amish Mom to bake 3 – 5 loaves for bread for the week. Most Amish Cooks still use lard in their recipes, although there has been a gradual shift towards other types of shortenings through the years. The first recipe featured in this chapter is that basic white bread recipe that is the cornerstone of an Amish kitchen. Once the basic is baked, Amish cooks turn their legendary culinary creativity to more "fun" types of breads, ranging from "pumpkin pie bread", to dilly bread, apple bread, broccoli bread, and many others. Baking bread is almost an art among the Amish. Hard-working hands punch down dough and patient persons watch it rise. Grandmothers bake bread in the kitchen beside their daughters, and grandchildren. Girls learn at a very early age how to gently work the dough into a tasty creation. It's not uncommon to see a 5-year-old girl, locks of hair pulled under a tiny bonnet, creating her own little loaf of bread under the watchful eye of her mother. In this way the recipes and methods pass safely to a new generation.

WHITE BREAD

6 cups all-purpose flour
2 1/4 cups milk
1 tablespoon shortening, margarine or butter

1 package active dry yeast
2 tablespoons sugar
1 teaspoon salt

In large mixing bowl combine 2 1/2 sups of flour, and the yeast. In a saucepan warm milk, sugar, margarine or butter and salt. Add to flour mixture. Beat with mixer on slow for 30 seconds, then on high for 3 minutes. Using a spoon, stir in as much remaining flour as you can. Turn out onto a lightly floured surface. Knead until moderately stiff. About 3 minutes, though in my experience, you can't knead it too long. Place in lightly greased bowl, turn over to grease both sides. Cover and let rest in a warm place till double (about 45 minutes). When double, punch dough down. Turn out onto lightly floured surface and divide dough in half. Cover and let rest for 10 minutes. Lightly grease 2 loaf pans. Shape each half into a loaf by patting or rolling. Place shaped dough into loaf pans. Cover and let rise in a warm place till nearly double (about 30-40 minutes). Bake at 375° for about 40 minutes or till top is brown. Remove from pans immediately, and cool on wire racks. Makes 2 loaves.

HONEY NUT BREAD

1/2 cup coarsely chopped nuts
1 egg, beaten
2 cups flour
1/2 cup honey

3 teaspoon baking powder
2 tablespoon melted butter
1/2 teaspoon salt
1/4 cup milk

Add nuts to the sifted dry ingredients. Combine the beaten egg, honey, milk, and melted butter. Add to the first mixture; stir until ingredients are just moistened. Bake in a greased bread pan in a 350° oven for 45-50 minutes.

POPPYSEED BREAD

3 cups flour
1 1/2 teaspoons salt
1 1/2 teaspoons baking powder
3 eggs, beaten
1 1/2 cups oil
2 1/2 cups sugar
1 1/2 cups milk
1 1/2 teaspoons vanilla
1 1/2 teaspoons almond flavor
1 1/2 teaspoons butter flavor
1 1/2 tablespoons poppy seeds
Glaze:
1/4 cup orange juice
3/4 cup sugar
1/2 teaspoon almond flavor
1/2 teaspoon butter flavor
1/2 teaspoon vanilla

Sift together first 3 ingredients. Add remaining ingredients. Mix and pour into 2 greased and floured bread pans. Bake 1 hour at 350° or until toothpick test comes out clean. Stir all glaze ingredients together. Prick bread loaves with a fork after baking and pour glaze over bread while it is still hot.

SUNFLOWER BREAD

1 3/4 cup unsifted all-purpose flour
2/3 cup sugar
1 t cup cinnamon
1/8 teaspoon nutmeg
3/4 teaspoon baking powder
1/2 teaspoon baking soda
1/2 teaspoon salt
3/4 cup unsalted sunflower kernels
1/2 cup golden raisins
3 teaspoons lemon zest
1/2 cup sunflower oil
2 eggs
1 1/2 teaspoon vanilla
2 cups shredded zucchini

Preheat oven to 350°. Grease & flour loaf pan. Mix flour, sugar, spices, baking soda, and salt. Mix in sunflower seeds, raisins & lemon zest. Blend oil, eggs & vanilla in small bowl; add zucchini. Stir into dry ingredients. Spread into pan. Garnish with sunflower seeds. Bake 55-60 minutes. Cool in pan 10 minutes. Cool on rack. Wrap & store overnight before slicing.

AMISH PUMPKIN PIE BREAD

3 cups sugar
1 cup oil
4 eggs, beaten
2/3 cup water
1 can pumpkin
3 1/2 teaspoons flour
2 teaspoons baking soda
2 teaspoons salt
1/2 teaspoon cinnamon
1/2 teaspoon allspice
1/2 teaspoon nutmeg
1 teaspoon pumpkin pie spice (optional)

Mix sugar, oil, eggs and add pumpkin and dry ingredients, then water stirring until just mixed. Pour into 2 well-greased and floured loaf pans. Bake at 350° approximately 1 hour. May be frozen.

BROCCOLI BREAD

2-1 pound loaves frozen bread dough
1 head fresh broccoli
1 small can pitted black olives, chopped
2 teaspoons garlic powder
4 tablespoons olive oil

1 1/2 teaspoons red pepper flakes
1 cup grated cheese
Salt and pepper to taste

Cook broccoli florets (no stems) until tender; drain and chop fine.

Thaw and flatten out bread loaves into individual long rectangles on floured surface.

Sprinkle cooled, chopped broccoli (squeeze out any excess water) on top of the flattened bread. Next, sprinkle chopped black olives and red pepper flakes over broccoli. Salt and pepper to taste. Drizzle olive oil sparingly over vegetables, then sprinkle with grated cheese to cover thinly. Sprinkle garlic powder on top of cheese.

Roll up lengthwise; tuck ends under and pinch short ends closed. Bake on parchment paper covered cookie sheet at 350° until golden brown. Remove and place on wire racks to cool. Cut into slices and serve with a salad or as an appetizer to your meal. Once baked, this bread can be frozen in freezer paper – simply reheat from frozen state until thawed.

AMISH DILLY BREAD

1 package dry yeast (dissolved in 1/2 cup warm water)
1 cup warmed cottage cheese
2 tablespoons granulated sugar
1 tablespoon dry onion flakes
1 tablespoon melted butter
2 teaspoons dill seed
1 teaspoon salt
1/4 teaspoon baking soda
1 unbeaten egg
2 1/4 to 2 1/2 cups all-purpose flour

Mix all ingredients, except flour, in a large bowl. Gradually stir in flour to make a soft dough. Put in a warm draft free place and cover with a damp cloth. Let rise until double in bulk (about 1 hour). Knead on a well floured board. Divide in half and place in 2 or 3 greased loaf pans. Cover with damp cloth and let rise until double in bulk. Bake 40 to 50 minutes at 350°. Brush with butter and sprinkle with salt while still hot.

COUNTRY CREAM CHEESE BRAIDS

Dough:
1 cup sour cream
1/2 cup granulated sugar
1 teaspoon salt
1/2 cup melted butter
2 packages dry yeast

1/2 cup warm water
2 beaten eggs
4 cups flour (may need more)

Filling:
16 ounces cream cheese
3/4 cup granulated sugar
1 beaten egg
1/8 teaspoon salt
2 teaspoons vanilla extract

Glaze:
2 cups confectioners' sugar
4 tablespoons milk
2 teaspoons vanilla extract

Heat sour cream. Stir in sugar, salt and butter. Cool.

Sprinkle yeast over warm water, stir until dissolved. Add sour cream mixture, egg and flour. Mix. Cover tightly and place in refrigerator overnight.

The next morning divide dough into 4 parts. Roll each on well floured board into 12 x 8-inch rectangle. Spread each with 1/4 of cream cheese filling.

Filling: Soften cream cheese and sugar. Mix well. Add egg, salt and vanilla extract and mix well. Roll up jellyroll style. Place seam side down on 2 greased baking sheets. Slit about 2/3 of the way through dough to resemble braids. Cover and rise until doubled (one hour).

Bake at 375°.

Yields 4 braids.

Glaze: Mix confectioners' sugar, milk and vanilla extract. Spread on braids while still warm.

These freeze well. Wrap in foil and place in oven straight from freezer. Heat at 325° for about 15 minutes.

OLD GERMAN SOUR CREAM PEAR BREAD

1/2 cup butter (1 stick)
1 cup sugar
2 eggs
3 tablespoons pear juice
4 to 5 canned pear halves, chopped

2 cups flour
1 teaspoon baking soda
1 teaspoon baking powder
1/4 teaspoon salt
1 cup sour cream

Filling:
1/2 cup walnuts, chopped
1/2 cup brown sugar
2 teaspoons cinnamon

Cream butter, sugar and eggs. Add pear juice and chopped pear halves. Mix dry ingredients and add alternately with sour cream. Pour not quite half the batter into a greased loaf pan. Mix together filling ingredients and generously heap the mixture into the center of the batter. Pour remaining batter on top. Bake at 350° for 50 to 60 minutes. Cool in pan on wire rack.

GINGER CAKES

1 cup shortening
1 cup brown sugar
2 eggs, well beaten
1 cup molasses
1 teaspoon baking soda dissolved in 1 tablespoon boiling water
4 cups flour
1 teaspoon ginger
Pinch salt

Cream the shortening and brown sugar together. Add eggs and beat thoroughly. Add the molasses and baking soda which has been dissolved in the boiling water. Sift the flour and ginger together and combine with other ingredients. Mix well. Pour into well-greased muffin pans and bake at 350° about 20 minutes.

ZUCCHINI AND CARROT MUFFINS

1 3/4 cups unsifted self-rising all-purpose flour*
3 tablespoons granulated sugar
1/2 teaspoon ground cinnamon
1 cup coarsely grated carrots
1 cup coarsely grated zucchini

2 teaspoons finely grated orange rind
2 eggs
1/4 cup vegetable oil
1/4 cup orange juice
Preheat oven to 400°. Place cast iron muffin pans or molds in oven to heat.
In a large bowl, combine self-rising flour, sugar, cinnamon, carrots, zucchini and orange rind. Add eggs, oil and orange juice to flour mixture; stir just until dry ingredients are moistened.
Carefully remove hot muffin pans or molds from oven. Grease with vegetable cooking spray or brush with vegetable shortening. Divide batter among muffin cups. Bake for 15 to 18 minutes or until tops spring back when gently touched. Serve warm.

* To use all-purpose flour which is not self-rising, add 2 1/2 teaspoons baking powder and 3/4 teaspoon salt to flour mixture.

SOUR CREAM MUFFINS

1 1/4 cups unbleached flour
3/4 cup cornmeal
3 teaspoons baking powder
1/2 teaspoon salt
1/4 cup granulated sugar
1/2 teaspoon baking soda
1 cup sour cream (at room temperature)
1 egg, beaten (at room temperature)

Preheat oven to 350°. Grease 12 muffin cups or line with paper liners.

Sift together flour, cornmeal, baking powder, salt and sugar.

In a separate bowl, add baking soda to sour cream. Stir until frothy. Add egg to sour cream mixture and mix well. Add sour cream mixture slowly to dry ingredients and stir just until mixed.

Fill muffin tins about 2/3 full. Bake 25 minutes or until browned

AMISH MUFFINS

5 cups flour
3 cups sugar
5 teaspoons soda
2 teaspoons salt
2 teaspoons allspice

15 ounces raisin bran
1 cup oil
1 quart buttermilk
2 teaspoons vanilla

Mix first six ingredients together. Add oil, buttermilk, and vanilla; mix well. Butter muffin tins well and fill 3/4 full. Bake at 375° for 20 minutes.

AMISH FRIENDSHIP MUFFINS

1 cup Amish Friendship Bread Starter
2 cups all-purpose flour
3/4 cup oil
1 teaspoon baking soda
2 teaspoons baking powder
1 cup granulated sugar
3 eggs
1 1/2 teaspoons cinnamon
2 teaspoons vanilla extract
1 cup nuts, chopped
1 cup apples, chopped (optional)
1 cup raisins or 1 cup blueberries (optional)

Topping:
1 cup brown sugar
1/2 cup all-purpose flour
1/4 cup margarine

Preheat oven to 350°. Liberally grease muffin tins if not using liners.

Combine starter, flour, oil, baking soda, baking powder, sugar, eggs, cinnamon and vanilla extract; stir well. Add nuts. Stir in optional ingredients, if desired. Mix well. Put into muffin tins. Sprinkle each muffin with Topping and bake for 15 to 20 minutes or until done.

TENNESSEE CHERRY MUFFINS

2 cups all-purpose flour
1/2 cup granulated sugar
1/2 cup firmly packed brown sugar
1 teaspoon baking soda
1 teaspoon baking powder

1/4 teaspoon salt
1 teaspoon cinnamon
1 cup crushed spoon-size Shredded Wheat-'n-Bran® cereal
1/2 cup finely chopped pecans or walnuts
2 large eggs
2/3 cup vegetable oil
1 teaspoon vanilla extract
1-21 ounces can cherry pie filling

In a large bowl, combine flour, sugars, baking soda, baking powder, salt and cinnamon; mix well. Stir in crushed cereal and nuts. In another bowl, lightly beat eggs. Add oil and vanilla extract. Add egg mixture and cherry pie filling to dry ingredients. Stir gently until mixed. Fill paper-lined or greased muffin cups about 2/3 full. Bake at 375° for 25 to 30 minutes or until wooden pick inserted in center comes out clean.

PEACH MUFFINS

2 cups all-purpose flour
2 teaspoons ground cinnamon
1 teaspoon powdered ginger
3/4 teaspoon salt
3/4 teaspoon baking soda
1/4 teaspoon baking powder
2 eggs
3/4 cup dark brown sugar
3/4 cup oil, preferably half walnut oil and half vegetable
1 teaspoon vanilla extract
1 3/4 cups (about 2 large) chopped peaches
2/3 cup chopped pecans, toasted
1/4 cup poppy seeds

Preheat oven to 400°. Grease muffin tins.
Sift together flour, cinnamon, ginger, salt, baking soda and baking powder into a bowl; set bowl aside.
In another larger bowl, beat together eggs with brown sugar, oil and vanilla extract. Fold in the peaches, pecans and poppy seeds. Add the flour mixture, and lightly stir to combine. Spoon the batter into the prepared muffin tins. Bake the muffins 22 to 25 minutes, or until they are browned and a wooden pick inserted in the center of a muffin comes out clean.

PUMPKIN CINNAMON ROLLS

1/3 cup milk
2 tablespoons butter
1/2 cup canned pumpkin or mashed cooked pumpkin
2 tablespoons sugar

1/2 teaspoons salt
1 egg, beaten
1 package dry yeast
1 cup unbleached all-purpose flour
1 cup bread flour
1/2 cup brown sugar, packed
1 teaspoon ground cinnamon
2 tablespoons melted butter

In small saucepan, heat milk and 2 tablespoons of butter just until warm (120° to 130°) and butter is almost melted, stirring constantly.

In large mixer bowl, combine pumpkin, sugar and salt. Add milk mixture and beat with electric mixer until well mixed. Beat in egg and yeast.

In separate mixing bowl, combine flours. Add half of flour mixture to pumpkin mixture. Beat mixture on low speed for 5 minutes, scraping sides of bowl frequently. Add remaining flour and mix thoroughly (dough will be very soft). Turn into lightly greased bowl then grease surface of dough lightly. Cover and let rise in warm place until doubled, about 1 hour.

Punch dough down. Turn onto floured surface. Knead a few turns to form a smooth dough, sprinkling with enough additional flour to make dough easy to handle. On lightly floured surface, roll dough into 12- by 10-inch rectangle.

In small bowl, combine brown sugar and cinnamon. Brush surface of dough with melted butter. Sprinkle with brown sugar mixture. Beginning with long side of dough, roll up jelly-roll style. Pinch seam to seal.

With sharp knife, cut roll into twelve 1-inch slices. Place rolls, cut side up, in greased 9-inch square baking pan. Cover and let rise until nearly doubled (30 to 45 minutes).

Bake rolls at 350° for about 20 minutes or until golden. Remove from pan to waxed paper-lined wire rack. Cool for 10 to 15 minutes. Makes 12 rolls.

BUTTERMILK BISCUITS

2 cups sifted all-purpose flour
1/2 teaspoon salt
3 level teaspoons baking powder
1/2 teaspoon baking soda
3 tablespoons butter
3/4 cup buttermilk

Sift flour, salt, baking powder and baking soda together. Cut in butter with 2 knives. Add buttermilk and knead lightly. Roll to 3/4 inch thickness and cut with cutter. Dab top with melted butter and bake 12 minutes at 450°.

Makes 12 to 16 biscuits, depending on size of cutter.

PARENTING TIPS & FAMILY LIFE

By Kevin Williams
The Amish Cook's Editor (kwilliams@oasisnewsfeatures.com)

Some of the most serene scenes I've witnessed have been of the Amish partaking in family suppers. A silent prayer is observed, adding a solemn start to a daily ritual that is vanishing in most of America: the family meal. Amish families still regard supper together as the cornerstone of their day. Stories are shared, laughs abound, and plenty of hearty comfort food soothes the soul. It's not uncommon to see several kinds of meats, cheeses, and fresh garden vegetables on the table, along with some mouth-melting desserts. It's a time when little else matters but the company of loved ones and the sampling of comforting food. Children of all ages help with the task of washing the dinner dishes when the meal is over. You'll rarely hear grumbling or complaining – such chores work are simply expected. The drudgery of doing dishes is erased by the joy of singing or yodeling as they work.

Amish children have a reputation has being well-behaved and well-mannered and this is probably deserved, although this writer has seen his fair share of unruly Amish youngsters. Kids are kids regardless of their culture. Amish childhood can be basically divided into three segments:

EARLY CHILDHOOD (newborn to age 3)

Amish babies are born into an environment of unlimited love. Babies are regarded as a blessing and excitement ripples through an Amish church district when news of a newborn circulates. Amish babies are almost always born at home under the watchful eye of an Amish midwife or country doctor. From birth the community embraces the newborn. One new Amish mom laughingly told me "The first time I took her to church I hardly saw her because all the ladies wanted to take turns holding her." Large families, in the eyes of the Amish, are a sign of closeness to God and of ample bounty. The largest family this writer has seen consisted of 14 children belonging to some Amish parents in a community near Reed City, Michigan, although I have heard of families as large as 18 children.

CHILDHOOD (age 3 to age 13)

The family is the most important social unit in the Amish culture. Large families with seven to ten children are the norm. Chores are clearly divided by gender role in the Amish home - the man usually works on the farm, while the wife does the washing, cleaning, cooking, and other household chores. Capitalism will sometimes trump these gender roles, though. There are cases when a mother will own a quilt business or bakery and find that her income outstrips her husband's. The matronly roles, however, of cooking, cleaning, and gardening would remain unchanged in these cases. The father would still be considered the head of the Amish household. A dialect of Swiss or German is spoken in the home, with English taught as a secondary necessity. The Amish school-children will begin school around age 5 and continue until the eighth grade. A common misconception is that most Amish kids attend "Little House on the Prairie"-style "one room schools." While Amish parochial schools are becoming more and more common, they are a relatively recently development in Amish society as they seek to shield their children from the increasingly intrusive outside word. A great many Amish kids still attend public schools, although the trend is towards parochial. Quality of instruction varies greatly in these private schools. The basics of reading, writing, and `rithmatic are emphasized and some Amish end the eighth grade so well grounded in these basic skills that they could pass a GED.

Amish kids attend school just until the eighth grade. The Amish culture emphasizes vocational skills that most parents want their children learning as soon as they are ready. It is during this childhood time when Amish youth enjoy a life that is full of hard work (many are already helping on the farm), but tempered by fun lazy summer days.

ADOLESENCE

Much has been written and broadcast about the period that many Amish youth experience as they become teenagers. The period is sometimes referred to by it's Pennsylvania German word "rumspringa." Rumspringa – roughly translated as "running around period" – is a time not too different from what non-Amish teens experience.. This is a time when Amish youth begin to turn their sights to their future. Some Amish teens glide through these years with little rebellion or wild ways, other teens – just like non-Amish teens – can seek to push the envelope and act out. Some Amish teens will drink, smoke, and wear non-Amish clothing, some will even move away. Amish parents, for the most part, will turn a blind eye to the antics of their adolescents. Amish parents want their children to see with their own eyes whether the world has anything to offer them, they want their kids to feel like the Amish lifestyle is one that they themselves have chosen and not been imposed upon them. Most Amish children eventually choose to stay in the faith. By around age 21, they'll have to decide whether they want to be baptized into the church and become full members.

An Amish woman in Ohio shared these parenting tips that are rather revealing:

- Answer all misbehavior with love. Children understand the language of love, and don't understand hate.

- Keep God at the center of family life, the old saying " a family that prays together stays together" is correct.

- Sometimes you have to let babies cry, they have to learn "no" at an early age.

- Let your children work out differences among themselves, this can teach them valuable lessons early on about working out conflicts without running to their parents.

- Always prepare hearty meals for your youngsters as they need plenty of nutrition to keep them growing!

Amish

Main Dishes

Large gatherings require easy-to-assemble and filling food. Amish cooking is perfect for such occasions. Casseroles, stews, meat loaves (using chicken, pork, or beef) are common centerpieces on the menu. Whether the occasion is the meal after a barn-raising, an after-church gathering, or simply a large Sunday supper for the family, Amish menus provide a little something for everyone.

POT PIE

Filling:
4 pounds of meat of your choice (beef roast, ham, turkey legs or whole chicken)
2 quarts water
1 teaspoon salt
5 potatoes, diced
2 cups corn
1 medium onion, diced
Pot Pie Dough:
2 cups flour
1/2 teaspoon salt
2 eggs
4 tablespoons water

Cook the meat of your choice in 2 quarts salted water until tender. Cool and debone, set aside. To the boiling broth add: potatoes, corn, and onion. Cook for 10 minutes.
To prepare pot pie dough: combine flour and salt. Beat eggs with the water. With a fork work into flour and salt until you have formed at stiff dough. Add more water if mix is too dry. Roll dough out onto a floured board until thin. Cut into 2 inch squares. Drop Pot Pie dough into boiling broth a few at a time and cook about 20 minutes. Stir meat pieces in and heat through.

WHOLE WHEAT CHICKEN POT PIE

1 1/2 cups whole wheat flour
1/2 cup white flour
1 teaspoon salt
5 tablespoons cold water
1 egg white
3/4 cup shortening
16 ounces frozen mixed vegetables
2 cups chicken and 1 cup chicken broth (boil a whole chicken and pick choice parts and save broth)

Preheat oven to 425°.
Combine flours in large mixing bowl. Mix in salt. Cut in shortening with pastry blender. Sprinkle with water, one tablespoon at a time, until dough will form a ball. Roll out half of the dough on floured surface. Roll dough for bottom crust 1 inch larger than pie dish.

Fold dough and unfold into pie dish. Bake bottom crust at 425° for 10 minutes so that the crust will not be soggy. Roll out other half of dough and cut with cookie cutters to form top crust. Thicken chicken broth to form a gravy. In large mixing bowl, combine gravy, chicken and vegetables. Stir to coat, salt and pepper to taste, pour into pie shell and top with cookie cutter shaped dough. Brush top crust with egg white. Bake at 425° for 30-35 minutes.

PENNSYLVANIA CHICKEN CASSEROLE

4 whole boneless skinless chicken
breast, halved
1/4 teaspoon salt
1/8 teaspoon pepper
2 cups sauerkraut, well drained
4 slices fat free Swiss cheese
1 1/4 cups Russian dressing
1 tablespoon chopped parsley

Spray pan with Pam. Place chicken in pan, salt and pepper. Cover chicken with sauerkraut. Pour dressing evenly over all and top with cheese. Cover with foil and bake 1 hour at 325° or until fork tender. Sprinkle with chopped chives to serve.

CHICKEN DRESSING CASSEROLE
1-14.5 ounces can chicken broth
1/2 cup (1 stick) unsalted butter
1 cup water
2 tablespoons chicken base or bouillon granules
1 cup milk
3 eggs
1-1 pound loaf white bread, cubed and toasted
2 cups cooked chicken, cubed
3 ribs celery, diced
3 carrots, shredded
1 onion, diced
1/4 cup chopped parsley
1 tablespoon seasoning salt
1/4 teaspoon freshly-ground pepper
1/4 teaspoon salt
1/4 teaspoon celery seed

Preheat oven to 350°.
Combine broth, butter, water and chicken base in small saucepan over medium heat until butter is melted and base is dissolved; cool completely. Whisk in milk and eggs.

Mix together bread, chicken, celery, carrots, onion, parsley, seasoning salt, pepper, salt and celery seed in large bowl. Pour broth mixture over; toss until liquid is absorbed. Place dressing into 13 x 9-inch baking pan. Bake until golden brown, about 1 hour and 20 minutes.

EMMA'S CHICKEN BROCCOLI CASSEROLE

3 chicken breasts, cooked and diced (or use 3 cans cooked, flaked chicken)
2 pounds fresh broccoli, steamed (or 2 packages frozen broccoli)
1 can cream of mushroom soup (can use reduced fat, reduced sodium, etc.)
1 can cream of celery soup
1 can cream of chicken soup
1 cup mayonnaise
1 teaspoon Lemon juice
1 1/2 cups grated cheese
2 cups dry bread cubes
2 tablespoons melted butter

Drain broccoli and place in 9x13 baking dish. Place chicken over top. In a bowl, combine all the soups, mayonnaise and lemon juice. Spread on top of chicken. Sprinkle with grated cheese. Mix dry bread cubes with melted butter and sprinkle on top of cheese. Bake at 350° for 35 minutes.

WORKING HUSBAND'S CASSEROLE

2 cup cooked diced chicken
2 cup noodles, cooked
2 cup milk
1 can cream of mushroom soup
1 can cream of celery soup
1/4 cup chopped onion
2 cup grated mild cheddar cheese
1 teaspoon Salt
1 can mushrooms

Preheat oven to 350°. Mix together all ingredients. Spoon into a 9 x 13 inch baking dish and bake for 45 minutes, until heated through.

OVEN FRIED CHICKEN

1/3 cup vegetable oil
1/3 cup butter
1 cup all-purpose flour
1 teaspoon salt
2 teaspoons black pepper
2 teaspoons paprika
1 teaspoon garlic salt
1 teaspoon dried marjoram
8 or 9 pieces chicken

Place oil and butter in a shallow cooking pan and place in 375° oven to melt butter, set aside. In a large paper sack combine dry ingredients. Roll the chicken pieces 3 at a time in butter and oil then drop into a sack and shake to cover. Place on a plate until all pieces are coated. Leave any excess butter and oil in pan. Place chicken in the pan skin side down or its just as good if you remove all the skin first. Bake at 375° for 45 minutes. With spatula or tongs, turn chicken pieces over and bake 5 to 10 minutes longer or until crust begins to bubble.

AMISH BAKED CHICKEN

1/2 cup all-purpose flour
2 teaspoons paprika
1 teaspoon pepper
1/4 teaspoon dry mustard
3 teaspoons salt
1 cut-up broiler or young chicken
1/4 pound butter

Preheat oven to 350°.
Mix the dry ingredients well in a plastic bag, then coat the cut up chicken parts with the mixture.
Melt the butter in a 13 x 9-inch baking pan. Place the chicken parts in the pan, but do not crowd them. Bake the chicken for 1 1/2 to 2 hours or until done.

ROAST CHICKEN

3 whole chickens
6 1/2 quarts bread crumbs
1 pound margarine, divided in half
3 cups celery
1 cup onion, chopped
6 eggs, beaten
2 teaspoons parsley flakes
6 teaspoons salt
1 teaspoon celery salt
1 teaspoon garlic or onion salt
1 teaspoon paprika
1 teaspoon pepper
3 teaspoons poultry seasoning

Cook chickens till tender. De-bone chicken and reserve broth. Brown bread crumbs in 1/2 lb margarine, stirring constantly. Combine bread crumbs and all other ingredients except chicken. Mix well and add chicken broth till mixture is very moist. Put chicken meat in bottom of a large roast pan. Cover with bread mixture. Roast in a slow oven till golden brown ($250°$ for 1 1/2 to 2 hours).

CHICKEN LOAF

1-5 pound chicken, cooked and cubed
2 cups chicken broth
2 cups uncooked rice
2 cups milk
2 cups bread cubes
4 eggs
Salt and pepper to taste
2 cups diced celery

Stir and mix all ingredients. Spoon into a greased baking dish. Bake at $350°$ for 1 hour or until a knife inserted in the center of the loaf comes out clean. Slice and serve.

DUTCH MEAT LOAF

2 pounds hamburger
1 medium onion
1 egg
1 cup bread crumbs
1 can tomato sauce
1/2 teaspoon salt
1/4 teaspoon pepper

Sauce:
1/2 cup tomato sauce
1 cup water
2 teaspoons molasses
2 tablespoons white vinegar
2 teaspoons yellow mustard

Mix meatloaf ingredients well and form into loaf.

Mix sauce ingredients and pour entire mixture over the meat loaf. Bake covered at 350°
for about 1 1/2 hours. Baste the meatloaf periodically with pan juices.

SWISS AMISH MEAT LOAF

1 egg
1/2 cup evaporated milk
1/2 teaspoon rubbed sage
1 teaspoon salt
1/2 teaspoon black pepper
1 1/2 pounds lean ground beef
1 cup Ritz cracker crumbs
3/4 cup grated Swiss cheese
1/4 cup finely chopped onion
2-3 strips bacon, cut into 1-inch pieces

Preheat oven to 350°.
Beat the egg in a large bowl. Add evaporated milk, sage, salt and pepper, and mix. Add
beef, crumbs, 1/2 cup of the cheese and the onion; blend. Form into a loaf and place in a
2-quart rectangular baking dish. Arrange bacon pieces on top of loaf. Bake 40 minutes.
Sprinkle remaining cheese on top and bake 10 minutes longer.

AMISH HAM LOAF

Loaf:
1 pound cured ground ham
1 pound pork or hamburger
2 eggs
2/3 cup cracker crumbs or oatmeal
1/3 cup minute tapioca
1/4 cup milk

Dressing:
1/4 cup vinegar
1/2 cup water
1/2 cup brown sugar
1 tablespoon mustard

Mix loaf ingredients and form into a loaf. Boil dressing a few minutes. Pour over loaf and bake in moderate oven (325-350°) about 2 hours basting occasionally. Dressing should become thick and syrupy.

HAM SALAD

1 pound ham, sliced
1 pound can of saurkraut, drained
1 pound coarsely grated Swiss cheese
1 tablespoon caraway seed
1 tablepoon finely chopped dill weed
Mayonnaise as a binder

Mix all ingredients together with a kitchen fork and add mayonnaise until it reaches consistency that you like. Serve on favorite rye bread. Can also be used on crackers as an appetizer.

HAM & POTATO DINNER

4 potatoes, boiled
1/4 cup oil
1/2 cup diced cooked ham
1/4 cup sliced onion
1/4 cup thinly sliced green pepper
2 tablespoons diced pimento
6 eggs, beaten
1/4 teaspoon salt
Dash of black pepper
1/2 cup finely diced Swiss or Jack cheese
Chopped parsley

Peel potatoes and cut into 1/2-inch cubes. In skillet,
heat oil over medium heat. Add potatoes and fry until evenly
browned. Add ham, onion, green pepper and pimento. Cook,
stirring occasionally, until vegetables are crisp-tender.
In bowl, combine eggs, salt and pepper and add to potato
mixture. Stir in diced cheese. Cook, stirring occasionally,
until eggs are cooked and cheese begins to melt. Sprinkle with chopped parsley.

HAM CASSEROLE

1/8 cup butter or margarine
3/8 cup flour
3 cups milk
1 1/2 cup cheese, grated
1 quartt green beans, drained
3 cups ham, diced and cooked
5 medium potatoes, cooked and mashed

Melt butter and stir in flour. Add milk, stirring constantly, until thickened. Add cheese and stir
until melted.
Arrange green beans in greased casserole dish and cover with 1/2 of the cheese sauce.
Add ham; pour remaining cheese sauce on top.

Spread mashed potatoes on top and sprinkle with cheese. Bake at 350° for 30 minutes or until slightly browned. Serve with ham gravy.

HAM & KRAUT BALLS

1 c cup diced ham
1 cupdiced swiss cheese
1/4 cup chopped onions
1/4 cup chopped green pepper
2 cups drained sauerkraut
1 egg
2 eateaspoonoon prepared mustard
1 teaspoon celery seed
1/4c grated parmasean cheese

Mix all ingredients
Shape into balls
Dip into beaten egg. Roll in bread crumbs
Fry in oil 3-5 minutes

HAM & STRING BEANS

5 potatoes cubed
Ham (cubed)
2 pounds string beans

Take some ham (however much you have or want to make) and potatoes. Boil in 2 quarts of water until done. Add 2 pounds of string beans (canned, frozen or fresh) cook an additional 10 minutes or until beans are tender. Serve.

SKILLETBURGERS

1 1/2 pounds ground beef
1 large. onion and 1 green pepper, chopped
3/4 cup ketchup
2 tablespoons sugar
1 teaspoon salt
2 tablespoons yellow mustard
1 tablespoon vinegar

Brown meat and add other ingredients. Simmer 30 minutes.

HAYSTACKS

When I think of dinner at The Amish Cook's house, my memories often take me to "haystack suppers." This is a favorite fun meal that is really easy to assemble and enjoyable to eat. Simply layer your favorite stuff and eat. It's also a favorite dish for gatherings of Amish teenagers. – Kevin Williams, Editor

1 pkg. saltine cracker crumbs
16 ounces white rice (instant or regular),
cooked
1 1/2 pound ground beef, browned and drained
1 jar spaghetti sauce
1 head lettuce, chopped
3 green peppers, chopped
4 tomatoes, chopped
1 can black olives, chopped
1 bag tortilla chips
1 can cheese sauce

Assemble "haystacks" by layering all ingredients in order of recipe, beginning with cracker crumbs and ending with cheese sauce. Serves 10

COUNTRY PEPPERS

6 medium green peppers
1 pound sausage or hamburger
1/4 cup chopped onion
1 1/2 cups cooked rice
1-8ouncesjar processed cheese spread (Cheez Whiz)
1/2 cup chopped tomatoes

Remove tops and seeds from peppers; parboil 5 minutes and drain. Brown meat; drain. Add onion and cook till tender. Stir in rice, 3/4 cup process cheese spread and tomatoes. Fill each pepper with mixture. Place in baking dish; cover and bake at 350° for 40 minutes. Top with remaining cheese spread.

SAUSAGE SKILLET AROMA

2 medium zucchini cut in halves lengthwise
1 pound bulk sausage
1 can cream of mushroom soup
½ cup thinly sliced onion
1 medium clove garlic, minced
1 medium tomato, cut in wedges

Cut zucchini into 1-inch pieces. Shape sausage into 16 meat balls. Brown in skillet; drain off grease. Add remaining ingredients, except tomato. Cover and simmer 15 minutes, stirring often. Add tomato; heat 2-3 minutes longer. Serve.

MUSHROOM CASSEROLE

32 ounces canned mushrooms, drained or use fresh mushrooms
2 tablespoon finely chopped onion
2 tablespoon butter
1/4 cup bread crumbs
2 teaspoon lemon juice
1/2 teaspoon salt
1/4 teaspoon Worcestershire sauce
1/2 cup light cream
1/2 cup shredded American cheese

Sauté mushrooms with onion and butter for 5 minutes, until tender. Add crumbs, cook 2 minutes. Stir in lemon juice, Worcestershire sauce and seasonings. Place in baking dish. Pour cream around mushrooms. Bake at 400° for 15 minutes. Top with cheese and continue baking 8-10 minutes, until cheese melts.

SAUSAGE POTATO CASSEROLE

4 cup scalloped potatoes
2 cup cheddar cheese, grated
1 1/2 cup milk
1 pound sausage, cooked & crumbled
1 can cream of mushroom soup
Mushrooms, onions, salt and pepper to taste

Make layers of potatoes, sausage, mushrooms, onion and cheese. Season to taste. Mix soup and milk and pour over layers. Cover and bake at 350° until potatoes are tender.

PORK PIE

1 1/2 pounds ground pork
4 potatoes
3/4 stick butter or margarine
1/4 cup milk
1 package onion mix
1 package brown gravy
1 cup water
Crust:
2 cups flour
3/4 cup Crisco shortening
1/2 cup ice water

Cook pork in frying pan. Drain fat. Cook potatoes with a little salt. Drain, cut and mash them. Add butter or margarine and milk to potatoes. Add 1 package dry onion mix. Mix 1 package brown gravy mix with 1 cup water into small sauce pan and stir until smooth. Add it to the pork & potato mixture and mix altogether.
Mix together crust ingredients. If dough is too dry add more water. Shape into ball and roll it out between 2 pieces of wax paper. Place in pie pan and pour in pork filling. Baste top of pie with milk. Bake at 350° for about 25-30 minutes, or until crust is browned to your desire.

HUSBAND'S DELIGHT

24 ounces tomato sauce
1/2 cup sour cream
1 cup cottage cheese
8 ounces cream cheese
1 medium onion, chopped fine

1 cup grated Cheddar cheese
1 1/2 pounds ground beef
1/4 cup bell pepper, chopped
2 cloves garlic, chopped
1-8 ounces package wide noodles, cooked as directed
1 tablespoon granulated sugar

Brown ground beef, bell peppers, and garlic together and drain. Cook noodles as directed then drain. Add noodles, sugar, and tomato sauce to ground beef. Stir.Combine sour cream, cottage cheese, cream cheese and onion. Put into a 3-quart casserole dish a layer of noodles and meat, then a layer of cheese mixture; add the remaining meat and noodle mixture. Add grated cheese. Bake at 350° for 25 to 30 minutes.

AMISH CORN BALLS

2 tablespoons butter or oil
1 cup celery, chopped
1/2 cup onion, chopped
1 large can or 2 small cans cream style corn (2 1/2 cups)
1/2 teaspoon salt
1/4 teaspoon pepper
1 1/2 teaspoon poultry seasoning or sage
1 cup water
1-16 ounces loaf bread
3 egg yolks
1/2 cup (1 stick) butter
1 can cooked chicken, drained
1 can cream of chicken soup

Brown celery and onion in butter. Add cream corn, salt, pepper, seasoning or sage, and water. Bring to a boil and then pour over sliced bread. Add 3 egg yolks and toss thoroughly. Let cool; form into balls (about 12) and place in a buttered 9 X 13 dish. Melt 1/2 cup butter and pour some over each ball. Bake at 375° for 25 minutes. Heat the drained, cooked chicken and cream of chicken soup and pour over corn balls.

BOLOGNA
(Editor's note: Unless you're feeding100 people, you'll probably want to cut this recipe down and only make a fraction of it)

60 pound beef
60 pound pork
2 pound salt
6 1/2 pound brown sugar

3 ounces pepper
3 ounces allspice
1 ounces salt

Grind and mix well. Put in cloth casings and smoke as often as desired.

AUNT EMMA'S HEARTY STEW

2 pounds cubed stew beef
1/4 cup flour
1/4 teaspoon celery seed
Sprinkle of Accent
1 1/4 teaspoon salt
1/8 teaspoon pepper
4 medium onions, sliced
6 medium potatoes, thinly sliced
2 medium carrots, thinly sliced
1 1/2 cups hot water
1 teaspoon Worcestershire sauce
Butter

Mix together the flour and the seasonings and dredge the meat cubes in the mixture.
In a large casserole with tight fitting cover arrange in layers the meat and vegetables.
Add the bouillon to the water and add the Worcestershire sauce. Pour it evenly over the
casserole.
Dot the surface with butter.

AMISH TURKEY LOAF

3 cups diced turkey
8 slices bread
1/2 cup diced celery
1 onion, chopped
1 egg, beaten
1/4 teaspoon poultry seasoning

Cover and bake at 325° for 3 hours.

Mix all the ingredients together and place in greased loaf pan. Bake at 350° for 55 minutes. Serve with gravy.

AMISH TURKEY CASSEROLE

2 or more cups white or dark cooked turkey, cut into chunks
4 to 6 cups prepared Bread Stuffing
1/2 cup Parsley
1 cup chopped onion
1 cup chopped celery
Pepper to taste
2 cups Turkey Broth from the pan (tastes better if a little turkey fat is left in) or Chicken stock
1/4 cup (1/2 stick) Margarine

Mix stuffing, parsley, cooked turkey and broth in roasting pan. Top with onion, celery and the margarine cut into chunks. Bake for 1 to 11/2 hours, covered, in preheated 325° oven. Add more broth if necessary. The top should be crisp, the interior soft.

AMISH POT ROAST

4 pound beef roast (sirloin tip, rump, English cut)
1 tablespoon oil
1/4 cup soy sauce
1 cup coffee
2 bay leaves
1 garlic clove, minced
1/2 teaspoon oregano
2 onions, sliced

Sear roast in 1 tablespoon oil on all sides in heavy Dutch oven. Pour sauce over meat. Put half of onions on meat, the other half in sauce. Cover and roast 4-5 hours at 325°.

AMISH PORK CHOPS

6 pork chops
1 can chicken rice soup
1/2 soup can of water
Spread mustard on both sides of 6 pork chops. Brown in skillet approximately 10 minutes, turning frequently. Add chicken rice soup and 1/2 can of water and simmer for 25 to 30 minutes. Serve with rice.

AMISH BARBECUE PORK CHOPS

8 pork chops
1 cup ketchup
1 can celery soup
1/2 cup chopped onion
1 or 2 tablespoons Worcestershire sauce

Mix the ketchup, celery soup, chopped onion and Worcestershire sauce together and pour over the pork chops in a buttered casserole dish. Bake at 275° for 2 hours.

AMISH SUNDAY PORK CHOPS

6-8 pork chops
1 cup catsup
1 tablespoon Worcestershire sauce
1/2 cup onion, diced

Preheat oven to 375°.
Brown pork chops and drain. Mix together catsup, Worcestershire sauce and onion. Pour over pork chops. Cover. Bake for 2 hours.

CORN PIE

1 dozen ears fresh white corn (or 2 packages frozen corn)
4 potatoes, diced
1 medium onion, chopped
2 cups water
1 cup celery, chopped
4 hard boiled eggs, chopped fine
1 cup milk (maybe a little more if needed)
4 tablespoons butter
1 unbaked pie crust

Cook corn, potatoes, onion and celery in water until done, drain water. Add eggs, milk and butter. Use a deep pie dish. Put into pie crust and scratch the lid. Bake on a cookie sheet at 375° for 30-45 minutes or until golden brown.

FRIED SALMON PATTIES

2 cups cracker crumbs
1 cup salmon
1 teaspoon salt
2 eggs, beaten
1 1/2 cups milk
Pepper to taste

Mix all ingredients together and shape into patties.
Fry in oil over medium heat until they start to brown.

PEPPERS STUFFED WITH SCRAPPLE

1 1/2 pounds Scrapple
6 large green peppers
3 tablespoons chopped onion
1 cup cheese sauce

Cube and soften scrapple over low heat. Add onion. Cut thin slice from stem end of peppers. Remove seeds and plunge in boiling water 5 minutes. Drain and stuff with scrapple. Set peppers in muffin pans. Bake in moderate oven (350°) 25 to 30 minutes. Top with cheese sauce. Serving suggestion: Fluffy buttered rice and honeyed carrots are a good addition to the meal. Stuff peppers with cooked sausage and rice for another

Amish

Salads/Sides

HOME CANNING

By Kevin Williams
Oasis Newsfeatures Editor (kwilliams@oasisnewsfeatures.com)

The first time I visited Elizabeth Coblentz, the original Amish Cook, she introduced me to "canning." This was rather jarring (pun intended!). I was expecting to see a bunch of tin cans in her basement sealed shut with vegetables – ah the ignorance of a city kid. Elizabeth never did explain to me why canning isn't called jarring. Most Amish women are avid home "canners", socking away a colorful kaleidoscope of fresh food in sealed clear glass Mason jars. I've always been amazed by what Amish women will can from meats, vegetables, fruits, and even desserts. If properly canned and sealed, foods will stay fresh in a cool cellar for years. On a cold winter's night with the icy fingers of an Arctic storm swirling outside, one can pick from a bounty of summer season vegetables to open and experience an immediate taste of July. Elizabeth told me the only thing that really doesn't can well, in her experience, is rhubarb, everything else is fair game. Elizabeth was typical of most Amish women: she would seal her foods in sterilized glass jars, label and date the top, and then store them in her cellar. Lovina, who now pens the Amish Cook, practices canning with the same enthusiasm.

Home canning is very much a part of the Amish tradition of self sufficiency and independence from the outside world. During the Y2K scare of 2000, had the nation's power grids gone awry, most Amish families would have fared far better than the rest of us with their ample stocks of food.

Amish women will delicately pass the tradition and technique of home canning on to their daughters who will pass it on to theirs and thus the tradition lives on from generation to generation.

Canning recipes rarely appear in the Amish Cook newspaper column anymore because every time we used to put those recipes in the paper I would field calls from legions of angry USDA and 4-H officials taking Elizabeth to task for her canning techniques. They said her canning methods were unsafe, not meeting USDA guidelines. Elizabeth bristled at the criticisms, saying that her way never made her sick.

Some home-canning tips:

1. Clean and sterilize tools and equipment. All tools should be thoroughly washed. All canning jars and bands should be sterilized in boiling water. This eliminates harmful germs and bacteria.
2. Keep jars warm (180-190 degrees) until ready to use.
3. Proper cleaning and preparation of food. Select fresh and healthy fruits and vegetables at their peak ripeness. Wash and prepare according to directions for canning.
4. In preparing food, precooking is recommended over raw packing.

5. Process high acid foods in a boiling water bath. Use a Pressure canner for low acid foods.
6. Make sure acidity is sufficient. The absence of high acidity in the food can promote harmful bacteria. Make sure you know the acid levels of the food you are canning and add acid if need be. A common way to increase acidity is to add a teaspoon of lemon or lime juice to each jar.
7. Remove air bubbles by pushing food down firmly into the jar. Then, take a knife around the inside edge of the jar to help air bubbles to escape.
8. Leave 1/4 to 1/2 inches of headroom for most fruits and vegetables. Consult your canning recipe for the particular food you are processing.
9. Wipe the top of the lids clean. It is important that they have no food or residue on them to assure a good seal.
10. Put lids and bands on the jars, fingertip tight.
11. Just before placing the jars in the canner, fill the sink with hot water and set the jars in it for a moment. This will raise the temperature of the glass and minimize breakage.
12. Process properly. Once the food is prepared, packed ,and sealed into sterilized jars, it is processed in a boiling water bath. Immerse the jars in boiling water for the time required in the particular recipe. You can process it longer, but do not cut short the process.

Did you Know? The USDA no longer recommends open kettle boiling water bath. They recommend only using pressure canning. Despite this recommendation, a large percentage of home canners still use boiling water bath.

CUCUMBERS IN SOUR CREAM

2 cup peeled and sliced cucumbers
1/2 teaspoon Salt
1/2 cup sour cream
2 teaspoon Vinegar
1/2 teaspoon Sugar
2 teaspoon instant minced onion
1/2 teaspoon dill weed
Dash of cayenne pepper
Cracked black pepper

Place cucumber slices in bowl. Sprinkle with salt and cover with cold water. Refrigerate

30 minutes. Drain well, combine remaining ingredients and toss lightly. Chill at least 1 hour before serving.

HAM SPREAD

3 cans sauerkraut drained and rinsed (use a hearty, unsweetened variety, like Stokley's)
2 cups ham scraps/leftovers
1 cup grated Swiss cheese
3/4 cup mayonnaise
2 tablespoons Dijon mustard
2 teaspoons caraway seeds

Chop ham or pulse briefly in food processor (don't over process). Turn out into a bowl and mix in other ingredients. Chill and serve on bread or crackers

SUMMER BLUEBERRY SALAD

2 small boxes grape Jell-O
2 cups boiling water
1 cup cold water
1-15 ounces can crushed pineapple
1 can blueberry pie filling
Topping:
1-8 ounces container sour cream
1-8 ounces package cream cheese, softened
1/3 cup sugar

Dissolve Jell-O in hot water. Add cold water. When Jell-O begins to set, add crushed pineapple with juice and the pie filling. When Jell-O is fully set, spread the topping on top. Add chopped pecans or walnuts, if desired.

AUTUMN COLORS SALAD

4 cups peeled, cubed sweet potatoes or yams
1 apple, skin on, chopped
2 stalks celery, sliced
1 small red onion, chopped
1 teaspoon fresh ginger, grated
Rice vinegar to taste

Steam potatoes till soft (not mushy). Toss all ingredients lightly.
Dress with vinegar to taste. Chill.

AMISH STUFFING

2 pounds homemade bread (or high quality store bought)
2 pounds poached chicken thighs
1/2 cup minced fresh parsley
3/4 cup chopped onions
1 cup chopped celery
1 cup shredded carrot
1 1/4 cup boiled potatoes
1 tablespoon rubbed sage
1 tablespoon celery seed
1 teaspoon dried thyme
1/2 teaspoon black pepper
1/2 tablespoon tumeric
5 fresh eggs
12 ounces evaporated milk
2 1/2 cups of homemade or canned chicken broth

Preheat oven to 350°. On 2 cookie sheets, toast the bread cubes for 15 minutes, or until the bread is golden brown. Transfer to a very large mixer bowl. Bone the chicken and very finely chop the meat, discarding the skin. (I do this in the food processor while chicken is still warm.) Add the chopped vegetables and chicken meat to the bread, along with the seasonings. Toss. In a medium bowl, beat the eggs; add the evaporated milk and broth. Pour over the bread mixture and blend. The mixture will be quite moist. Allow to stand 1 hour. Preheat oven to 350°. Transfer the dressing to an oiled 3 quart glass casserole that is 10 inches in diameter and 3 inches deep (at this point, dressing can be frozen for future use; thaw before baking.) Bake dressing for 2 hours, or until the center of the dressing puffs up and is golden brown on top

STUFFING CASSEROLE

1 1/2 cups chopped onions
1 1/2 cups chopped celery
1 1/2 cups chopped carrots
2 tablespoons canola or olive oil
6 ounces fresh mushrooms sliced
3 tablespoons chopped parsley
2 tablespoons dry butter buds
2 large yellow summer squash cut in 1/2" cubes
1 medium red potato cut in 1/4" cubes
3/4 cup bulgar wheat (cracked wheat)
1 cup boiling water

2 cups cooked meat cut in 1/4" pieces
4 cups pre-seasoned stuffing crumbs
3 cups broth (same flavor as meat used)

Put bulgar wheat in small bowl and cover with 1 cup boiling water. Cover and let stand for 15 minutes while cooking vegetables below. Add butter buds while stirring in boiling water. In VERY large frying pan saute onion, celery and carrots in oil until onions are transparent. Add mushrooms and saute until they give up their liquid. Add squash and potato and 1/2 cup water, cover and cook for 5 minutes. Add salt and freshly ground pepper to taste. Add parsley and sage (or marjoram if you prefer), meat and soaked bulgar wheat. Mix thoroughly then add stuffing crumbs and mix thoroughly. Add broth and mix until the whole mixture is gooey. Put in large greased casserole dish, cover, and bake at 350° for 1 hour. Reheats well over low heat in a frying pan.

SAUERKRAUT

1 head cabbage
2-4 teaspoons salt
2-4 tablespoons vinegar

Shred cabbage and place in jars. Add 1 teaspoon of salt and 1 tablespoon of vinegar to each jar. (The number of jars depends on what size you use.) Fill jars with cold water. Put lids on tightly and store in basement for 2 months.

BAKED MAC & CHEESE

2 cups elbow macaroni
2 teaspoons salt
2 1 /2 cups milk
8 ounces Velveeta cheese
Salt and pepper
1/8 cup butter
3/4 cup bread crumbs

Cook macaroni according to package directions, drain, pour into casserole dish and add milk, cheese, salt-pepper, mix thoroughly. Dot with butter, top with bread crumbs. Bake 30 minutes at 350°.

GREEN BEANS DELUXE

2-10ounces pkg. frozen cut green beans, cooked and drained
1/4 cup margarine or butter
1/4 cup chopped onion
2 tablespoons flour
1 teaspoon salt
1/4 teaspoon Dijon mustard
1/8 teaspoon pepper
1 cup dairy sour cream
1/2 cup shredded Swiss cheese
1/4 cup fine bread crumbs

Melt half the margarine in medium saucepan. Melt 1/4 cup margarine. Sauté onion in margarine until crisp and tender. Blend in flour, salt, mustard and pepper. Cook until thickened, stirring constantly. Remove from heat. Stir in sour cream, cheese, and beans and mix well. Pour into increased 1 1/2 quart casserole. Combine bread crumbs and remaining margarine; sprinkle over beans. Bake at 350° for 20-25 minutes or until crumbs are a light golden brown.

HEARTY POTATO SALAD

8 boiled potatoes
3 stalks celery, diced
3 hard cooked eggs, diced
1 onion, minced
1 tablespoon parsley, minced
2 eggs, well beaten
1 cup sugar
1/2 cup vinegar, diluted with 1/2 cup cold water
1/4 teaspoon dry mustard
1/2 teaspoon salt
1/4 teaspoon pepper
6 slices bacon

Boil potatoes in their skins. When soft, peel and dice. Add the celery, diced eggs, parsley, and onion. Meanwhile, fry bacon until crisp and brown. Remove the bacon and keep the fat hot. When bacon cools a little, break it up over the potato mixture. Beat the eggs, add the sugar, spices and vinegar and water. Mix well. Pour egg mixture into the hot bacon fat and stir until mixture thickens (about 10 minutes). Pour over the potato mixture and mix lightly. Can be served warm, or let stand in cool place several hours before serving.

delicious meal.

FRENCH FRIED ONION RINGS

4 medium onions
1 cup pancake ready mix
1 cup milk

Combine pancake mix and milk. Dip onions into mixture and fry in deep, hot fat until lightly browned.

PENNSYLVANIA DUTCH PEPPER CABBAGE

1/2 head cabbage or 1 small head cabbage
1 red pepper
1 yellow pepper
1 green pepper
1 onion
1 stalk celery
3 tablespoons salt
1/2 teaspoon mustard seed
1/2 teaspoon celery seed
1 tablespoon sugar
Vinegar

Chop vegetables medium coarse, not fine. Sprinkle with salt and drain in colander overnight. Stir in mustard seed, celery seed, sugar, and enough vinegar to cover. Let stand at least 3 hours before serving, or seal in jars. Makes 3 pints.

PENNSYLVANIA DUTCH RED PEPPER CABBAGE

2 tablespoons cooking oil
4 cups shredded red cabbage
2 cup unpared cubed apples
1/4 cup brown sugar
1/4 cup vinegar
1/4 cup water
1/2 teaspoon caraway seed
Salt and pepper

In skillet, heat oil. Add remaining ingredients. Cover and cook over low heat, stirring occasionally. Cook 25-30 minutes. Garnish with apple wedges, if desired.

TURNIP CASSEROLE

1 1/2 cups turnip, grated
1 1/2 cups potato, grated
3/4 cup milk
1/2 cup low fat yogurt
1/2 cup whole grain bread crumbs
1/4 cup sunflower, olive, or safflower oil
2 medium onions, chopped
1 tablespoon dried parsely
1/2 teaspoon ground black pepper

Thoroughly mix all the ingredients, except the bread crumbs, in a large bowl.

Pour the mixture into a lightly greased shallow casserole (approx. 9 inches). Then sprinkle the bread crumbs over the top. Bake for 40 minutes at 375°. Serve immediately

GREEN BEAN DISH

3 strips bacon
1 small onion, sliced
1/4 teaspoon dry mustard
2 teaspoons cornstarch
1/4 teaspoon salt
1-16 ounces can green beans, reserve 1/2 cup bean liquid
1 tablespoon brown sugar
1 tablespoon vinegar
1 hard cooked egg, sliced (opt.)

Fry bacon in skillet until crisp. Remove bacon, crumble; set aside. Drain off all but 1 tablespoon fat. Add onion and brown lightly. Stir in mustard, cornstarch and salt. Add liquid from beans. Stir until mixture boils. Blend in brown sugar and vinegar. Add beans and bacon, stir and heat thoroughly. Turn into serving dish. May garnish with sliced egg and crumbled bacon.

CHEESE BALL

2-8 ounces packages cream cheese
2 tablespoons Worcestershire sauce
1 tablespoon A.1. steak sauce
1 package dried beef, diced fine
1 tablespoon garlic, minced, or onion

Mix together and chill. Form into ball and serve with crackers.

PARTY CHEESE BALL

2-8ounces packages cream cheese
2 cups (8 ounces) Velveeta cheese

1 tablespoon chopped onion
2 teaspoons Worcestershire sauce
1 teaspoon milk
3/4 cup finely chopped pecans (optional)

Combine softened cream cheese and Velveeta cheese, mixing will. Add onions, Worcestershire sauce and milk; mix well. Chill for 1-2 hours. Shape into ball and roll in pecans (optional).

CARROT DISH

5 cups peeled and sliced carrots
1 raw pepper, diced
1 onion, diced
1 cup sugar
1/2 cup vinegar
1 can tomato soup
3 teaspoons Worcestershire sauce
1 teaspoon salt
1 teaspoon pepper
2 teaspoons prepared mustard

Cook carrots until almost tender. Add remaining ingredients to carrots. Mix together and refrigerate for at least 12 hours.

AMISH CORN FRITTERS

2 eggs, separated
2 tablespoons flour
1 tablespoon sugar
2 cups grated fresh corn

Beat the egg yolks and add the flour, 1 teaspoon salt and a little pepper. Add the corn and fold in the stiffly beaten egg whites. Drop small spoonfuls on a greased griddle or frying pan. Do not cook too long.

PENNSYLVANIA PICKLED EGGS AND BEETS

8 eggs
2-15 ounces cans pickled beets, juice reserved
1 cup white sugar
1 onion, chopped
1/2 teaspoon salt
3/4 cup cider vinegar
2 bay leaves
1 pinch ground black pepper
12 whole cloves

Place eggs in saucepan and cover with water. Bring to boil. Cover, remove from heat, and let eggs sit in hot water for 10 to 12 minutes. Remove from hot water, cool, and peel.

Place beets, onion, and peeled eggs in a non-reactive glass or plastic container. Set aside. In a medium-size, non-reactive saucepan, combine sugar, 1 cup reserved beet juice, vinegar, salt, pepper, bay leaves, and cloves. Bring to a boil, lower heat, and simmer 5 minutes. Pour hot liquid over beets and eggs. Cover, and refrigerate 48 hours before using.

Amish

Cookies

BOILED COOKIES

These are also called "funeral cookies" by the Amish because they can be made up quickly to take to the grieving family.

1/2 cup butter
1/2 cup milk
2 cups granulated sugar
3 tablespoons unsweetened cocoa
1/2 cup peanut butter
1 teaspoon vanilla extract
1/4 teaspoon salt
3 cups quick cooking oats (not instant)
1/2 cup coarsely chopped pecans

In a small saucepan over medium heat bring butter, milk, sugar, and cocoa to a boil and cook for 1 minute longer. Remove from heat and stir in peanut butter, salt and vanilla extract. Mix in oats and pecans. Drop by teaspoon onto wax paper and allow the cookies to stand unrefrigerated for 1 hour.

Store in an airtight container with wax paper between the layers.

LITTLE HONEY CAKES

1 1/2 cups lard
2 cups granulated sugar
4 eggs, beaten
1 cup molasses
1 cup honey
1 cup hot water
2 teaspoons cinnamon
1 teaspoon ginger
2 teaspoons baking soda
2 teaspoons baking powder
About 5 cups all-purpose flour to stiffen

Cream together the shortening and sugar. Add the eggs. Blend in the molasses, honey and hot water. Add the spices, baking soda, baking powder and flour. Chill the dough overnight.

Roll dough out and cut into cookies shapes or drop it by spoonful onto a cookie sheet. Bake at 350° until brown.

OATMEAL CINNAMON KRISPIES

2 1/2 cups butter, softened
5 cups granulated sugar
1/3 cup dark molasses
4 eggs
4 teaspoons baking powder
1 teaspoon baking soda
2 teaspoons salt
1 tablespoon cinnamon
1 tablespoon vanilla extract
4 1/3 cups flour
4 1/3 cups old fashioned oats
2 cups finely chopped pecans

Preheat oven 375°. Grease 2 cookie sheets.
Cream first 3 ingredients. Beat in eggs one at a time. Add baking powder, baking soda, salt, cinnamon and vanilla extract. Beat until well blended. Add flour and blend. Add oats and nuts and blend. Drop by rounded spoonfuls onto prepared cookie sheets. Bake for 5 to 9 minutes. It may be necessary to turn sheets around so they brown evenly. Let cool on sheets for about 3 minutes. Remove to wire rack till cool. Cookies will look medium brown and crinkly.

JACOB'S PUFF COOKIES

1 cup shortening, rounded
1 cup brown sugar, firmly packed
1/2 cup granulated sugar
2 eggs
2 3/4 cups flour
1 teaspoon cream of tartar
1 teaspoon baking soda
1 teaspoon baking powder
1/2 teaspoon salt

Preheat oven to 375°.
Mix shortening, sugar and eggs thoroughly. Mix dry ingredients and add to sugar mixture. Chill.
Form into balls and roll into mixture of 2 tablespoons sugar and 2 teaspoons cinnamon or add chocolate chips. Press balls with hand. Bake on an ungreased cookie sheet for 8 to 10 minutes. Cookies will puff, then settle down and look crinkly.

AMISH WHOOPIE PIE COOKIES

Cookie:
1 cup shortening
2 cups sugar
2 whole eggs
2 egg yolks
1/2 teaspoon salt
1 cup cocoa
1 cup sour milk
1 cup hot water
2 teaspoons baking soda
1 teaspoons baking powder

4 cups flour

Filling:
2 egg whites
2 teaspoons vanilla
3 1/2 cups, plus 2 tablespoons powdered sugar
4 tablespoons milk
4 tablespoons flour
1 1/2 cups shortening

Mix dry *cookie* ingredients. Set aside. Mix moist *cookie* ingredients;
slowly add dry and mix well. Drop by tablespoonfuls on
greased cookie sheet. Bake 10 minutes at 400°.

For filling, cream shortening, egg whites, vanilla, 2 tablespoons powdered sugar, milk,
and flour. Add an additional 3 1/2 cups of powder sugar.
When cookies are cooled, spread with filling and place another
cookie on top to make like a sandwich.

PUMPKIN WHOOPIE PIES

2 cup brown sugar
1 cup oil
1 1/2 cup pumpkin, cooked and mashed
2 eggs
3 cups flour
1 teaspoon salt
1 teaspoon baking powder
1 teaspoon vanilla
1 1/2 tablespoons cinnamon
1/2 tablespoons ginger

1/2 tablespoons ground cloves

Filling:
1 egg white, beaten
2 tablespoons milk
1 teaspoon vanilla
2 cups confectioners sugar
3/4 shortening

Cream the sugar and oil. Add the pumpkin and eggs. Add the flour,
salt, baking powder, soda, vanilla, and spices. Mix well. Drop by
heaping teaspoons onto a greased cookie sheet. Bake at 350° for 10
to 12 minutes.
For filling mix egg white, milk, vanilla and 1 cup confectioners sugar. Then add one
more cup of confectioners sugar and 3/4 cup of shortening. Make sandwiches from two
cookies filled with the filling.

Variation:
Adding half a cup of ground black walnuts gives these cookies a
special delicious flavor.

IOWA AMISH SNICKERDOODLES

1/2 cup margarine
1/2 cup Crisco solid shortening
2 eggs
1 1/2 cups sugar
2 3/4 cups flour
2 teaspoons cream of tarter
1 teaspoon baking soda
1/4 teaspoon salt
2 tablespoons sugar
2 teaspoons cinnamon

Mix first four ingredients thoroughly. Pre-sift the next 4 ingredients together. Add to the first mixture. Form balls (walnut size). Roll into mixture of sugar and cinnamon. Place about 2 inches apart on ungreased cookie sheet. Bake 8 to 10 minutes at 375°. Cookies will flatten into circles as they cook. May top with red hots or leave unadorned. Store well in covered container and can be frozen.

AMISH DROP COOKIES

2 cups sugar
1 1/2 cups lard
2 eggs
2 teaspoon vanilla
1 1/2 teaspoon nutmeg
6 cups flour
1 1/2 cups milk
3 teaspoons baking powder
1 teaspoon baking soda

Cream together sugar and lard. Add eggs, nutmeg, vanilla, and baking powder. Mix soda with milk and add alternately with flour and drop with soup spoon on ungreased cookie sheet. Bake plain or sprinkle with sugar and cinnamon or sugar. Bake at 350° for 10 to 12 minutes or until lightly browned. Make 5 dozen.

AMISH ANGEL FOOD COOKIES

1 cup shortening
1/2 cup granulated sugar
1 egg
1 teaspoon vanilla extract
1/4 teaspoon salt
1 teaspoon baking soda
 2 cups all-purpose flour
1 teaspoon cream of tartar
1 cup coconut
1/2 cup brown sugar

Mix shortening and sugar until creamy; add egg and vanilla. Sift all dry ingredients, except brown sugar, together. Add to shortening mixture. Roll dough into small balls and dip into water, then into brown sugar. Put on cookie sheet. Bake at 375° for 15 minutes.

SOUR CREAM COOKIES

4 cups brown sugar
2 cups sour cream
1 teaspoon vanilla extract
4 teaspoons baking powder
1 1/2 cups margarine or Crisco
6 eggs, well beaten
4 teaspoons baking soda
Enough flour to make a soft dough

Frosting:
1/2 cup margarine or butter, softened
1 teaspoon vanilla
4 tablespoons milk
4 cups powdered sugar
Mix all ingredients together. Put in refrigerator overnight, then roll out or drop by spoon onto baking sheet and flatten with a glass. Bake at 350o for 8-12 minutes or until done. Let cool. Combine frosting ingredients (add more milk or powdered sugar to desired consistency) and frost cookies.

AMISH BUSHEL OF COOKIES

5 pounds granulated sugar
2 1/2 pounds lard
12 eggs
1 cup pure maple syrup
3 tablespoons baking soda
2 pounds quick oats
3 tablespoons baking powder
1 quart sweet milk
6 pounds flour

1 pound salted peanuts, coarsely ground
1 pound raisins, coarsely ground

Mix all ingredients. Drop by heaping teaspoons onto ungreased cookie sheet. Bake at 350° for 8-11 minutes or until done.

WALNUT KISSES

1 pound walnuts
2 cups granulated sugar
5 tablespoons all-purpose flour
6 egg whites
1 teaspoon vanilla extract

Beat egg whites until stiff but not dry. Gradually add sugar and continue to beat until blended. Sift flour lightly over beaten whites and fold in with a wire whisk. Blend in vanilla extract and nuts. Drop by teaspoon onto greased cookie sheet about 2 inches apart. Bake at 325° for 10 minutes.

AMISH CHRISTMAS SUGAR COOKIES

Part One Ingredients:
1 cup sugar
1 cup powdered sugar
1 cup butter
1 cup oil
2 eggs

Part Two Ingredients:
1 teaspoon vanilla
1 teaspoon salt
1 teaspoon cream of tartar

1 teaspoon baking soda
4 ½ cups flour

Combine all ingredients from part one together. Mix well.

Combine all ingredients from part two together.Mix well.
Blend part one and part two together, then chill at least an hour or overnight. Flour hands, then shape into balls and place on cookie sheet. Flatten with bottom of chilled glass dipped in sugar. Sprinkle with colored sugar. Bake at 350° until edges are golden.

WHITE DROPPED AMISH CHURCH SUGAR COOKIES

1 1/2 cups margarine
2 cups granulated sugar
2 eggs
1 teaspoon vanilla extract
1 teaspoon almond extract
3 1/2 cups all-purpose flour
1 teaspoon baking powder
1 teaspoon baking soda
1 teaspoon grated nutmeg
1 teaspoon salt
1 cup sour cream
Sugar and raisins (for garnish)

Preheat oven to 375°.
In a large mixer bowl, cream together the margarine and sugar for 3 minutes. Add the eggs and extracts and beat until well combined.
In a large mixing bowl, whisk together the flour, baking powder, baking soda, nutmeg and salt. Add to creamed mixture and combine until mixture is moistened. Stir in sour cream by hand and blend well. Using a tablespoon, drop the dough onto a parchment-lined or nonstick baking sheet. Top each with sugar and one to three raisins. Bake for 10 to 12 minutes or until the bottoms are lightly browned. Remove the cookies to a rack to cool. Store in airtight containers or freeze.

Amish
Cakes

BLUEBERRY CAKE

3/4 cup granulated sugar
1/4 cup vegetable oil
1 egg
1/2 cup milk
2 cups flour
2 teaspoons baking powder
1/2 teaspoon salt
2 cups blueberries, well drained

Topping:
1/4 cup butter
1/2 cup granulated sugar
1/3 cup flour
1/2 teaspoon cinnamon

Preheat oven to 375º. Cream together sugar, oil and egg until lemon colored. Stir in milk. Sift together flour, baking powder and salt and stir into creamed mixture. Gently fold in blueberries. Spread batter into greased and floured 9-inch square pan. Combine topping ingredients and sprinkle over batter. Bake for 45 to 50 minutes.

SUGARLESS APPLE CAKE

2 1/2 cup apples, peeled and diced
2/3 cups oil
3 eggs
1 tablespoon tapioca

3 teaspoon vanilla
1/4 cup raisins
2 cups flour
1 tablespoon baking soda
1 teaspoon cinnamon
1/2 teaspoon salt

Stir together the first 6 ingredients, then add the last 4 ingredients. Bake in greased loaf pan at 350° for 45-50 minutes or until toothpick inserted in middle comes out clean.

PENNSYLVANIA APPLE DUMPLING CAKE

3 pounds apples, peeled, cored and sliced
2 cups all-purpose flour
1 1/2 cups white sugar
2 teaspoons baking powder
1 teaspoon salt
2 eggs, beaten
1 cup vegetable oil
1 teaspoon ground cinnamon

Preheat oven to 350o. Lightly grease a 9x13 inch baking dish. Place sliced apples in baking dish. In a medium bowl, mix together the flour, sugar, baking powder, and salt. Stir in eggs and oil; pack on top of apples. Sprinkle with cinnamon. Bake in preheated oven for 40-45 minutes, or until topping is puffed and golden brown.

BUTTERSCOTCH APPLE CAKE

3 eggs
1 1/4 cups vegetable oil
1 teaspoon vanilla
2 1/2 cups flour
2 cups sugar
2 teaspoons baking powder
1 teaspoon salt
1 teaspoon baking soda
1 teaspoon cinnamon
4 medium tart apples, peeled and chopped

1 cup chopped pecans
1-11 ounces package butterscotch chips

In a mixing bowl, beat the eggs, oil and vanilla. Combine flour, sugar, baking powder, salt, baking soda and cinnamon; add to egg mixture and mix well. Stir in apples and pecans. Pour into an ungreased 13-by-9-inch baking pan. Sprinkle with butterscotch chips. Bake at 325° for 40 to 45 minutes or until toothpick inserted in center comes out clean. Cool on wire rack.

CHEESE POUND CAKE

1 1/2 cups (3 sticks) butter
1-8 ounces package cream cheese
3 cups sugar
6 eggs
3 cups flour
1 teaspoon vanilla

Preheat oven to 300°. Cream butter, cream cheese and sugar together. Add eggs one at a time. Add vanilla and then flour slowly. Pour into a Bunt or Tube pan. Bake for 1 1/2 hours.

CREAM CHEESE CAKE

4 eggs, separated
2 pounds cream cheese
1 cup sugar
1 grated lemon rind
Juice of 1 lemon
1 teaspoon vanilla

Heat oven to 325°.
Mix egg yolks, cream cheese, sugar, lemon rind, lemon juice and vanilla in a large bowl until smooth. Add egg whites and beat until smooth. Place inch warm water in cake pan. Pour batter in another cake pan and put on top of cake pan with water. Bake for 1 hour. Turn oven off and let cake stand in oven for 1 hour.

INDIANA CARROT CAKE

2 cups sugar
1 1/2 cups oil
4 whole eggs
3 cups raw grated carrots
2 cups flour
2 teaspoons baking powder
2 teaspoons soda
2 teaspoons cinnamon
1 teaspoon salt
1 - 2 cups raisins
Preheat oven to 350°.
Mix sugar and oil thoroughly. Add eggs, one at a time and beat well. Add carrots. Add dry ingredients to mixture and beat 3 minutes. Stir in raisins. Bake in three greased 8 or 9 inch layer pans for 35 minutes in 350 degree oven.

Icing:
1 8-oz package cream cheese
1/2 stick butter or margarine
2 teaspoons vanilla
1 box confectioners' sugar

Combine cream cheese, butter, vanilla and confectioners' sugar. Mix well and spread on cake.

WISCONSIN CARROT CAKE

2 cups flour
2 teaspoons cinnamon
1 teaspoon baking powder
1/4 teaspoon salt
2/3 cup butter, softened
1 cup sugar
3 large eggs
2/3 cups milk
3 medium carrots, grated
1/2 cup coarsely chopped pecans
Icing:
1/2 cup butter, softened
8 ounces cream cheese, softened
1 teaspoon vanilla
21/2 cups powdered sugar
Topping:
1/4 cup finely chopped pecans
2 tablespoons firmly packed brown sugar

Preheat oven to 350°. Grease a 9-inch round cake pan. Dust with small amount of flour; tap out excess. In a large bowl, mix together 2 cups flour, cinnamon, baking powder and salt. In a separate bowl, beat together butter and sugar at medium speed until light and fluffy. Add eggs to the butter mixture, one at a time, beating well after each addition. At low speed, alternately beat flour mixture and milk into butter mixture. Stir in carrots and nuts.

WACKY CAKE

1 1/2 cups flour
1 cup sugar
4 tablespoons unsweetened cocoa powder
1 teaspoon baking soda
1/2 teaspoon salt
1 teaspoon vanilla
1 tablespoon cider vinegar
6 tablespoons vegetable oil
1 cup cold water

Sift flour, sugar, salt, baking soda, and cocoa together into an ungreased, 8" x 8" square cake pan. Make 3 indentations in the dry ingredient mixture. Pour vegetable oil into one of them. Put vinegar in the second one. Put vanilla into the third one. Pour a cup of water over all ingredients and mix well with a fork, Bake at 350 ° for 30 to 40 minutes, or until a tooth pick inserted in the center of the cake comes out clean. Remove from oven and place on a wire rack to cool. Leave cake in pan. Spread your favorite frosting on top.

AMISH OATMEAL CAKE

1 cup uncooked 1 min quick oatmeal
1 1/2 cups boiling water
1 1/2 cups flour
1/2 teaspoon baking soda
1/2 teaspoon cinnamon
1/2 teaspoon nutmeg
1/2 teaspoon salt
1/2 cup butter -- softened
1 teaspoon vanilla extract
1 cup brown sugar
1 cup granulated sugar
2 eggs

Place the oats in a small bowl and pour the boiling water over them. Let stand for 20 minutes. Preheat the oven to 350 F. Sift together the flour, baking soda, cinnamon, nutmeg, and salt on wax paper. Set aside. In a large mixing bowl, beat the butter until creamy. Add the vanilla and gradually add the sugars, beating until fluffy. Beat the eggs into the mixture one at a time. Add the oatmeal mixture and blend. Add the flour mixture and blend again. Pour the batter into an oiled 13 x 8-inch pan. Bake for 35 minutes, or until the top of the cake springs back when touched with fingertip. Frost cake with icing of your choice.

SMEARCASE CAKE

Cake:
2 cups flour

1/2 cup cooking oil
2 eggs
2 teaspoons baking powder
1 cup granulated sugar

Filling:
2-8 ounces packages cream cheese
3/4 cup granulated sugar
1 1/2 teaspoons flour
3 eggs
1 1/2 teaspoons vanilla extract
1-13 ounces can evaporated milk or 1 1/2 cups sweet milk
Cinnamon

For cake: sift dry cake ingredients together. Add eggs, oil and sugar; stir well. Press into a 13 x 9-inch pan. Spread cake dough in bottom and on sides.
For filling: Mix all filling ingredients together; beat until smooth. Pour filling into shell. Sprinkle with cinnamon and bake at 325° for 1 1/4 to 1 1/2 hours

MOLASSES SPICE CAKE
1 cup butter, room temperature
1 cup sugar
3 eggs
1 cup molasses
2 cups all-purpose flour
1/2 teaspoon salt
1/2 teaspoon ground allspice
1/2 teaspoon ground ginger
1 teaspoon ground cinnamon
1 cup boiling water
2 teaspoons baking soda
Sifted confectioners' sugar (optional)

In a mixing bowl cream butter with sugar, beating with a wooden spoon (or an electric mixer) until light and fluffy. Add eggs, one at a time, beating well after each addition. Add molasses and blend well.
In another bowl combine flour, salt, allspice, ginger, and cinnamon; stir well. Slowly add flour and spice mixture to creamed mixture, beating at low speed. Combine water with baking soda and add to cake batter, mixing well. Batter will be thin.
Pour batter in to a greased and floured 13x9x2 baking pan. Bake at 350° for 45 minutes, or until a wooden pick or cake tester inserted in center comes out clean. Cool in pan and cut into serving-size squares. If desired, sprinkle with sifted confectioners' sugar.

AMISH BUTTER CAKE

3 eggs

1-18.5 oz package yellow cake mix
1/2 cup butter, softened
1/2 pound cream cheese, softened
1-16 ounces package confectioners' sugar, sifted
1 teaspoon vanilla extract

Preheat oven to 350°. Grease and flour one 9x13 inch pan.
Beat together 1 egg, cake mix, and butter for 5 minutes. Mix and pour into prepared pan.
Cream together cream cheese, confectioners' sugar, 2 eggs, and vanilla extract. Pour this on top of the cake base.
Bake 35-40 minutes until browned. Cool 15 minutes and cut into squares.

LEMON SPONGE CAKE

1 cup milk
Lemon juice of 1 lemon
1 tablespoon butter melted
1 cup sugar
3 tablespoons flour
Lemon rind of 1 lemon, grated
1 pinch Salt

Mix the sugar and flour together and add the lemon juice and rind, slightly beaten egg yolks, butter and salt. Stir in the milk and mix well. Beat the whites until stiff and fold into the first mixture. Pour into custard cups.
Set the cups in pan with hot water and bake at 350° about 40 minutes. The sponge may also be baked in pie shell.

AMISH ROLLED OATS CAKE

1 cup oats
1 1/4 cups boiling water
1/2 cup shortening
1 cup white sugar
1 cup brown sugar
2 eggs
1 teaspoon vanilla
1 1/2 cups flour
1 teaspoon baking soda
1/2 teaspoon baking powder
1 teaspoon cinnamon

1/2 teaspoon nutmeg

Topping: (use on warm cake)
5 tablespoon oleo
1/2 cup brown sugar
1/2 cup milk
1/2 cup coconut
1/2 cup chopped nuts

Mix oats and boiling water; let stand 20 minutes. Cream together shortening, sugars, eggs, and vanilla. In a separate bowl, combine flour, baking soda, baking powder, cinnamon and nutmeg. Once the cooked oats have cooled, add them to the dry ingredients. Add shortening mixture next and stir well. Pour into 9x13 pan and bake at 350° for 35 minutes. For the topping, melt oleo in small saucepan and add sugar and milk. Bring to boil. Boil 7 minutes then add 1/2 cup coconut and 1/2 cup nuts. Pour over warm cake.

AMISH SHOO FLY PIE CUPCAKES

1 1/2 cups boiling water
1 teaspoon baking soda
1 cup molasses (dark preferred)
3 cups flour
1 cup brown sugar, packed
1/4 cup (1/2 stick) butter or margarine
1/4 cup unsweetened applesauce

Add baking soda to boiling water; then add molasses and set mixture aside. Make a crumb mixture from flour, brown sugar and margarine, and reserve 1 cup of this. Mix together the remaining crumb mixture, liquid mixture and applesauce. Pour (batter is thin) into cupcake pan lined with cupcake papers, and sprinkle reserved crumb mixture on top of each cupcake. Bake at 350° for 20-25 minutes. Yield: 2 dozen cupcakes.

SHOO FLY CAKE

4 cups flour
2 cups brown sugar
1 cups butter
2 cups boiling water
2 teaspoons baking soda
1 cup molasses

Work flour, sugar, and butter into fine crumbs. Set aside 1 1/2 cups for topping. Mix water, molasses, and baking soda together, and add remaining crumbs. Pour into greased and floured 9 x 13 inch cake pan. Bake at 350° for 35 minutes.

HUMMINGBIRD CAKE

3 cups flour
1 1/2 cups sugar
1 teaspoon salt
1 teaspoon baking soda
1 teaspoon cinnamon
3 eggs, beaten
1 1/2 cups salad oil
1 1/2 teaspoons vanilla
1-8ounces can crushed pineapple
1 cup pecans, crushed
2 cups bananas, chopped

Icing:
1-8ounces package cream cheese, softened
½ cup butter
1-2pound package powdered sugar
1 teaspoon vanilla
1/4 cup pecans, chopped (optional)

Combine dry ingredients; add eggs and oil and stir until dry ingredients are moistened.
Do not beat or cake will be tough. Stir in vanilla, pineapple, pecans and bananas.
Put in well-greased 9 x 13 pan; bake at 350° for 55 minutes. Cool at least 15 minutes.
Beat icing ingredients together and spread over cake. Sprinkle pecans on top (optional).

AMISH ANGEL FOOD CAKE

11 egg whites
1 1/2 cup sugar
1 cup flour
1 teaspoon cream of tartar
1/8 teaspoon salt
1 teaspoon vanilla

Sift the dry ingredients together 4 times. Beat the egg whites until dry. Fold dry ingredients into eggs. Add flavoring and bake at 250° for 1 hour in an ungreased angel food cake pan.

DUTCH CHOCOLATE MARBLE CAKE

1 /3 cup butter
2 eggs, well beaten
1 cup sugar
1 1 /2 cups flour
2 teaspoons baking powder
1 /2 cup milk
1 square of unsweetened chocolate
1 tablespoon butter
1 teaspoon vanilla

Cream the 1/3 cup butter and sugar together, add the well beaten eggs and mix well. Sift flour and baking powder and add alternately with the milk to the first mixture. Put 1/3 of mixture into a bowl and add the 1 tablespoon of butter and chocolate which have been melted together. To the white batter, add the vanilla. Drop white batter, then chocolate, by spoonfuls into a well-greased, deep cake pan and bake at 350° about 40 minutes.

Amish

Pies

PIE CRUST

1 cup flour
1/4 teaspoon salt
1 teaspoon sugar
1/2 cup shortening
2 tablespoons water

Mix all ingredients together and form into ball. Use flour to roll out.

OATMEAL PIE CRUST

1 cup quick-cooking oats
1/3 cup sifted flour
1/3 cup brown sugar
1/2 teaspoon salt
1/3 cup butter

Combine all ingredients except butter. Cut in butter until crumbly. Press firmly on bottom and sides of 9-inch pie plate. Bake in moderate oven (375°) about 15 minutes. Cool crust completely and fill with desired cream filling.

CHERRY CUSTARD PIE

1-16 ounces can sour cherries or pitted fresh sour cherries
3 eggs
2 cups milk
1/2 cup sugar
2 tablespoons flour

1 teaspoon vanilla
1 unbaked pie shell

Drain cherries thoroughly and place them in the bottom of an unbaked pie shell. Scald milk. Beat eggs and add to scalded milk. Mix flour and sugar together. Combine with milk & eggs and beat well. Stir in vanilla. Pour mixture into an unbaked pie shell over cherries. Bake at 400° for 35 to 40 minutes or until custard is set in center.

FRESH STRAWBERRY PIE

1 baked 9 inch pie shell
1 tablespoon powdered sugar
3 ounces cream cheese
1 1/2 quarts whole strawberries 1 cup sugar
2 tablespoons white Karo syrup
1 cup water
3 tablespoon strawberry Jell-O mix
Pinch of salt
3 tablespoon Cornstarch

Beat cream cheese with sugar, spread carefully over bottom of baked pie shell. Arrange berries in shell. Berries should be cut and sliced. Pour the following glaze over the berries: Cook all ingredients except Jell-O mix over medium heat until thick. Add Jell-O mix. Cool before pouring over berries. Enjoy.

GLAZED STRAWBERRY-RHUBARB PIE

Pastry for 2 crust pie
1/8 teaspoon salt
2 cups fresh strawberries
2 tablespoons butter or margarine
1 1/4 cup sugar, plus 1 tablespoon sugar
1/3 cup flour
2 cup fresh rhubarb cut into 1 inch pieces

Combine 1 1/4 cup sugar, salt and flour. Arrange half of the strawberries and rhubarb in pastry lined 9 inch pie pan. Sprinkle with half of the sugar mixture. Repeat with the remaining fruit and sugar mixture, dot with butter. Cover with other 1/2 of the pastry. Bake at 425° for 40-50 minutes, or until rhubarb is tender and crust is browned.

PENNSYLVANIA RHUBARB CREAM PIE

1 1/2 cup sugar
3 tablespoons flour
1/2 teaspoon ground nutmeg
1 tablespoon butter
2 beaten eggs
3 1/2 cup rhubarb, cut and sliced
1 (9 inch) unbaked pie crust

Place rhubarb in crust. Blend sugar, flour, nutmeg and butter. Add the eggs and beat smooth. Pour over rhubarb. Bake at 450° for 10 minutes, then at 350° for 30 minutes.

RHUBARB CRUMB PIE

Filling:
2 cups sugar
3 TABLESPOON flour
2 eggs, well beaten
5 1/2 cups coarsely chopped rhubarb
2 - 8" unbaked pie shells

Crumbs:
1 cup flour
1/2 cup brown sugar
1/4 cup shortening
1/2 teaspoon baking soda
1/2 teaspoon baking powder
Preparation -
Combine sugar, flour and eggs. Fold in chopped rhubarb. Spoon mixture into 2 unbaked pie shells. To prepare crumbs, combine all ingredients and mix well. Sprinkle evenly over rhubarb. Bake at 400° for 10 minutes; reduce heat to 350° and bake 40 to 50 more minutes

SUGAR PIE

1 unbaked 8" pie shell
1 cup brown sugar
3 tablespoons flour
1/4 teaspoon salt
1 1/2 cups evaporated milk
3 tablespoons butter
cinnamon to taste
Preheat oven to 350F. In a small bowl blend together the sugar, flour and salt. Spread in bottom of pie shell. Pour the milk over the sugar mixture, but do not stir. Dot with the butter and sprinkle cinnamon over all. Bake for 50-60 minutes, or until filling bubbles in center.

AMISH PIE

1 cup granulated sugar
1 cup packed brown sugar
2 tablespoons all-purpose flour
1/2 teaspoon ground cinnamon
1/4 teaspoon ground cloves
1/4 teaspoon salt
3 eggs, beaten
2 cups milk
1 tablespoon melted butter
1 teaspoon vanilla extract
1 (9 inch) unbaked pie shell

Preheat oven to 350°.
Beat all ingredients together very well and pour into pie shell. Bake for 45 minutes. The pie will quiver. The top will be puffed up when you remove the pie from the oven and will fall down after it has cooled. The pie is best served at room temperature.

KANSAS SUGAR CREAM PIE

1 (9 inch) unbaked pie shell
3 tablespoons flour
1 1/2 cups sugar
1/ 8 teaspoon salt
1 1/2 cups heavy cream
1 tablespoon butter, melted
1/2 teaspoon vanilla

1 dash nutmeg
1 dash cinnamon

Preheat oven to 425°. Combine ingredients. Pour into pie shell. Bake at 425° for 10 minutes, reduce heat to 325°, and bake for 45-55 minutes more, wrapping edges of pie crust with foil, if necessary, to prevent excessive browning. Cool completely before cutting.

AMISH BOB AND ANDY PIE

This pie is named for an Amish farmer's two prize gelding work horses. The farmer came in from working in the fields. He took a bite of this pie and said, "This pie is as good as Bob and Andy." The name has evolved through the years to where most Amish simply refer to it as "Bob Andy Pie."

1 cup granulated sugar
1 cup packed brown sugar
2 tablespoons all-purpose flour
1/2 teaspoon ground cinnamon
1/4 teaspoon ground cloves
1/4 teaspoon salt
3 eggs, beaten
2 cups milk
1 tablespoon melted butter
1 teaspoon vanilla extract

Preheat oven to 350°.Beat all ingredients together very well and pour into a 9-inch unbaked pie shell. Bake for 45 minutes. The pie will quiver. The top will be puffed up when you remove the pie from the oven and will fall down after it has cooled. The pie is best served at room temperature.

SUGAR CREAM PIE

3/4 cup granulated sugar
1/8 teaspoon salt
2 1/2 cups half-and-half

1/4 cup brown sugar
1/4 cup cornstarch
1/2 cup butter
1 teaspoon vanilla extract
1 (9-inch) lightly baked pie crust
1/8 teaspoon ground cinnamon

Preheat oven to 325°.
In medium saucepan, combine granulated sugar, salt and half-and-half. Bring to boil, stirring occasionally.
Meanwhile, in another saucepan, combine brown sugar and cornstarch. Gradually whisk in half-and-half mixture. Add butter. Cook mixture over medium heat, whisking constantly, for 5 minutes, or until boiling and thickened. Simmer 1 minute. Stir in vanilla extract.
Pour mixture into prepared pie crust; sprinkle with cinnamon. Bake 20 minutes or until top of pie is golden.
Place on wire rack. Pie filling will be very loose, but will thicken on cooling. Cool completely before slicing.

PECAN MAPLE SYRUP PIE

3 egg yolks
1 /4 cup flour
1 /2 cup water
1 cup maple syrup
1 pinch salt
1 teaspoon vanilla
2 tablespoons oleo
1 cup pecans

Beat egg yolks, adding flour and water alternately. Add syrup, salt, and vanilla. Pour into double boiler and add oleo. Cook until smooth, stirring as it boils. Add pecans. Place in 9-inch baked pie crust. Top with meringue consisting of: 3 egg whites and 3 tablespoons of maple syrup. Add syrup slowing while beating egg whites. Bake at 400° for 8-10 minutes.

AMISH THICK MILK PIE

3 eggs
1 cup molasses
1 cup granulated sugar
1/2 cup flour
1 teaspoon baking soda
3 cups thick sour milk
2 (9 inch) unbaked pie shells

Preheat oven to 400°.

Beat eggs. Add molasses. Combine sugar, flour and baking soda and add to egg mixture. Add thick milk. Pour into unbaked pie shells. Bake for 10 minutes; then reduce oven temperature to 325° and bake for 40 to 45 minutes.

Sprinkle top of pie with cinnamon, if desired.

SOUR CREAM APPLE PIE

1/3 cup flour
6 apples, sliced or cubed
2/3 cup sour cream
1 cup dark brown sugar
3/4 teaspoon cinnamon
3 tablespoons butter
2 to 3 teaspoons cider vinegar to taste
1-9" pie crust

Mix butter, sugar, and flour into crumbs. Sprinkle 1/4 of the crumbs into bottom of shell. Put in apples and heap them in the pie. Sprinkle apples with cider vinegar. Mix 1/2 the remaining crumbs with sour cream and pour mixture over apples. Mix the remaining crumbs with cinnamon and sprinkle on top. Bake in a 425° oven for 10 minutes until the crust browns slightly and then reduce heat to 350° for another 30 minutes. Serve with ice cream or old nippy cheese.

BLUEBERRY PIE

1 cup sugar
1 tablespoon cornstarch
1 /8 teaspoon salt
1 cup water
1 quart blueberries (washed and drained)
1 (9 inch) baked pie shell

Mix together sugar, cornstarch, salt, water, and 1 cup blueberries in saucepan. Cook and stir over low heat until thick. Add remaining blueberries and 1 tablespoon butter. Mix well. Let cool. Just before serving, pour into a pie shell. Garnish with whipped cream.

HOMEMADE MINCEMEAT PIE

1 1/4 pounds lean beef stew meat
2 3/4 pounds Granny Smith apples, cored and chopped
1/3 pound beef suet, coarsely ground
3/4 pound dark raisins
1/2 pound dark brown sugar
1/2 cup white vinegar
1/4 cup molasses
1/2 pound currants
1/2 cup apple juice
3/4 teaspoon ground cloves
3/4 teaspoon ground nutmeg
3/4 teaspoon ground allspice
3/4 teaspoon ground cinnamon
1/2 cup bourbon or rum, divided
2 crust recipe pie crust (homemade or store bought)
Water

Place stew meat in 2-quart pot; add water to cover. Bring to boil, cover, and simmer for 1 hour or until tender. Drain and put through meat grinder or processor until coarse.
Place the ground meat and the remaining ingredients, using only 1/4 cup of the liquor, in a 4-quart pot. Stir and bring to a simmer. Cover and cook gently for 1 hour, stirring occasionally. Remove from heat and allow to cool. (*Can be made several days ahead and refrigerated or frozen.*)

Prepare a 9-inch pie crust or use a store bought one. Place one-half of meat mixture into pie shell. Place remaining crust on top and seal to bottom crust. Put hole in center of top crust to allow steam to escape while baking. You can make a second pie or freeze remaining meat mixture for later use. Bake in preheated 375° oven for 1 hour or until brown and bubbly. Place pie on rack. Using a funnel, pour in the remaining 1/4 cup bourbon or rum. Tilt pie back and forth to incorporate. Serve warm.

HOLMES COUNTY PUMPKIN PIE

5 eggs
3/4 cup brown sugar
3/4 cup granulated sugar
3 cups milk
2 tablespoons flour
1 teaspoon cinnamon
1 teaspoon allspice
1 1/2 cups pumpkin
1 unbaked pie crust

Separate 2 eggs and beat the whites; set aside. Beat the two yolks and the remaining eggs well. Mix flour, spices, and sugars together and add to eggs. Add pumpkin, part of milk, and eggs whites to mixture, then add rest of milk. Pour into unbaked pie crust. Bake at 450° for 10 minutes to brown, then reduce heat to 325° and bake for 30 to 35 minutes.

INDIANA PUMPKIN PIE

1 cup cooked pumpkin
1 cup brown sugar
1/4 teaspoon nutmeg
2 tablespoons flour
2 eggs, beaten
1/3 cup water
1/4 teaspoon cinnamon

1/2 teaspoon salt
1 2/3 cup rich milk
 1 (9 inch) unbaked pie shell

Mix together sugar, spices, salt, flour & stir into pumpkin. Add milk and eggs. Pour into 9-inch unbaked pie shell. Bake in 450° oven for 10 minutes. Reduce heat to 350° and bake for another 25 to 30 minutes, or until filling is firm.

DELUXE PUMPKIN PIE

2 large eggs
3/4 cup sugar
1/2 teaspoon salt
1/2 teaspoon nutmeg
1/2 teaspoon ginger
1/2 teaspoon ground cloves
1 3/4 cups canned pumpkin
1 1/2 cups milk
1 unbaked pie shell

Beat the eggs; add sugar, spices, salt, and pumpkin. Mix well. Add milk; stir until mixture is smooth. Pour into pie shell and bake at 450° for 10 minutes. Reduce heat to 350° and bake for 45 minutes.

CORN MEAL PIE

1 egg, beaten
3/4 cup brown sugar
2 tablespoons butter

4 tablespoons cream
2 tablespoons corn meal
1/2 cup chopped nuts
1 unbaked pie shell

Mix all ingredients together and pour into pie shell. Bake at 325° for 35-40 minutes.

BUTTERMILK PIE

1 unbaked pie shell
1 1/2 cup of sugar
1/4 cup of flour
1/2 cup (1 stick) of butter or margarine, melted
3 eggs
1/2 cup buttermilk
1 teaspoon vanilla

Mix sugar and flour together. Cream the melted butter and sugar/flour mixture together. Add eggs, buttermilk and vanilla. Mix well and pour in pie shell. Bake at 350° for 45 minutes or until knife inserted in center comes out clean.

COTTAGE CHEESE PIE

2 cups cottage cheese
3/4 cup white sugar
1/4 teaspoon salt
2 egg yolks
1-3 ounces package egg custard mix
2 teaspoons vanilla extract
3/4 cup milk
1/4 teaspoon lemon juice
3 tablespoons butter, melted
2 egg whites, stiffly beaten

1 (9 inch) pie shell

Preheat oven to 325°. Blend together the cottage cheese, sugar, salt and egg yolks. Beat in the custard mix, vanilla, milk, lemon and butter or margarine. Fold in the egg whites. Pour the mixture into the pastry shell and bake in the preheated oven for 30 minutes. Let cool and serve.

MOLASSES PIE

3/4 cup flour
1 tablespoon butter
1/4 cup water, boiling
1 each pie crust
1/2 cup sugar
1/4 cup molasses
1/4 teaspoon baking soda

Mix the first 3 ingredients together, using the hands, and pinching mixture until very fine. Add the water to the molasses and soda and beat until this is foamy and rises. Pour this into a pie pan lined with pastry (a deep pan is preferable) and taking a spoon, mix the above crumbs well into the molasses filling. Bake at 325° for 30 minutes.

PENNSYLVANIA PIE

1 (9 inch) unbaked pie shell

Filling:
1/2 cup molasses
1/2 cup sugar
1 egg
1 cup water
2 tablespoons flour

Juice and rind of 1/2 lemon

Topping:
2/3 cup sugar
1/4 cup butter
1 egg
1/2 teaspoon baking soda
1/2 cup sour milk
1 1/4 cup flour

Combine ingredients for pie filling. Pour into unbaked pie shell. For topping, combine butter and sugar. Add egg and beat thoroughly. Add milk and sifted dry ingredients alternately. Spread topping over mixture in pie shell. Bake at 375° for 35-40 minutes.

MAPLE CREAM PIE

1 can sweetened condensed milk
2/3 cup real maple syrup
1 pinch of salt
1 cooked pie shell
1 whipped cream

Combine sweetened condensed milk, syrup, and salt in saucepan. Heat (stirring CONSTANTLY) over very low heat until bubbles in center. Cool slightly. Pour into precooked shell. Top with whipped cream when ready to serve.

BLACK WALNUT OATMEAL PIE

3 eggs, lightly beaten
1 cup brown sugar, packed
1/2 cup dark corn syrup
1/2 cup evaporated milk
1/2 cup quick-cooking rolled oats
1/2 cup coarsely chopped black walnuts
1/4 cup (4 tablespoons) butter, melted
1 teaspoon vanilla
1/8 teaspoon salt
1 unbaked pastry for single-crust pie

In large mixing bowl, combine eggs, sugar, syrup, milk, oats, nuts, butter, vanilla and salt, mixing well.

Line 9-inch pie plate with pastry, trim and flute edge. Place plate on oven rack and pour in filling. Protect edge of pie with foil to prevent over browning. Bake at 350° for 25 minutes. Remove foil. Bake for about 25 minutes more or until top is deep golden brown and slightly puffy. Filling with be slightly soft, but will firm up as it cools.

AMISH SAUERKRAUT SURPRISE CUSTARD PIE

Editor's Note: the sauerkraut ends up with a coconut-consistency; it's a very sur – pie – sing pie!

2 1/2 cups milk
3/4 cup sauerkraut, drained and chopped
1/2 cup white sugar
3 eggs
1 teaspoon vanilla extract
1/4 teaspoon salt
1 (9 inch) unbaked pie shell

Preheat oven to 425°.
In a large bowl, combine milk, sauerkraut, sugar, eggs, vanilla and salt.
Pour mixture into pie shell. Bake at 425° for 35 minutes, or until a knife inserted into filling 1 inch from the edge comes out clean.

WISCONSIN SOUR CREAM APPLE PIE

1 cup sour cream
1 large egg
3/4 cup sugar
2 tablespoons all-purpose flour
1/4 teaspoon salt
1 teaspoon vanilla extract
2 1/2 cups apples, peeled and diced

1 (9 inch) unbaked pie shell

Crumb Topping:
1/2 cup packed brown sugar
1/3 cup all-purpose flour
1/4 cup butter
1 teaspoon ground cinnamon

Beat sour cream and egg together; add sugar, flour, salt and vanilla. Mix until smooth. Stir in apples. Place mixture into pie shell and bake in preheated 400° oven for 25 minutes. Meanwhile mix topping ingredients until crumbly. Remove pie from oven and sprinkle the crumb topping over pie. Bake 20 minutes longer.

WET BOTTOM SHOO-FLY PIE
1 1/2 cups all-purpose flour
1/2 cup brown sugar
2 tablespoons shortening
1 teaspoon baking soda
1/2 cup dark corn syrup
1/2 cup molasses
1 cup boiling water
1/4 teaspoon salt
1 egg
1 (9 inch) unbaked pie crust

Crumbs:
In a medium bowl, combine flour, brown sugar and shortening. Cut ingredients together with a pastry cutter or rub with fingers until dough forms fine crumbs. Set aside. Preheat oven to 375°.
Filling:
In a medium bowl, dissolve the baking soda in the boiling water. Stir in corn syrup, molasses, salt and egg being sure to stir well. Pour 1/3 of mixture into unbaked pie crust. Sprinkle 1/3 of the crumbs over mixture. Continue this process until crumb mixture and liquid mixture have been used up. Bake for 10 minutes, then lower oven temperature to 350° and bake for an additional 30 minutes

CHOCOLATE PIE

2 cups water
1 tablespoon butter
1 tablespoon cocoa
2 1/2 heaping tablespoons cornstarch
3/4cup sugar

1 teaspoon vanilla
Pinch of salt
1 baked pie crust.

Heat water to boiling; add butter.
In separate bowl, while water is coming to a boil, mix together cocoa, cornstarch, sugar, and salt. Add enough water to make a paste, add to boiling water and stir. Remove from heat and add the vanilla. Stir and pour into baked pie crust.

RASPBERRY CREAM PIES

2 baked 9-inch pie crusts

Raspberry Filling:
1 cup water
1/2 cup granulated sugar
1 tablespoons cornstarch
1/8 teaspoon salt
1/4 cup water
1/4 teaspoon lemon juice
1 (3 ounce) box raspberry Jell-O
3 cups fresh raspberries

Vanilla Filling:
3 cups milk, scalded
1/2 cup cornstarch
1 1/3 cups granulated sugar
1/8 teaspoon salt
1 teaspoon vanilla extract
3 egg yolks
1 cup milk

Raspberry Filling directions: Heat the 1 cup water and sugar. Mix the cornstarch, salt, 1/4 cup water and lemon juice. Add to water-sugar mixture. Boil until mixture is clear. Add Jell-O and stir to dissolve. Add raspberries and let cool.

Vanilla Filling directions: Place 3 cups milk in kettle. Heat to scalding.
Mix cornstarch, sugar, salt, vanilla, egg yolks and 1 cup milk. Slowly blend into hot milk, stirring constantly until thick. Cool.
Pour 2 cups Vanilla Filling into each pie shell. Place 2 cups Raspberry Filling on top of Vanilla Filling.

Serve topped with whipped cream.

MARSHMALLOW PIE

1/2 cup milk
1/4 cup butter
1-10ounces package miniature marshmallows
1 cup whipped topping
Chopped chocolate pieces or chocolate syrup
1 graham cracker pie crust

Melt marshmallows in pan over low heat with milk and butter. Let cool, then mix in whipped topping. Spoon into pie crust and top with chocolate pieces or chocolate syrup to your taste.

PEAR PIE

2 cups pears, chopped
1/2 cup sugar, plus 1/3 cup sugar
1 egg, beaten
1 cup cream
2/3 cup plus 1 tablespoon flour
1 teaspoon vanilla
1/8 teaspoon salt
1/4 cup margarine
1 unbaked pie shell

Combine pears, 1/2 cup sugar and egg. Add cream, 1 tablespoon flour, vanilla, and salt. Mix together and pour into pie shell. Bake at 350° for 15 minutes.
Combine remaining flour, 1/3 cup sugar and margarine. Sprinkle crumb mixture on top of pie. Return to oven; continue baking for 30 minutes or until browned on top.

AMISH PEACH PIE

4 cups peaches, peeled and sliced
1/2 cup granulated sugar
1/4 teaspoon salt
2 1/2 tablespoons tapioca
1 (9-inch) unbaked pie shell

Crumbs:
2 1/2 tablespoons butter or margarine; melted
1/4 cup all-purpose flour
1/2 teaspoon cinnamon
1/3 cup brown sugar

Preheat oven to 425°.
Gently mix together peaches, sugar, salt and tapioca. Let blend for 5 minutes before spooning into pie shell.
Mix *crumb* ingredients well and sprinkle over pie shell. Bake for 45 to 50 minutes.

AMISH CARAMEL PIE

3 cups brown sugar
3 cups water
2 tablespoons butter
1 cup all-purpose flour
3 cups milk
6 egg yolks

Boil brown sugar, water and butter together for 2 or 3 minutes for a good strong caramel flavor. Slowly stir eggs, milk and flour into boiling syrup, stirring constantly until it comes back to a boil. Remove from heat; cool 5 minutes and stir once. Pour into 2 baked pie shells. Top with either a meringue from the egg whites or allow pie to cool and top with whipped cream.

BROWNED BUTTERSCOTCH PIE

1 baked 9-inch pie shell
1 1/4 cups brown sugar, firmly packed
1/3 cup all-purpose flour
2 cups milk, scalded
1/2 teaspoon salt
1/4 cup butter
1 tablespoon vanilla extract
Whipped cream topping

Set out pie shell. In a large, heavy saucepan, combine 1/2 cup of the brown sugar and the water. Bring to a boil over medium heat, and continue cooking for about 3 1/2 minutes or until mixture is thick and bubbly. Set aside.

In a large mixing bowl, beat the eggs until frothy. Add the flour and blend until smooth. Add the remaining 1/4 cup brown sugar and blend. Slowly add the scalded milk, stirring constantly.

Over low heat, reheat the sugar-water mixture until liquefied again. Pour the milk-egg mixture into it, stirring with a rubber spatula all the time. Cook over medium-low heat until mixture bubbles up and is very thick, 3 to 5 minutes.

Add salt, butter and vanilla extract. Let cool for 15 minutes, then pour into baked shell. Chill in refrigerator. Frost the pie with the whipped cream topping.

PENNSYLVANIA PLAIN RAISIN PIE

2 cups raisins
1 cup water
1/2 cup light corn syrup
1/4 cup cornstarch
1/4 teaspoon salt
Pastry for 2-crust, 9-inch pie

Place raisins in a saucepan; add water. Simmer raisins for 10 minutes. Mix corn syrup with cornstarch and salt; add to raisin mixture in saucepan. Continue cooking, stirring constantly, for about 3 minutes. Cool; pour into pastry-lined pie pan. Arrange top crust over top or cut into strips and arrange strips in a lattice fashion. Bake at 425° for 20 minutes, until pie is done and pastry is golden brown.

AMISH SOUR CREAM RAISIN PIE
2 cups sour cream
4 egg yolks
1 3/4 cups sugar
4 teaspoons flour (heaping)
1 1/2 cups raisins
1 baked 10-inch single-crust pie shell

Meringue:
Place 12 medium egg whites in bowl. Add 1/4 heaping teaspoon cream of tartar. Stir until stiff (for those with electricity, a high-speed mixer can work here). Add 2 cups powdered sugar and beat until meringue forms soft peaks. Using a rubber spatula, spread onto pie filling, making a good seal with the edge of the crust.

Stir the sour cream and yolks in a heavy medium saucepan. Add the sugar. Dump in the flour, then raisins, and mix using a wooden spoon. Cook over medium heat until the raisins are plump and the filling is glossy (about 5 minutes after a full boil, or just a little longer, depending on your burner).

Cool the filling slightly, then pour into the crust. Preheat oven to 400°. Prepare meringue and spread onto pie. Put the pie in the oven. Watch closely for 15 to 20 minutes, then take it out when the peaks are golden brown. Let it cool. Serve immediately.

SOUR CREAM RAISIN PIE

1 baked pie shell

Filling:
1 cup raisins
1 cup water
1 cup sour cream
1 cup sugar
1/4 teaspoon salt
2 rounded tablespoons flour
3 egg yolks, beaten
1 teaspoon vanilla

Meringue:
3 eggs whites
1/2 teaspoon cream of tartar
5 tablespoons sugar

Stew raisins in water until tender. Save 1/2 cup of juice. Add sour cream, sugar, salt, flour and cook for 5 minutes. Add beaten egg yolks and cook until thick. Remove from heat and add vanilla. Pour filling into pie shell and top with meringue.

For meringue: Beat egg whites with cream of tartar until stiff. Add sugar and beat in. Pour in pie shell. Bake at 350° for 12 to 15 minutes or until meringue is golden brown.

SIMPLE SOUR CREAM RAISIN PIE

2 tablespoons cornstarch
3/4 cup sugar

1/4 teaspoon salt
1 teaspoon cinnamon
1/2 teaspoon nutmeg
1/4 teaspoon cloves
2 egg yolks, beaten
1 cup sour cream
1 cup raisins
2 egg whites
4 tablespoons sugar
Baked pastry shell

Mix together in top of double boiler the first six ingredients, add well beaten egg yolks and mix well. Add sour cream and raisins and cook over hot water till thick. Cool and pour into baked pastry shell. Top with meringue made by beating 2 egg whites until stiff, and adding 4 tablespoons sugar gradually. Brown in a 325° oven for 15 minutes.

CHEESE CUSTARD PIE

1-16ounces container of small curd cottage cheese
1 cup sugar
3 tablespoons flour
1 teaspoon lemon juice
3 egg yolks
3 egg whites
1-12 ounces can evaporated milk
1 1/2 cups milk
1/4 teaspoon nutmeg
1/8 teaspoon salt
Pastry for 2 - 9" pie shells

Empty cottage cheese into a colander or strainer to drain the watery liquid for about an hour. Meanwhile, line 2 - 9" pie pans with pie dough; Flute the top edges. Mix sugar, flour, salt and nutmeg together; set aside. Place drained cottage cheese in a separate bowl. Add lemon juice, egg yolks, milk, and evaporated milk. Combine with the dry ingredients. and mix well. In another bowl, beat egg whites until stiff. Fold beaten egg whites into pie mixture. Pour into 2 unbaked pie shells. Bake at 400° for 15 minutes. Reduce oven temp to 350° and bake for 25 minutes or until a knife inserted in the center comes out clean. Cool on wire racks at room temperature, then refrigerate.

Amish

Other Desserts

PEANUT BUTTER BALLS

2 cups peanut butter
1 cup honey
1/2 cup raisins (optional)
1/2 cup *date sugar

Mix and roll first 3 ingredients into 1 inch balls. Roll in date sugar and put into freezer.
* Date Sugar is found in most health food stores - or you can replace it with regular granulated sugar.

CARAMEL CORN

1 cup Butter
2 cup brown sugar
1/2 cup light corn syrup
1 teaspoon salt
t teaspoon baking soda
1 teaspoon vanilla
1 cup popping corn (un-popped)

Make Pop Corn in Air Popper. Set aside.
Then melt butter, stir in brown sugar, corn syrup and salt. Bring to a boil, stirring constantly!! Then boil without stirring for 5 minutes. Remove from heat and stir in the baking soda and vanilla. Gradually pour syrup over the pop corn.
Mix well and pour into large baking pans.   Bake in a 250° oven for 45 minutes.
Stir every 15 minutes. Let cool.

CHOCOLATE SANDWICH GOBS

2 cups granulated sugar
1 cup margarine, melted
2 eggs, beaten
4 cups all-purpose flour
1/2 cup unsweetened cocoa
2 teaspoons salt
1 cup soured milk or buttermilk
2 teaspoons baking soda
Your favorite chocolate icing

Combine sugar, margarine and beaten eggs in large mixing bowl; mix well. Sift together flour, cocoa and salt; add dry ingredients to creamed mixture, alternately with sour milk or buttermilk. Add baking soda to dough, mixing until combined.
Drop by spoonfuls on greased cookie sheets. Bake at 350° for 10 minutes. Remove to cooling rack. Form sandwich cookies by placing icing between two cookics.

APPLE DUMPLINGS

6 apples, peeled and cored
Lemon juice
1/2 cup white sugar
1 teaspoon cinnamon
Brown sugar
Butter
2 pie pastries for 2 pies

Sugar sauce:
3 cups water
4 1/4 cups sugar
2 teaspoons vanilla
2 tablespoons butter
1/4 teaspoon nutmeg

Roll out pastry and cut into squares enough to cover each apple completely. Peel and core apples. Roll in lemon juice. Then roll in white sugar and cinnamon combined. Place on pastry square. Stuff core cavity with brown sugar, butter, brown sugar, butter, etc., in equal parts. (The amount depends on size of core cavity, just stuff full.) Fold pastry up around apple to completely enclose it. Place in pan. Prepare sugar sauce by mixing water, sugar, vanilla, butter, and nutmeg, and boiling for 1 minute. Let cool slightly. Pour over apples. Bake in 375° oven for 1 hour. Serve warm. Note: Apples may be frozen in pastry (before sugar sauce is added) if desired. Good dessert to make ahead if you have lots of apples around. Just let thaw when ready to use, and cover with sugar sauce and bake.

BAKED ORCHARD APPLES

8 to 10 apples, cored, peeled & halved
3/4 cup white sugar
3/4 cup brown sugar
1/2 cup flour
1 teaspoon cinnamon
2 teaspoon butter, melted
1 cup water

Place apples in greased 9 x 13 dish. Mix remaining ingredients together in the order given, in a pan, and bring to a boil. Simmer, stirring until thick. Pour syrup over apples and bake at 350° for 35 to 40 minutes or until tender.

AMISH APPLE GRUNT

1/2 cup sugar
2 tablespoons butter
1 egg
1 cup flour
1/2 teaspoon salt
1/2 teaspoon baking powder
2 cups sliced apples
1/2 cup sour milk or buttermilk
1/2 teaspoon vanilla
6 tablespoon brown sugar
1 1/2 tablespoon flour
1/2 teaspoon cinnamon
1 1/2 tablespoons butter

Cream sugar and butter together; add eggs and mix. Blend flour, salt and baking powder together and add to mixture. Mix soda with milk and vanilla; mix all together. Add apple slices and pour batter into a buttered baking dish. In a separate bowl, combine brown sugar, flour, cinnamon and butter, mixing until crumb texture; sprinkle over apple batter. Bake at 375° for 35 to 40 minutes. Serve hot with rich milk, Half and Half or a scoop of ice cream. Makes 4 generous servings

PENNSYLVANIA APPLE POT PIE

6 baking apples
4 cups flour
1 teaspoon cinnamon
Water
1/4 pound lard
1/4 teaspoon salt
1/8 pound butter

Make a dough of the lard, flour and salt, adding enough water to moisten and hold together. Roll out like pie dough and cut into 2 inch squares.
Wash and peel apples and cut into eighths. Put alternate layers of apples and dough into kettle, sprinkling each layer of apples generously with sugar and adding a little cinnamon. Have top layer of dough, dot with butter and fill kettle half full of water, cover and cook over a low flame until apples are soft. Serve with milk or cream

GERMAN BROWNIES

2 eggs
3/4 cup sugar
1 teaspoon vanilla
1/2 cup butter or margarine, melted
3/4 cup chocolate, ground
2/3 cup unsifted flour
1/4 teaspoon baking powder
1/4 teaspoon salt
1/2 cup walnuts, chopped

Heat oven to 350°. Using a spoon, stir eggs with sugar and vanilla; add
butter. Sift ground chocolate with flour, baking powder and salt. Stir into egg mixture;
add nuts. Spread into greased 8 or 9" square pan. Bake at 350° for 20-30 minutes. For
extra chewy brownies, use 8" pan and less baking time. For cake like brownies use 9" pan
and longer baking time. Cut into squares and serve

MINT BROWNIES

Brownie:
1 Stick oleo
1 cup sugar
4 eggs
1 can chocolate syrup
1 cup plus 2 tablespoons flour
½ cup chopped nuts

Mint Layer:
1 stick oleo, melted
4 cups powdered sugar
¼ cup milk
green food coloring
1 teaspoon crème de menthe flavoring

Chocolate Layer:
1 stick oleo
1-12ounces package chocolate chips

Mix together brownie ingredients and pour into 9 x 13 greased and floured pan. Bake at
350° for 20-25 minutes (just until center is set). Cool completely.
Mix together ingredients for mint layer and spread on cooled brownies. Melt and mix
two ingredients for chocolate layer and pour over mint layer. Do not cut brownies until
chocolate layer is set. They can cool in the refrigerator

SIMPLE CHOCOLATE FUDGE

1 1/4 cups margarine or butter
3-1 ounces squares unsweetened chocolate
1/2 cup Karo light or dark corn syrup
1 tablespoon water
1 teaspoon vanilla extract
1 pound confectioners' sugar
1 cup chopped nuts

Grease an 8-inch square baking pan.

Melt margarine or butter and chocolate over low heat. Stir in corn syrup, water and vanilla extract. Remove from heat. Stir in remaining ingredients until smooth. Turn into pan. Cool and cut into squares.

AMISH COOKED FRUIT DRESSING

2 eggs
2 tablespoons flour
1 cup brown sugar, firmly packed
1 cup water
2 tablespoons vinegar
1/4 teaspoon salt

Combine all ingredients together in a saucepan and bring to a boil. Remove from heat.

When cool, pour over sliced apples, sliced bananas or grapes. Nuts and dates may be optionally added.

CHOCOLATE TAPIOCA

1 tablespoon cocoa
1/2 cup granulated sugar
1 teaspoon vanilla extract
1/2 teaspoon salt
1 cup tapioca
1/2 cup brown sugar
Boiling water (4-6 cups)

Place cocoa and granulated sugar in a large bowl. Add 2 cups boiling water and stir. Add vanilla, salt, tapioca and brown sugar. Add more water until right consistency.

Serve hot with cold milk.

AMISH CHURCH SPREAD

1/2 cup creamy peanut butter
1/4 cup marshmallow cream
1 cup light corn syrup

In a mixing bowl, stir all the ingredients together until combined. Place in a covered container. Store in refrigerator.
Bring to room temperature to serve as a bread spread or ice cream topper. Makes about 1 1/2 cups.

BUTTERY BLUEBERRY COBBLER

1/2 cup butter or oleo
1 cup flour
1 1/4 cups sugar
1 teaspoon baking powder
1/2 cup blueberries
1 tablespoon lemon juice

Melt butter or oleo and pour into approx. 9 x 9 baking dish. Sift together 1 cup sugar, flour, and baking powder; stir in milk. Pour batter evenly over butter; do not stir. Mix berries, ¼ cup sugar and lemon juice in saucepan. Heat to boiling, stirring continuously. Pour evenly over batter; do not stir. Bake at 350° for 45 minutes. Serve with whipped cream, ice cream or plain.

HOMEMADE CHERRY PUDDING

1 can sour pitted cherries
1 cup sugar
1 tablespoon butter
2 cups flour
2 teaspoons baking powder
3/4 cup milt
Pinch of salt

Drain cherries. Add sugar and butter. Stir well. Add flour and baking powder, sifted together. Put aside 2 tablespoons for crumb topping. Add milk and salt. Stir by hand. Batter will be a little stiff. Layer batter and cherries in greased, floured 9x13 pan. Bake at 350° for approximately 40 minutes, until lightly browned. Check with toothpick. Serve warm in a bowl, with sugar and milk (as shortcake).

FLORIDA PUDDING

Crust:
1 cup flour
1/2 cup chopped nuts
1/2 cup oleo or butter

Second layer:
1 cup powdered sugar
1 cup whipped topping
1-8oz package cream cheese

Third Layer:
1 box butterscotch pudding, prepared by directions on the box
or 2-21/2 cups homemade butterscotch pudding

Mix crust ingredients and press in bottom of 9 x 9 pan; bake until browned, then cool.
For the second layer, cream the cheese and powdered sugar together and fold in whipped topping. Spread over cooled crust.
Top with butterscotch pudding and serve.

MOTHER'S BUTTERSCOTCH PUDDING

1 cup brown sugar
2 tablespoons butter
1 tablespoon flour
1 egg yolk
1 cup milk
1 teaspoon vanilla
1/4 teaspoon salt

(Best done in a cast iron skillet.) Boil sugar and butter together until soft. Beat the egg yolk well and add it to the flour, milk, vanilla, and salt. Carefully stir a little at a time into the sugar mixture. Cook, stirring constantly, until thick and bubbly.

RHUBARB CRUNCH

2 cups flour
1 1/2 cups uncooked oatmeal
2 cups brown sugar
1 cup (2 sticks) melted butter
2 teaspoon cinnamon
8 cups fresh rhubarb, diced
2 cups sugar
4 tablespoon cornstarch
2 cups water
2 teaspoon vanilla

Combine flour, oatmeal, brown sugar, melted butter and cinnamon together until crumbly. Press half of the crumb mixture into a greased 9 x 13 cake pan; set remaining crumb mixture aside to be used a topping.. Combine sugar, cornstarch, water and vanilla in a saucepan. Cook on medium-high heat until thick and clear (about 10-12 minutes), stirring constantly. Place diced rhubarb on top of crumb mixture in cake pan. Pour cooked mixture over top rhubarb, then top with remaining crumb mixture. Bake at 350° for 1 hour.

GRAHAM CRACKERS

2 cups brown sugar
1 cup butter or shortening
1 cup milk or cream
1 teaspoon baking soda
4 cups whole wheat flour
2 cups pastry flour
1 teaspoon salt
1 teaspoon baking powder
1 teaspoon vanilla

Mix all ingredients together; roll out thin. Prick with a fork and cut into squares. Bake at 375o until golden brown.

BROWN SUGAR DUMPLINGS

Dumplings:
1 cup sugar
2 teaspoons butter
2/3 cup milk
2 teaspoons baking powder
2 cups flour
1 cup raisins
1 cup dates, chopped

Syrup for boiling:
2 cups brown sugar
½ cup butter
2 cups water

Mix all dumpling ingredients together. Mix syrup ingredients together in a saucepan and bring to a boil. Drop dumplings by heaping teaspoon into boiling syrup and cook till firm.

AMISH OATMEAL CANDY

2 cups sugar
1/2 cup (1 stick) margarine
1/2 cup milk
2 tablespoons cocoa
2 1/2 cups oatmeal

Mix sugar and cocoa together. Add milk. Bring to a boil and boil for 1 1/2 minutes. Remove from heat and add oatmeal. Drop on wax paper with a teaspoon and allow to cool.

AMISH CRACKER JACK

3 quarts popped corn
1 cup brown sugar
1/4 cup butter
1/4 cup corn syrup
1/2 teaspoon salt
1/8 teaspoon cream of tartar

Cook all except popped corn over medium heat to 260° or "hard ball" on a candy thermometer. Boil slowly. Pour over popped corn in baking dish . Put in 200 degree oven for 1 hour. Stir 3 times during the hour. Let cool.

CASHEW CRUNCH

1 cup butter
1 cup granulated sugar
1 tablespoon light corn syrup
2 cups cashews

Lightly butter a cookie sheet.

In a heavy pan cook and stir the butter, sugar and corn syrup over low heat until the butter is melted and the mixture comes to a boil. Cook until it starts to turn golden brown (290° on a candy thermometer) like peanut brittle.

Remove from heat. Quickly stir in cashews. Pour out onto prepared cookie sheet. Cool completely.

Break up into pieces. Store in airtight container.

Amish

Miscellaneous

PARTY MIX

2 1/2 cups Cheerios
2 1/2 cups Rice Chex
2 1/2 cups Corn Chex
2 1/2 cups Wheat Chex
1/2 bag pretzel sticks
1 pound peanuts
1 teaspoon onion salt
3/4 cup salad oil
1 teaspoon celery salt
Pinch of garlic salt
2 tablespoons Worcestershire sauce

Mix cereals, pretzels and nuts in a large bowl. Add spices and sauces. Pour salad oil over mixture. Bake at 250° until hot.

SLUSH PUNCH

2 pints orange sherbet
1 pint lemon sherbet
1-2 liter bottle ginger ale
1-2 liter bottle 7-UP

Mix all ingredients together until sherbet is dissolved.

HOT CHOCOLATE MIX
This mix makes a great Christmas gift.

1-8qt. box Carnation instant milk
2 pounds powdered sugar
1-12ounces jar Cremora
2 pounds Nestle Quik

Mix all ingredients together. Use 1/3 cup mix per cup of hot water.

LEMON SAUCE

1 tablespoon butter
1/2 cup sugar
4 teaspoons all-purpose flour
1/4 teaspoon nutmeg
1 cup water
1 teaspoon grated lemon peel
2 tablespoons lemon juice

In a small saucepan, melt butter. Stir in sugar, flour and nutmeg to blend. Add water, lemon peel and juice. Cook and stir over medium at until thickened and bubbly. Cook for 1 minute longer.

PENNSYLVANIA TOMATO KETCHUP

1 gallon ripe tomatoes
1 large onion
1 tablespoon pickling spice
2 teaspoons salt
1 cup sugar
1 cup vinegar

Chop tomatoes; cook with onion and pickling spice in saucepan until well done. Drain off juice. (You can save juice for use in chili or vegetable soup.) Put tomato and onion through sieve. Return to saucepan and add salt, sugar and vinegar. Cook until thick. Makes 3-4 14ounces bottles.

HOMEMADE SALAD DRESSING

1/2 cup cream
1 tablespoon brown sugar
2 tablespoon white sugar
1 tablespoon vinegar

Put cream in bowl. Add sugars and vinegar. Mix well. Use over lettuce, endive, escarole or spinach.

AMISH SALAD DRESSING

1 1/4 cups sugar
1/3 cup vinegar
3/4 teaspoon celery seed
3/4 teaspoon pepper
1/2 teaspoon salt
1/4 cup mayonnaise
1/4 cup mild prepared mustard
1 tablespoon onion
3/4 cup chilled salad oil

Put all ingredients in blender, except salad oil, and blend until well mixed. Gradually add well chilled salad oil and blend until thick and emulsified. This is good both on lettuce and cabbage salads. Yields: 1 pint.

THE GERMAN BAPTISTS

COOKING & CULTURE

By Kevin Williams
Editor

Close cousins of the Amish live quietly in pockets of rural America. Their population is smaller, but their devotion is no less. Clustered in communities along the Ohio-Indiana border, central Kansas and near Modesto, California live a Plain People known as the German Baptists. Sometimes called "Dunkers" because of their triple-immersion adult baptismal rites, German Baptists share similarities with the Amish. Their dress is considered "plain", with simple hand-sewn dresses being the apparel of choice for women and solid-colored shirts and trousers for men. Women will occasionally be seen wearing floral print dresses. Colors are more ample and pronounced in German Baptist clothing than in the Amish and Old Order Mennonites.

As an 18-year-old cub reporter with the *Middletown Journal* in Middletown, Ohio, I found myself intrigued with the "Amish-looking" people I would sometimes see at area malls. I decided to find out more about them and ended up discovering a wonderfully rich culture: the German Baptists.

The German Baptists come from the same Anabaptist religious movement that the Amish and Mennonites grew from. The German Baptists, however, embrace more technology than their Amish cousins. German Baptists drive cars and work in modern occupations (I met one German Baptist truck driver in Preble County, Ohio). As noted, they dress plainly. Another big difference between the German Baptists and the Amish: the German Baptists have church buildings (called meeting-houses). Old Order Amish hold church services in their homes. A German Baptist minister once told me that their church is always trying to strike a balance between the old ways and "too much acculturation."

German Baptist cooking is similar to Amish fare, although I have found it to be a touch more elegant with more of the old European touches.   While plenty of Amish restaurants abound, I'm unaware of any German Baptist restaurants. Beare's Restaurant and Catering in West Alexandria, Ohio (77. E Dayton St.) has a menu that seems to have been influenced by regional German Baptist cuisine. West Alexandria sits in the heart of Ohio's Dunker country and this restaurant is a great way to experience of taste of the culture. Tiny farm markets throughout southwest Ohio (check Barn N Bunk, 3677 Wayne Madison Road Trenton, OH) occasionally sell pies, breads, and butters produced by local German Baptist bakeries. Or head to your kitchen and whip up one of these old German Baptist creations:

German Baptist

Breakfasts

DUTCH MOLASSES CRUMB CAKE

2 cups brown sugar
1 cup butter and lard mixed
3/4 cup New Orleans molasses

3 cups flour
1 cup hot water
1 teaspoon baking soda

Mix sugar, flour, and butter together using the hands to make into crumbs. Put molasses into separate vessel, and into it stir the soda and hot water. Put soda into cup and dissolve with a little water, before filling the cup to the full. Put pastry into deep pans, pour in the molasses mixture, and sprinkle the crumbs over the top, and bake at once in a moderate oven (375° for 30 minutes or until done).

CORN MEAL DOUGHNUTS

2 cups cornmeal, sifted
1 cup white flour
1 tablespoon sugar
1 teaspoon salt
3 eggs
1 teaspoon baking soda dissolved in: 1 cup buttermilk (or sour milk)

Mix all ingredients together. Drop from a spoon into hot grease/oil.
Fry brown on both sides and serve hot with molasses.

FRIED BREAD

1 egg
Pinch of baking soda stirred into 1 tea-cup of sour cream
Salt
Slices of bread not less than 3-4 days old

Butter for frying

Beat egg to a froth. Pour in sour cream and baking soda mixture. Add a little salt and stir together well. Put a lump of butter into the frying pan and let it get real hot. Take slices of bread or bread scraps and dip into the batter. Fry a nice brown on both sides and serve with hot syrup.

MILK TOAST

3 slices bread, lightly toasted
3 pints fresh milk
Pinch of salt
1 tablespoon sugar

Heat milk until it comes to a boil, add salt and sugar. Put the bread into a dish, pour the hot milk over it. Let it stand for a few minutes before serving.

OMELET

4 eggs, well-beaten
1 cup of rolled cracker crumbs
2 1/2 cups of milk
Salt to taste

Mix well and pour into buttered skillet. Cook on low heat slowly until brown on both sides.

BREAKFAST DISH

2-3 cups cold leftover meat, chopped fine
2 medium potatoes, boiled with skins on, then chopped
1 1/2 cups bread crumbs, moistened with hot water
2 eggs

Salt and pepper
Butter for frying

Beat all ingredients together. Form batter into patties and fry until brown.

BREAD PANCAKES

Week old, dry bread
Sour milk or buttermilk
1 egg
1 teaspoon baking soda
1 cup flour
Salt

Cut dry bread into small pieces and place in dish. Cover bread with buttermilk and let stand overnight. Mash up fine in the morning. Stir in egg, baking soda, flour, and salt. Add more flour if needed to make a thick pancake batter. Fry in butter until bubbly on top. Flip over and fry till done.

German Baptist

Soups

GERMAN TOMATO SOUP

1 medium sized onion, chopped fine
2 medium sized carrots
1 small head of celery
1 quart of tomatoes

salt and pepper to taste
butter
2 tablespoons of flour moistened with water

Stew all of the above gently over low heat for one hour. Remove from heat and put through a sieve to remove seeds, pressing as much of the pulp through as possible. Return to heat, adding salt and pepper to taste and a generous amount of butter. When boiling stir in 2 tablespoonfuls of flour moistened with water, to the consistency of cream.

BEAN SOUP

1 pint beans
1 1/2 quarts ham broth
1 tablespoon flour mixes into 1 cup of cream
Pinch of parsley

Stew beans until tender. Mash through a colander to remove shells. Add to ham broth and bring to a boil. Add flour and cream mixture. Add a pinch of parsley and serve with toasted bread squares.

CARROT SOUP

3 medium carrots, peeled and sliced very thin
3 Irish potatoes, sliced thin
1 onion, shredded fine
2 tablespoons butter
Red pepper and salt for seasoning
Water
1 1/2 pints rich milk or cream
Small bunch of minced parley

Boil carrots in clear water for 20 minutes; drain carefully. Add potatoes and onion. Add butter, pepper, and salt. Cover with enough water to boil all tender. Add rich milk and parsley. Serve very hot.

CORN SOUP

3-4 pints sweet milk
2 tablespoons butter
Pepper

1 can corn
Salt

Put a small amount of water into a pan

German Baptist

Salad/Sides

ONION SALAD

6 large onions, chopped
3/4 cup apple cider vinegar

3 tablespoons sugar
1/2 teaspoon salt
1/2 teaspoon cinnamon
1/2 teaspoon Allspice

Soak onions in cold water for 10 minutes. Drain. Pour rest of ingredients over onions and stir several times so the ingredients are evenly distributed. Let stand 30 minutes and serve.

COLD BEAN SALAD

1 can kidney beans
4 hard boiled eggs
1 large sour pickle
1 large onion
2 tablespoons vinegar
Salt and pepper

Drain kidney beans and place in bowl. Chop eggs (reserve one to garnish), pickle and onions separately and add to beans. Add vinegar, salt and pepper and chill.

RED BEET SALAD

3-4 red beets
Butter or ham gravy for frying
1/2 cup cream
1 egg
Vinegar
Salt and pepper

Wash and cook red beats until tender (pierce with a fork before cooking). Peel and chop not to fine. Before frying, prepare a dressing with the cream, egg, enough vinegar to suit your taste, and salt and pepper. Then fry beets lightly in butter or gravy, stirring all the time. Pour dressing over beets in pan and cook for a few minutes, stirring briskly. Turn out in a dish at once and serve.

CABBAGE SALAD

1 large red pepper, chopped
12 ears sweet corn, cut off cob
1 head cabbage, shredded

3 onions, chopped
4 tablespoons ground mustard
2 tablespoons salt
1/2 cup sugar
Vinegar to season

Mix all ingredients into a pan and cook about 15 minutes or until corn is tender. Can while hot. Makes 3 quarts.

GREEN TOMATO PIE

1 pint green tomatoes, sliced
5 tablespoons sugar
1 tablespoon vinegar
1 teaspoon flour
1 1/8 teaspoon allspice
1 1/8 teaspoon ground cloves
1 1/8 teaspoon cinnamon
2 unbaked pie shells

Line a pie pan with first pie shell and add 1 pint of green tomatoes, pared and sliced. Sprinkle sugar over pie then add vinegar. Next add flour, allspice, cloves, and cinnamon. Place second pie shell on top and bake in a moderately hot oven (400°) for about 30 minutes.

POTATO CAKES

1 cup potatoes, mashed
1/2 cup lard
1/2 cup butter
1 yeast cake

1 teaspoon salt
3 eggs
2 cups sugar

Mix 1 cup of sugar and the hot mashed potatoes; after cooling add 1 cup flour and dissolved yeast; beat and let rise 3 hours. Mix lard, butter, 1 cup of sugar, eggs and salt; mix this with the sponge and beat vigorously. Stir until stiff. Let rise overnight; roll out, cut, place biscuits in pans, spread with melted butter, and sift brown sugar over them. Bake 20-30 minutes in moderate oven (375°).

OLD GERMAN POTATO BALLS

1 1/2 cups fresh bread crumbs
1/2 medium onion, diced
2 stalks celery, diced
2 tablespoons fresh parsley, chopped
1 cup milk
2 large eggs, beaten
2 1/2 cups mashed potatoes, plain
1 stick plus 4 tablespoons butter, melted, divided
Salt and pepper to taste

Use a blender or food processor to make soft bread
crumbs. Mix all of the above together except 4 tablespoons
butter. Use an ice cream scoop to form mixture into balls. Place balls side by side in a buttered casserole
dish. Drizzle the 4 tablespoons of melted butter
over formed balls. Bake, uncovered, for 20 minutes, or until lightly browned in a pre-heated 350° oven.
During last 2-3 minutes, place under broiler.

SCALLOPED POTATOES

3 large potatoes, sliced
2 cups bread crumbs
Butter
Milk

Salt and pepper

Bake bread crumbs in oven till light brown. Pare and slice potatoes. Layer ingredients into a casserole dish as follows: potatoes, bread crumbs, salt and pepper, butter, then keep layering until all is used. Then cover with milk and bake at 400° for 30 minutes.

BREAD BALLS

2 cups bread crumbs
Salt and pepper
2 tablespoons butter
1 egg
Enough sweet milk to moisten above ingredients

Mix together first four ingredients. Add sweet milk till moistened. Roll in balls and bake with a roast or other meat. Serve with gravy.

ASPARAGUS FRITTERS

1 teaspoon butter
1 teaspoon baking powder
2 eggs
Pinch of salt
1/4 cup sweet milk
Enough flour to make drop batter
Cooked asparagus, cooled and chopped

Make a batter out of everything but the asparagus. Stir in asparagus and drop by heaping spoonfuls into hot lard.

MACARONI AND TOMATOES

1 can tomatoes
Pinch baking soda
1 cups cooked macaroni noodles
Flour and milk for thickening

Salt, pepper, and sugar for seasoning

Cook tomatoes and baking soda in a large pot for 3-4 minutes. Stir in macaroni. Add thickening made from flour and milk. Season with salt, pepper, and sugar and serve.

FRIED CARROTS

3-4 carrots
Water
Salt
Flour
2 tablespoons lard
Sugar

Peel and slice carrots 1/4 inch thick. Cover with salted water and boil until tender. Drain; roll each slice in flour and fry in lard, browning both sides. Place in a hot dish and sprinkle each layer with sugar.

BAKED ONIONS

6 large onions, peeled
Salt
Butter
Rich Milk
Bread crumbs

Cover whole onions with water and boil 10 minutes. Drain and cover again with water and add 1/2 teaspoon salt and boil till tender, but firm. Drain; place in a baking dish and sprinkle with salt. Put a lump of butter on each and cover with rich milk. Give a heavy sprinkling with bread crumbs and bake till light brown.

German Baptist

Breads

COFFEE BREAD

2 eggs
3 cups of flour
1 cup warm milk
1/2 teaspoon salt

1 cup blanched chopped almonds
1/2 cup sugar
1/2 cup butter
1 yeast cake
1 cup bread crumbs
2 1/2 tablespoons brown sugar
1 teaspoon cinnamon

Mix sugar and butter, pour warm milk over it, add yeast, dissolved. Add the eggs well beaten, flour and salt, to make a batter. Beat, and let rise for 1 1/2 hours. Beat again, place in deep pie pans, sprinkled with flour. Make crumb topping by mixing bread crumbs, brown sugar, cinnamon, 1 teaspoon salt, almonds and 2 tablespoons nearly melted butter. Mix in bowl, spread over cakes, let rise for 15 minutes and bake in medium hot (400°) oven till toothpick inserted in center comes out clean.

SWEET POTATO ROLLS

2 pounds boiled sweet potatoes, mashed very fine
2 tablespoons butter
1/2 cup of yeast
1 pint sweet milk
Pinch of salt
Enough flour to make a stiff dough.

Mix all ingredients together and make a soft dough. Let rise till light, then cut into any shape desired. Place on pan and let rise one hour. Bake a 350° for until no golden and no longer doughy.

BISCUITS

1 cup sweet cream
2 cups sweet milk
3 teaspoons baking powder
Pinch of salt
Flour

Mix baking powder and salt into a little flour in a large bowl. Add cream and milk, then enough flour to make a dough stiff enough to roll out on a board. Mix only until combined (don't over mix). Roll and cut out then bake 400° till light brown.

CORN MUFFINS

1 egg
1 tablespoon sugar
1 pint sour cream
1 cup corn meal
1/2 cup wheat flour
1 scant teaspoon baking soda

Mix all ingredients together and pour into muffin pans. Bake at 375° until golden brown.

CORN PONE

2 cups corn meal
1 cup flour
1/2 cup sugar
1/2 cup melted lard
1 egg
1 teaspoon baking soda
Pinch of salt
Sweet milk

Mix all ingredients except the sweet milk. Once mixed add enough sweet milk to make a thin batter. Pour into loaf pan and bake at 375° for 20-30 minutes

PERFECT WHITE BREAD

1 cake compressed yeast (or dry yeast), dissolved in 1 cup of lukewarm water
1 1/2 pints milk
1 1/2 pints water

1 teaspoon salt
1 teaspoon sugar
Flour

Mix milk and water together in a large pan. Add salt and sugar, then dissolved yeast. Add enough flour to make dough stiff enough to knead. Knead until it will not stick to the board. Set to rise. As soon as light, knead again for about 30 minutes. Divide into loaves and put into pans to rise again. After they are light, bake at 375° until they start to brown.

COMMUNION BREAD

1 1/2 pints of sweet cream
1/4 pound (1 stick) of butter
1/2 cup sugar
Enough flour to make pie-like dough

Mix all ingredients together and roll thin like pie dough. Mark bread into 1 inch strips and perforate strips with a fork at each end before baking. Bake on flat cookie sheets.

German Bapitst

Main Dishes

BAKED BEEF LOAF

2 pounds of chopped beef
1/2 cup bread crumbs
2 eggs, well beaten
1/2 cup milk

Pepper and salt
Pinch of cayenne pepper
1/2 teaspoon celery salt
2 cups fresh sliced tomatoes
2-3 tablespoons butter

Mix everything except for the tomatoes well and shape into a loaf. Put 1 cup of tomatoes into the bottom of a baking pan and then place the beef loaf on this and then pour over the remaining tomatoes. Dot with small bits of butter. Bake for at 350° for 45 minutes.

BEEF ROLL

5 pound round steak
3 cups sweet milk
2 cups cracker crumbs, rolled fine
Salt and pepper to taste
1-2 tablespoons butter

Ground the round steak fine. Add milk, crumbs, and seasonings. Mix with your hand. When well mixed put into a covered roaster. Dot with bits of butter and bake at 350° for 2 1/2 hours.

CHICKEN WITH NOODLES

1 fat chicken, cut in pieces
1 pint of rich, sweet milk
2 eggs
Enough flour to make a stiff dough
Pinch of salt

Cook chicken in broth until tender. Make a dough with milk, eggs, flour and salt. Roll thin and let dry until dough becomes brittle (3-6 hours). Cut up in fine strips and drop into broth. Boil only about 15 minutes.

HAM SOUFFLE

3 tablespoons butter
2 tablespoons flour

3/4 cup milk
1 1/2 teaspoons salt
Sprinkle of pepper
3/4 cup cooked ham, chopped fine (may substitute beef or chicken if preferred.)
3 eggs yolks
3 egg whites, beaten stiff

Make a sauce out of butter, flour, milk, salt and pepper. Add cooked meat and 3 egg yolks. Lastly, fold in 3 beaten egg whites. Turn into buttered dish and bake in slow oven (325°) until firm and lightly browned.

POTTED HAM

2 parts boiled lean ham to one part fat
Salt and pepper
1 heaping teaspoon gelatin dissolved in a small amount of water

Chop ham and fat very fine. Season with salt and pepper. Add gelatin dissolved in water to the meat. Put in pan and bake at 375° for 30 minutes. Press and set away to cool.

FRIED BACON AND POTATOES

1/2 pound bacon
3 medium potatoes
1/2 cup sweet milk
Pepper

Cut bacon into slices 3 inches long and layer them in the bottom of a frying pan. Place pared potatoes with skins, on top of bacon. Pour a little water over bacon and potatoes and fry slowly until soft and brown. Pour off all the fat and add sweet milk. Sprinkle with pepper and serve.

BAKED POT PIE

1 1/2 pints sour milk

1/2 teaspoon baking soda
1/2 teaspoon salt
1 teacupful of lard
Enough flour to make a stiff dough
1 1/2 cups cooked beef or pork, cut in small pieces
Pepper
Broth

Mix milk, soda, salt, lard, and flour to make a stiff dough. Divide into two equal parts and roll each 1/2 inch thick. Place one half into a pie pan. Place meat into pie crust and add a little pepper and broth. Cover with remaining dough and seal the edges. Make a small slit in center of top crust and bake at 350° till light brown. Serve with broth.

FRIED FISH

1 (2-3 pound) fish
Salt
Flour
Lard

Split fish lengthwise; remove backbone and cut in squares for frying. Wash thoroughly and dry with a towel. Salt fish, then roll in flour. Fry briskly in at least 1/2 inch deep hot lard till browned on both sides. Serve immediately.

German Baptist

Cookies

DEACON COOKIES

3 eggs
1 cup sugar
1/2 cup butter

1/2 cup lard
2 teaspoons baking powder
Enough flour to make a soft dough.

Combine all ingredients and mix well. Roll thin (1/4 inch or less), cut out and bake in a quick oven (425° for10 minutes).

OLD DUTCH MOLASSES CAKES

1 quart molasses
2 pounds brown sugar
2 ounces baking soda
1 ounces cream of tartar
1 pound lard
1 quart sour milk
4 pounds flour
1 egg

Make the dough 12 hours before the baking. Mix all ingredients except egg together and place in refrigerator. Roll dough out (must be very cold to roll) and cut with round cutter. Brush with egg before baking. Bake in slow oven (325° till golden).

COCONUT CRACKERS

2 1/2 pounds sugar
1 pint New Orleans molasses
6 ounces butter
1 1/2 pounds flour
1 large coconut freshly grated
1 teaspoon baking soda

Mix well to form stiff batter. Cut in cookie form and transfer to greased baking sheets. Bake in hot oven (425°) until brown, then remove from oven and allow to cool thoroughly on baking sheet.

CHOCOLATE JUMBLES

2 cups sugar
1 cup butter
1 cup grated milk chocolate

4 eggs
1 teaspoon soda
1 teaspoon cream of tartar
3 cups of flour

Mix together well and chill. Roll thin and cut into any pattern desired. Bake at 350° for 8-11 minutes or until they start to turn golden.

CREAM COOKIES

3 cups sour cream
3 eggs
Pinch of salt
3 cups white sugar
3 small teaspoons baking soda
Enough flour to make a soft dough.

Mix all ingredients together, adding the flour last. Roll dough out and cut into shapes. Sprinkle with granulated sugar and bake at 350° till golden.

SOFT GINGER COOKIES

1 pint molasses
1 cup lard
1 cup sour or buttermilk
1 tablespoon baking soda
1 teaspoon ginger
Flour

Stir baking soda and ginger into the molasses. Add milk/buttermilk and lard, then stir in enough flour to make a soft dough. Roll our a little thicker than sugar cookies. Cut into cakes and bake at 375° until done.

German Baptist

Cakes, Pies, and Puddings

DUTCH RAISIN CAKE

5 eggs, seperated
3/4 cup butter
2 cups sugar
1 cup raisins
1 cup milk

3 cups flour
2 teaspoons baking powder
1 cup hickory-nut meats

Mix all the ingredients except egg whites. Beat egg whites until stiff, and bake in layers. Ice with vanilla icing.

APPLE BUTTER CAKE

1 cup sugar
1/2 cup butter
4 eggs, well beaten
4 tablespoons sour milk
1 teaspoon baking soda
1 teaspoon ground cinnamon
1 teaspoon cloves
2 cups flour
1 cup apple butter

Cream together sugar, butter and eggs. Add sour milk, baking soda, cinnamon and cloves, then stir. Add flour and beat well. Add apple butter and bake at 350° until done.

A CHEAP CAKE

1/2 cup butter
2 cups sugar
3 eggs, well beaten
1 cup sweet milk
3 cups flour
3 teaspoons baking powder

Mix all ingredients together and bake at 350° until done.

WHIPPED CREAM PIE

1 1/2 cups thick sweet cream, chilled
1/2 cup of granulated sugar, chilled
1/2 teaspoon lemon extract
1 baked and cooled pie crust

Beat cream and sugar together until it stands stiff. Add lemon extract and pour the cream into a baked and cooled pie crust. Chill and serve.

SYRUP PIE

3 eggs
2 cups maple syrup
1 cup milk
1 rounded teaspoon baking powder
1 unbaked pie crust

Stir first four ingredients well and pour into pie crust. Bake at 350° for 30-45 minutes, or until golden.

BANANA PIE

1 baked pie crust
2 bananas
1 cup milk
1/2 cup sugar
2 egg yolks
2 tablespoons flour
2 egg whites

Slice bananas into cooked and cooled crust. Boil milk, sugar, egg yolks, and flour until thick. Let it cool then pour over bananas. Beat the egg whites and spread over the top. Bake at 400° only until the top starts to brown. Do not let the pie get hot throughout.

COTTAGE PUDDING

1 egg
1 cup sugar
1/2 cup milk and cream
1 tablespoon butter
2 teaspoons of baking powder
Flour enough to stiffen

Stir all ingredients in a bowl until well mixed. Pour into a loaf pan and bake at 350° until done. Serve with the following sauce:

1 1/2 cups of water
1/2 cup of sugar
2 tablespoons butter
1 tablespoon corn starch

Let water and sugar come to a boil. Add butter and corn starch. Serve over cottage pudding.

EGG PUDDING

5 tablespoons flour
1 teaspoon salt
1 1/2 pints sweet milk
5 eggs
Butter and lard for frying

Mix flour and salt in large bowl. Add half the milk to it and stir. Add eggs and stir again. Add remaining milk and stir well. Heat butter and lard in skillet till melted. Pour in batter. As it bakes, lift the edges of it with a knife until all of it is set. Place it in a hot oven to finish baking.

BREAD PUDDING

1/2 loaf of bread
Enough milk to cover
1 egg, beaten well
1/2 cup sugar
1/2 cup raisins
Pinch of salt
1/3 cup butter
1 teaspoon ground cinnamon

Break bread fine and cover with milk. Set on the stove where it is not too hot, to soften. Remove from stove and stir in egg, sugar, raisins, salt, butter, and cinnamon. Bake at 350° for 1 hour. Serve with sugar and cream.

MENNONITE

The Mennonite movement evolved during the Anabaptist reaction to the Reformation some 500 years ago. This movement is when the Amish church and German Baptist church also planted their roots. The term "Mennonite" comes from the church's founder, Menno Simons. The Mennonite church is much more diverse than the Amish or German Baptist. Most Mennonites fully embrace technology, while still embracing the pacificism and family values that the Old Order Amish still adhere to.

Mennonite cooking is much like Amish and German Baptist cooking. Nestled in rural Missouri along meandering State Route 5 near the tiny town of Versailles is a rare restaurant offering up mouth-melting Mennonite meals. The sign says "Traditional Mennonite Cooking," and Lehman's restaurant delivers plenty of that and more.

Barbara Lehman (a common Mennonite surname) and her family serve delicious home-cooked meals to hungry customers. Using traditional German recipes handed down from generation to generation, Barbara uses only the freshest ingredients like homegrown vegetables, farm-raised meat, and desserts made from scratch.

Entrees of baked ham and chicken, roast beef, deep-fried catfish, fried chicken, and breaded or boiled shrimp are menu regulars. Meals are served family-style with ample helpings of salad, vegetables, homemade noodles, mashed potatoes and gravy, and homemade bread all refillable on request. You can top off the feast with homemade pie or cake. Must-try desserts are the coconut-cream pie, Mom Lehman's caramel pudding, or Barbara's apple pie with its extra-flaky crust.

The atmosphere is so relaxcd; a young waitress might take your dessert order with her toddler perched on her hip. Don't be surprised if another worker comes up to your table and says, "Hi, I'm Mom." That would be Barbara's mother, who, along with her father, is usually helping out. Other family workers are Barbara's sister and niece and nephew. Barbara's husband is the official barbecue chicken chef in the summer.

Lehman's will move to its new location, eighteen miles south of Tipton, on Route 5 just north of Versailles before Easter. Hours are Thursday, Friday, and Saturday from 4:30 to 8:30 p.m. Entrees range in price from $7.95 to $12.95, with side dishes served family style.

Mennonite

Breakfasts

PENNSYLVANIA DUTCH APPLE MUFFIN CAKE

2 cups all-purpose flour
1 tablespoon baking powder
1/2 teaspoon baking soda
1 teaspoon salt
1 1/4 teaspoons cinnamon
1/2 teaspoon allspice
1/4 teaspoon ground cloves
1 1/4 cups sugar
1/4 cup butter
1 cup sour cream
2 eggs
1 cup apple, peeled and diced (Granny Smith are great for this recipe)
Butter and fine dry bread crumbs for coating the pan

Crumb topping:
In a small bowl blend: 3 tablespoon all-purpose flour 1/4 teaspoon cinnamon Cut in 2 tablespoons cold butter until mixture forms coarse crumbs.

Heat oven to 350°. Butter an 8 cup charlotte pan or souffle dish and sprinkle with bread crumbs, tapping out excess. Combine dry ingredients and blend well. Set aside. Melt butter in saucepan. Remove from heat; stir in sour cream and beat in eggs. Beat the butter mixture into the dry ingredients and blend until batter is smooth and satiny. Stir in apple. Scrape batter into prepared pan, smoothing batter until it is level. Sprinkle with crumb topping. Bake at 350° 1 hour until crumbs are crisp and lightly browned and a skewer inserted in center of cake comes out clean. If tube pan is used, start testing for doneness after 45 minutes. Cool cake in pan 20 minutes. Run a butter knife between cake and sides of pan and gently invert cake onto counter. Re-invert on a rack and cool thoroughly before serving.

MENNONITE FASTNACHTS

Fastnacht Day is a special Pennsylvania Dutch celebration that falls on Shrove Tuesday, the day before Ash Wednesday. It actually translates to "Fast Night". The tradition is to eat the very best, and lots of it, before the Lenten fast. Fastnachts (pronounced fost-nokts) are doughnuts. There are three types of fastnachts, one made with yeast, one made with potatoes and yeast, and the other without either. All are crispy on the outside and not as sweet as standard doughnuts

2 cups milk
3/4 cup sugar
1 teaspoon salt
2 eggs
1 yeast cake/package
About 7 cups flour
1/3 cup lard
Powdered sugar

Bring milk and lard to boiling point, but do not boil. Stir in sugar and salt and let cool to lukewarm. Beat eggs and add to the milk. Soak yeast cake in 2 tablespoon warm water and keep warm. Sift and measure the flour. Combine yeast and liquid. Add enough flour until able to handle easily. Knead well by punching, stretching, and folding it over itself. Knead well until snappy. Let rise overnight.

In the morning, roll out to 1/4 inch thickness. Cut into 2" squares and make a slit in the center. Cover and let rise 3/4 of an hour. Fry in deep fat until brown. While warm, roll in powdered sugar. **Note:** If desired, the cruellers can be rolled in granulated or powdered sugar while still warm. A common Pennsylvania Dutch tradition is to cut the crueller down the center and drizzle some molasses or corn syrup on each half.

DUTCH COFFEECAKE

1 1/2 cup sugar
1/2 cup oil or Crisco
2 1/2 cups flour
1 cup milk
1 egg
1 teaspoon vanilla
1/4 teaspoon salt
1 teaspoon cinnamon
1/2 teaspoon cloves
1 teaspoon baking soda

Cream oil and sugar; mix thoroughly. Add remaining ingredients, beat well. Pour into 2 (8 inch) pans. Bake at 350° for 30 minutes. If you want to have a crumb topping (not needed) hold out 3/4 cup of creamed mixture before adding rest of ingredients, then add to top of cake just prior to baking. Cake is great hot or cold; served with or without butter or margarine.

BREAKFAST CAKES

3 1/2 cups flour
2 cups sugar
1 cup shortening
2 eggs
1 cup milk

Mix flour, sugar and shortening together until "crumbly." Set half of mixture aside for later. Add to other half: 2 eggs and 1 cup milk. Mix together well and pour into three 8-inch pie pans. Sprinkle mixture that was set aside on top of each pan. Bake at 350° for 30 to 35 minutes.

BREAKFAST CAKE (variation)

4 cups flour
3 cups sugar
2 teaspoons baking powder
1 cup butter or margarine
2 eggs
1 cup milk

1 teaspoon vanilla

Mix flour, sugar, baking powder and butter until fine crumbs. Take out 1 cup of crumbs for top of cakes, and set aside. Then add to remaining crumbs, 2 eggs, milk and vanilla. Mix until smooth. Pour into 3 (8-inch) greased cake pans. Sprinkle crumbs over batter. Bake at 350° for 40 minutes or until done.

FARMERS BREAKFAST

4 Potatoes; medium
3 Eggs; large
1/2 pound bacon
2 medium tomatoes, peeled
3 tablespoons milk
1 cup cooked ham, cubed
1 tablespoon chives, chopped

Boil unpeeled potatoes 30 minutes. Rinse under cold water, peel and set aside to cool. Slice potatoes. In a large frying pan cook bacon. Add the potato slices; cook until lightly browned. Meanwhile blend eggs with milk and salt. Stir in the cubed ham. Cut the tomatoes into thin wedges; add to the egg mixture. Pour the egg mixture over the potatoes in the frying pan. Cook until the eggs are set. Sprinkle with chopped chives and serve.

CINNAMON FLOP

1 cup sugar
2 cup flour
2 teaspoon baking powder
1 tablespoon melted butter
1 cup milk
1 cup brown Sugar
1 teaspoon cinnamon
1/4 cup butter

Sift sugar, flour, and baking powder together. Add 1 tablespoon melted butter and milk and stir until well blended. Divide mixture between 2-9" well greased pie or cake pans. Sprinkle tops with brown sugar, then cinnamon and then push chunks of butter into the dough. This makes holes and later gets very soft as it bakes. Bake at 350° for 30 minutes.

CINNAMON FLOP COFFEE CAKE

1 cup brown sugar, firmly packed
1/2 teaspoon ground cinnamon
1/4 cup butter or margarine
2 cups sifted all-purpose flour
2 teaspoons baking powder
1/4 teaspoon salt
2 tablespoons shortening
1 1/2 cups sugar
1 egg
1 teaspoon vanilla
1 cup milk

Combine brown sugar and cinnamon; cut in butter until crumbly and well
mixed. Set aside. Sift together flour, baking powder and salt. Cream shortening in mixer
bowl. Gradually beat in sugar. Add egg and vanilla; beat well. Stir in flour mixture by
hand alternately with milk, starting and ending with flour mixture. Pour into greased 8"
square pan. Sprinkle with brown mixture. Bake at 425° for 30 to 35 minutes or until
golden brown.
Serve warm.

APPLESAUCE COFFEE CAKE

1 3/4 cups flour
1/2 cup sugar
1/2 cup margarine or butter
1/4 cup chopped nuts
1/2 teaspoon ground cinnamon
1 1/2 teaspoons baking powder
1/2 teaspoon baking soda
2 beaten eggs
1 cup Applesauce or Apple Butter
1 teaspoon vanilla
1/2 cup raisins (optional)

In a bowl combine 3/4 cup of flour and the sugar; cut in margarine. For topping, remove
1/2 cup of the crumb mixture; stir in nuts and cinnamon. Set aside.
To remaining crumb mixture add remaining flour, baking powder, and baking soda. Add
eggs, applesauce or apple butter, and vanilla; beat till well blended. If desired, stir in
raisins. Pour into a greased 9x9x2-inch baking pan. Sprinkle topping on batter. Bake at

375° for 30-35 minutes or until toothpick inserted in middle comes out clean. Serve warm.

PEANUT BUTTER COFFEE CAKE

Topping:
1/2 cup packed brown sugar
1/2 cup flour
1/4 cup peanut butter
3 tablespoons butter

Batter:
2 cups flour
1 cup packed brown sugar
2 teaspoon baking powder
1/2 teaspoon baking soda
1/2 teaspoon salt
1 cup milk
1/2 cup peanut butter
2 eggs
1/4 cup butter

For topping, combine 1/2 cup of brown sugar and 1/2 cup flour; cut in 1/4 cup peanut butter and 3 tablespoons butter till crumbly. Set aside.

For batter, combine 2 cups flour, 1 cup brown sugar, baking powder, baking soda, and salt. Add milk, 1/2 cup peanut butter, eggs, and 1/4 cup butter. Beat with mixer on low speed till blended. Beat on high for 3 minutes, scraping bowl frequently. Pour into a greased 13x9x2-inch pan, spreading evenly. Sprinkle with topping mixture. Bake at 375° for about 30 minutes, or until toothpick inserted in center comes out clean.

STICKY BUNS

1 package active dry yeast
1 /4 cup water
1 cup milk, scalded
3 tablespoons sugar
1 /2 teaspoon salt
3 1/ 4 cups sifted flour
3 tablespoons butter
1 /2 cup chopped raisins
2 tablespoons currants
2 tablespoons finely chopped citron
1 /4 cup firmly packed brown sugar
1 /2 teaspoon cinnamon
3 tablespoons brown sugar

Soften the yeast in the warm water and let stand 5-10 minutes. Add milk to sugar and salt; mix thoroughly and cook to lukewarm. Add 1 cup of flour and mix until smooth. Stir in yeast mixture and add remaining flour gradually, mixing well. On a floured surface, knead dough until smooth. Place in a greased bowl and grease top of dough. Cover and let rise in a warm place until dough is doubled; about 45 minutes. Punch down and roll into a rectangle about 1/4 inch thick. Brush with softened butter and spread evenly with raisins, currants and citron, 1/4 cup brown sugar and cinnamon. Roll up, starting at long side, as for a jelly roll and cut into 1 inch slices. Place slices cut side down in a buttered 13x9x2 inch pan. Cover and let rise until doubled. Sprinkle top with remaining brown sugar and bake at 375* for 20 to 30 minutes.

Mennonite

Soups

PEA POD SOUP

2 quarts green English Peas (pods only)
1 onion
2 cups milk
1 teaspoon sugar
3 tablespoons butter
1 teaspoon salt
1 teaspoon pepper
3 tablespoons flour
Nutmeg to taste

Wash the pea pods, cut into 1" long pieces, boil in water with onion 1 1/2 hours. Strain through colander; add pepper, salt, sugar, nutmeg and milk (scalded). Bring to a boil and thicken with butter and flour mixture.

MENNONITE CHICKEN SOUP

3-4 pounds chicken
3 quarts of water
2 teaspoons salt
1 teaspoon peppercorns
1 small onion, chopped
1 bay leaf
1 spring parsley

Cut chicken into serving pieces, add to water and bring to a boil. Take off scum as it forms. Allow to simmer for 3 hrs, covered, adding more water as required. Half hour before serving skim off fat. Add pepper, salt, onion, bay leaf and parsley.

MENNONITE POTATO SOUP

2 cups diced potatoes
1 cup water
2 tablespoons onion
2 tablespoons butter
2 tablespoons flour
2 cups milk
1/8 teaspoon celery salt
Salt & pepper to taste

Boil potatoes in water until tender. Brown onion in butter; blend in flour. Add ½ cup milk. Add flour mixture, seasonings and remaining milk to potatoes and cook in saucepan until hot.

GARDEN GREEN BEAN SOUP

1 ham bone
1 cup cut green beans
2 medium potatoes
1 bunch of summer savory

Cover ham bone with water and boil for 45 minutes. Skim the scum as it cooks. Remove the bone and add vegetables. Simmer for about 1 hour. About 15 minutes before serving, add the summer savory and parsley. Remove when serving. Just before serving, add the butter. Stir and serve hot.

MENNONITE BROWN FLOUR POTATO SOUP

6 medium potatoes
3 cups water
3 cups milk
3 tablespoons flour
3 tablespoons butter
Salt, pepper and parsley

Peel and cut potatoes in slices. Boil them in salted water until tender. Add the milk and let simmer. Meanwhile brown the flour in the melted butter, stirring all the time over low heat; add it to the soup, stirring until the mixture thickens. Sprinkle with parsley and pepper. Serve with buttered crumbs, or squares of fried bread, or pretzels on top. You might boil some sliced onion with the potatoes, if you like.

PENNSYLVANIA DUTCH ASPARAGUS SOUP

2 cups asparagus
1 tablespoon diced onion
1 cup milk
1 can of crushed cream corn
2 hard boiled eggs, diced
Processed heese (optional)
Salt & pepper to taste

Clean and cut asparagus into small pieces and place in pan. Add just enough water to cover. Add salt, pepper and diced onion. Bring to a boil and then reduce heat and simmer till asparagus is tender. DO NOT drain. Add milk and crushed cream corn. Simmer a few minutes till corn is heated then add diced hard boiled eggs and serve.

(Optional: Sprinkle with shredded cheese)

Mennonite

Breads/Muffins

MENNONITE OATMEAL BREAD

2 cups rolled oats (preferably coarse)
2 cups boiling water
1/2 cup molasses
1/2 cup brown sugar
1 teaspoon salt
1/2 cup shortening
1/2 cup lukewarm water
1 teaspoon sugar
2 packets yeast (about 2 tablespoons)
Chopped nuts (optional)
Raisins (optional)
About 6 cups all-purpose flour

Pour boiling water over rolled oats; stir and add molasses, salt, brown sugar and shortening. Let stand until lukewarm, then add yeast dissolved in warm water with sugar. Mix in flour until it requires muscle. Then knead a few minutes on well floured surface. Put dough in bowl; cover. Let rise 1 to 2 hours in warm, draft-free place to double its size (rolls and buns take less time to rise and bake; can make various shapes). It has risen enough if the dent stays when you press your fingers deeply into it. Nuts or raisins may be added with flour or kneaded into individual loaves. Divide into 2 loaves; let rise again until smooth and round over tops of pans, about an hour or more. Bake at 400° F for 35 minutes.

Rye Bread variation
Use half rye and half all-purpose flour and use 1/2 cup molasses as part of the liquid.

MENNONITE WHITE BREAD

1 package dry yeast
2 1/2 cups warm water
1/3 cup sugar
2 teaspoons salt
2 1/2 tablespoon shortening, melted
6 cups flour, more as needed to 7 cups

Dissolve yeast in the 1/2 cup of the water. In a large bowl, combine the
sugar, salt, remaining water and shortening. Add the yeast mixture. Gradually
add flour to form a soft dough. Turn dough out onto a floured board
and knead until smooth. Place dough in a greased bowl; turn to
grease the top. Cover with a clean tea towel and let rise until
doubled (about two hours).

Punch dough down; then divide in half and form loaves. Place in
previously greased bread pans. Prick the tops with a fork. Let rise
until higher than the pans (about two hours).

Bake in a 375° oven for 25 to 30 minutes. Cool for 10 minutes.
Butter the tops of the loaves. Place pans on their sides until bread
loosens. Remove loaves and cool completely on racks.

PULL-APART MONKEY BREAD

3 packages of regular refrigerated biscuits
3/4 cup butter
3/4 cup brown sugar
2 teaspoons cinnamon
4 tablespoons sugar

Grease a tube pan.
Mix together cinnamon and sugar. Cut biscuits into quarters and roll in the cinnamon and
sugar mixture. Just drop them into the greased tube pan as you coat them.   After all
biscuits are coated, melt the butter and brown sugar together in a small saucepan. When
sugar is dissolved, pour over the biscuits. Bake at 350° for approximately 40 minutes or
until golden brown.
Turn out onto a large platter and just pull apart.

ORANGE MUFFINS

2 cups all-purpose flour
2 teaspoons baking powder
1/4 teaspoon baking soda
1 teaspoon salt
1/2 cup white sugar
1 tablespoon grated orange zest
2/3 cup orange juice
1/2 cup melted butter
2 eggs
1/2 cup ground walnuts (optional)
1 tablespoon melted butter
1/4 cup packed brown sugar
1/2 teaspoon ground cinnamon

Combine flour, baking powder, baking soda, salt, white sugar and grated orange peel. Stir in orange juice, 1/2 cup melted butter, eggs and chopped nuts. Pour into 12 muffin cups. Blend 1 tablespoon melted margarine, 1/4 cup brown sugar, 1/2 teaspoon cinnamon and sprinkle on top of each muffin. Bake in a preheated 350° oven for 20-25 minutes. Serve hot.

Mennonite

Main Dishes

PENNSYLVANIA DUTCH MEAT LOAF

1 1/2 pounds ground beef
1 cup fresh bread crumbs
1 medium onion, chopped
1 medium green bell pepper, chopped
1-8 ounces can Hunt's tomato sauce
1 egg
1 1/2 teaspoons salt
1/4 teaspoon pepper
3/4 cup water
2 tablespoons brown sugar, packed
2 tablespoons prepared mustard
1 tablespoon vinegar

In a medium bowl, lightly mix beef, bread crumbs, onion, green pepper, 1/2 can tomato sauce, egg, salt and pepper. Shape into a loaf in a shallow baking pan.
Combine remaining tomato sauce with the rest of the ingredients. Pour over loaf. Bake at 350° for 1 hour and 15 minutes Baste the loaf several times during baking.

DUTCH MEAT LOAF

2 pounds hamburger
1 medium onion
1 egg
1 cup bread crumbs
1 can tomato sauce
1/2 teaspoon salt
1/4 teaspoon pepper

Sauce:
1/2 cup tomato sauce
1 cup water
2 teaspoons molasses
2 tablespoons white vinegar
2 teaspoons yellow mustard

Mix first 7 ingredients together well and form into loaf.
Mix sauce ingredients and pour entire mixture over the meat loaf. Bake covered at 350° for about 1 1/2 hours. Baste the meatloaf periodically with pan juices.

LANCASTER COUNTY CHICKEN LOAF

2 cups cooked chicken
1 cup soft bread crumbs
2 tablespoons parsley
2 tablespoons celery
1 teaspoon salt
2 eggs
1 cup milk
3 tablespoons melted butter

Mix all ingredients. Pour into buttered loaf pan. Bake at 375° for 30 minutes. Serve warm or cold.

MENNONITE MEAT BALLS (FLEISCHBÄLLE)

3/4 pound ground pork
3/4 pound ground beef
1 onion, finely chopped
Salt and pepper
3/4 cup rice, soaked in water
2 eggs
1 cup breadcrumbs
1 cup catsup or tomato sauce

Mix all but the catsup and form into balls; brown balls in a pan, put in an oven dish and cover with blended catsup and 1 quart of boiling water. Let simmer in oven for 3 to 4 hours, making sure the balls haven't gone dry.

BUWE SCHENKEL

1 1/2 pounds beef
5 potatoes
2 1/2 teaspoon salt
dash of pepper
1 tablespoon butter
1 tablespoon minced parsley
1 small onion, chopped
2 eggs, slightly beaten

Add about 2 teaspoon of salt to the beef, cover with water and stew for several hours. (One alternative would be to use 1 1/2 pounds of ground beef. In this case, brown the ground beef and put in a pot with the potatoes as outlined below.)

For the dough:
1 1/4 cups flour
1 teaspoon baking powder
1 tablespoon shortening (like Crisco)
1 tablespoon butter
5 or 6 tablespoon water

Make the dough by cutting shortening and butter into the flour and baking powder. Add the water to this mixture. Divide the dough into six parts and roll each into a 6-inch circle. Set aside while you prepare the filling. Peel the potatoes and slice thin. Add the potatoes, salt and pepper to the beef and add water to almost cover the potatoes. Cook until the potatoes are almost soft. Turn off the heat and add the butter, onion, parsley and the two beaten eggs to the potatoes. Let stand for ten minutes. Put a spoonful of the mixture in the middle of a dough circle. Fold the dough over to form half moons, pinching edges tightly together. Lift carefully into the kettle of boiling broth. Cover tightly and cook for about 30 minutes. This yields 6-8 generous servings.

PENNSYLVANIA DUTCH SCRAPPLE RECIPE

Pigs' knuckles
Pork
Onion
Water
1 1/2 teaspoons salt
1 teaspoon pepper
1 teaspoon ground sage
3 cups cornmeal
All-purpose flour, for dredging sliced scrapple
Butter, back fat or vegetable oil for frying

Place pigs' knuckles in a large pot; add pork, onion, and water. Cook
slowly, covered, for 2 1/2 hours; drain, reserve broth.
Chill meat and remove fat; separate meat from bones. Chop meat.
Place meat in a kettle with 2 quarts of the reserved broth. Add salt, pepper
and sage; bring to a boil. Combine cornmeal with remaining 1 quart of reserved
broth and stir into boiling mixture. Cook over medium heat until
thickened, stirring constantly. Cover and cook over very low heat; stir
again after 20 minutes.

TRADITIONAL SCRAPPLE

1 pound boneless cooked pork loin, chopped
1 cup cornmeal
1-14 1/2 ounces can chicken broth
1/4 teaspoon dried thyme
1/4 teaspoon salt
1/2 cup all-purpose flour
1/4 teaspoon pepper
2 tablespoons vegetable oil, or as needed

In a large saucepan combine pork, cornmeal, chicken broth, thyme and salt. Bring to a
boil, stirring often. Reduce heat and simmer about 2 minutes or until mixture is very
thick, stirring constantly. Line an 8x8x2-inch baking pan or a 9x5x3-inch loaf pan with
waxed paper, letting paper extend 3-4 inches above top of pan. Spoon pork mixture into
pan. Cover and chill in the refrigerator 4 hours or overnight. Unmold; cut scrapple into
squares. Combine flour and pepper; dust squares with flour mixture. In large skillet
brown scrapple on both sides in a small amount of hot oil.

PLAIN SCRAPPLE

1 medium onion, chopped
1/2 pound chopped raw pork
1 1/4 teaspoon salt
1/8 teaspoon pepper
1 1/4 quarts water
1 cup corn meal

Brown onion slowly in a little fat. Add meat, seasoning and water. Cook at simmering point 20 minutes. Add corn meal and cook over medium heat for one hour. Turn into loaf pan and cool.

BEEF PAPRIKA

1/4 cup shortening or oil
2 pounds beef cubes
1 cup onion, chopped
1/8 teaspoon garlic powder
2 tablespoons Worcestershire Sauce
3/4 cup ketchup
1 tablespoons brown sugar
2 teaspoons salt
2 teaspoons paprika
1/2 teaspoon mustard
1/8 teaspoon pepper
1 1/2 cup water
2 tablespoons flour or cornstarch
1/4 cup water

Brown meat, onion, and garlic powder in oil. Stir in next 8 ingredients.
Simmer for 2+ hours. After beef has simmered and is done, blend 2
tablespoons flour or cornstarch with 1/4 cup water and add to the beef mixture to make a sauce.
Serve over Buttered Noodles

DRIED BEEF CASSEROLE

5 ounces of dried beef, cut into bite-sized pieces
1/2 pound Monterey jack cheese, shredded
1 medium onion, chopped
4 hard-boiled eggs, diced
2-10.5 ounces cans cream of mushroom soup
1 3/4 cups milk
2 cups uncooked elbow macaroni

Combine all ingredients and pour into a greased casserole dish. Chill overnight covered. Bake at 350 for one hour. Serve and enjoy.

DUTCH BAKED CHICKEN

3 young, fresh chickens, salted
3 pounds of lard for frying
1/8 pound of flour
1 lemon for garnishing
2 1/2 cups of bread crumbs
1-2 eggs

The chickens are killed, dressed, washed, dried and prepared at once. Cut the chickens in half, salt them, dip them first into flour, then in beaten egg and then in bread crumbs. The lard is heated in an iron pot or kettle and the pieces of chicken placed into it carefully, one at a time, so as not to cool the fat too much and that the crumbs may not fall off. Bake them to a nice brown color. After the crust is hard, let them cook more slowly until well done. Then put on paper to drain, strew fine salt over the pieces and put on a platter after which they may be garnished with lemon slices.

DUTCH CHICKEN FLOATS

Remains of chicken meal
1 cup butter
1 cup mushrooms
3 eggs
2 tablespoons flour
1 teaspoon onion juice
1 green pepper
lemon juice
2 cups cream
1/2 teaspoon paprika

Melt two tablespoons of the butter in a saucepan and fry the green pepper (chopped and seeds removed). Then add the mushrooms (peeled and diced). Add the flour and cook until smooth, but not brown. Then add the cream and cook until thick. Pour in the chicken remains, boned and diced, and stand the pan in hot water. Meanwhile beat the rest of the butter to a cream, add the egg yolks one at a time while heating. Stir this into the hot chicken mixture until the egg thickens, and cook slowly. Add onion and lemon juice, salt, paprika. Serve on buttered slices of bread toasted on one side, the toasted side down.

HAMBURGER BARBECUE

1/2 cup Ketchup
2 tablespoons brown sugar
1 tablespoon vinegar
1 teaspoon yellow mustard (prepared)
2 tablespoons Worcestershire sauce
1 pound ground chuck
1/2 cup chopped green bell pepper
1/2 cup finely chopped celery
1/4 cup chopped onion

In saucepan, combine first 5 ingredients. Bring to boil, reduce heat and simmer 1/2 hour. In a frying pan, brown ground chuck. Add vegetables and sauté until vegetables are tender. Pour sauce over meat mixture. Simmer for 5 minutes.
Serve on hamburger buns.

MENNONITE OLD-FASHION BEEF POT PIE

2 pounds stewing beef
6 cups water
1 1/2 teaspoons salt
6 medium size potatoes
2 cups all-purpose flour
1 egg
3 tablespoons milk or water
1 teaspoon minced onion
1 teaspoon minced parsley

Cook meat in salt water until it is tender. Remove meat from broth; add minced onion and parsley to broth. Bring to the boiling point and add alternate layers of cubed potatoes and squares of dough.
To make dough, beat egg and add milk. Add flour to make stiff dough. Roll out paper thin and cut into 1-inch squares. Keep broth boiling while adding dough squares in order to keep them from packing together. Cover and cook for 20 minutes, adding more water if needed. Add meat and stir through pot pie.

DUTCH BARBECUE CHICKEN

1 broiler chicken
1 teaspoon onion juice
1/3 cup cider vinegar
1/2 teaspoon salt, pepper, paprika, garlic
1 teaspoon kitchen bouquet
1 teaspoon Worcestershire sauce
1 tablespoon tomato paste
1/2 cup melted butter

Cut tile broiler in half down the back. Get the broiling pan hot and grease well. Lay the chicken on rack and put immediately under hot fire. Sear on both sides. Have ready a barbecue sauce made of all the other ingredients listed above. Have also a new, clean paint brush, and during the broiling process paint the chicken on all sides at least three times with sauce. Make a gravy out of the drippings.

DUTCH LIVER-OYSTERS

1 pound Calf's liver
2 eggs, beaten
1/2 cup bread crumbs

Place the liver in salt and boil for 30 minutes. Then cut pieces about the size of an oyster, and dip in crumbs, then egg, then crumbs, and fry like oysters. Serve with lemon, and perhaps with squares of fried corn meal mush, or fried tomatoes.

LIVER DUMPLINGS (Leber Kloese)

1 soup bone
1 loaf bread
1 pound beef liver
2 eggs
4 onions
1 cup milk
2 cups flour

To make stock for boiling dumplings, cook the soup bone in plenty of water for 2 to 3 hours, spicing it with salt, pepper, celery tops, and parsley. Strain and keep hot while preparing dumplings. Scrape the liver, dice the onions and fry in butter. Season the milk and add the eggs, beaten, and make a soft paste with the flour. Shape into balls about 2 inches in diameter. Bring soup stock to boil, drop in dumplings, and boil without cover for 20 minutes. Serve on platter, and pour stock over it.

KONIGSBERGER KLOPS
1 1/2 pounds finely chopped raw beef
1/4 pound fat pork, chopped
1 1/2 roll-the crust cut off
1/8 pound butter
3 eggs
1 teaspoon minced onion
Salt
1 pinch of pepper
Some Flour
juice of 1 lemon

Mix the beef and pork well with the butter; soak a roll, press out and add together with other ingredients. Make small dumplings, rolled in flour, and boil slowly in bouillon or salt water for 15 or 20 minutes; or broil or fry them. Serve in deep dish with white fricassee gravy over them, add sauerkraut as a side dish.

DUTCH BEEF WITH ONIONS

1 1/2 pounds boiled beef
1 onion
2 tablespoons butter or suet
2 tablespoons flour
1 tablespoon vinegar
1 teaspoon meat extract
1/2 quart bouillon
1 pinch of pepper
Salt

Mince the onion, simmer it in butter or suet until soft; add flour, simmer until brown. Pour on the bouillon, vinegar, salt, pepper and meat extract and let come to a boil. Cut meat in slices and serve hot in gravy.

AUNT MAGGIE'S PENNSYLAVANIA DUTCH SAUERKRAUT CASSEROLE

1 pound sauerkraut
1 cup sugar
6 slices bacon
1 teaspoon black pepper

Mix sauerkraut, pepper and sugar together in 1 1/2 quart dish. Cut bacon slices in 1 inch picces and mix slightly with sauerkraut. Bake at 325° for 2 1/4 hours.

VARENIKY (pierogies)

1 pound dry curd cottage cheese
1 1/2 tablespoons finely chopped onion
1/2 teaspoon salt
3 egg yolks
3 egg whites, beaten
1 cup milk
3 1/2 teaspoons salt
3 1/2 cups flour
Water for boiling

Cream Gravy: (optional)
2 tablespoons butter or margarine
1 small onion, finely chopped
1 cup cream or evaporated milk
Salt and Pepper to taste.

In a bowl, combine cottage cheese, onions (if using), 1/2 teaspoon salt, and egg yolks.
Mix well with your hands until the cottage cheese is in fine curds. In a separate bowl,
combine egg whites, milk, 2 teaspoons salt and flour. Mix together, adding flour as
necessary until dough is stiff enough to roll out. Roll out dough (you may need to do this
in two batches) to 1/8 inch thick on floured surface. Cut into circles 5 inches in diameter
(use a cookie cutter, small bowl, or water glass to do this or use a pierogy maker.) Place 1
rounded tablespoon of the cottage cheese mixture on each circle. If you put too much
filling they won't stay together properly. Fold the circle over to make a half circle. Pinch
the edges tightly.
In a saucepan, bring 4-6 cups of water to a boil. Add 1 teaspoon salt. Drop vareniky into
boiling water a few at a time. Cook for five minutes. Remove with a slotted spoon and
drain. Keep hot.
Traditionally served with Cream Gravy: Melt the 2 tablespoons margarine or butter in a
skillet. Sauté the finely chopped onion. Add 1 cup of cream (or evaporated milk, to keep
the fat content lower) and salt and pepper to taste. Heat slowly, but do not boil.

DUTCH TURKEY SCALLOP

Leftover turkey
Leftover stuffing
Bread crumbs
Butter
3 hard boiled eggs, sliced
Salt and pepper

Dice leftover turkey. Place a layer of bread crumbs in the bottom of a buttered baking dish; add a layer of turkey, together with any cold dressing/stuffing left over. Slice three or four hard boiled eggs, placing some with each layer of turkey. Alternate the layers of meat and crumbs, adding bits of butter and seasoning to each. The last layer must be of crumbs. Place bits of butter over the top. Thin with milk any gravy that may he left, and pour over it. Cover the dish and bake at 350° for 30 minutes. A few minutes before serving remove the cover and let the scallop brown.

POTATOES AND GERMAN FRANKFURTERS

6 medium potatoes, about 2 pounds cooked, peeled, diced
2 green onions, chopped
6 dinner size frankfurters, sliced 1/2 inch thick
2 tablespoons salad oil
2 tablespoons sugar
1 teaspoon flour
1 teaspoon salt
1/4 cup vinegar
1 tablespoon chopped parsley
1/2 cup water

In a large serving bowl, combine hot potatoes and onions; cover and keep warm. Meanwhile, in large skillet over medium heat, brown frankfurters in oil. With slotted spoon, remove frankfurters, reserving drippings. Add frankfurters to potatoes and onions; keep warm. Into hot drippings, stir sugar, flour and salt until smooth and bubbly. Gradually stir in vinegar and ½ cup water; cook stirring constantly, until sauce thickens and boils. Pour sauce over potatoes and frankfurters. Sprinkle with parsley

POTATO BAKE

6-8 slices bacon
2 tablespoon flour
1/4 cup vinegar
1/2 teaspoon celery salt
1-10 3/4 ounces can Campbells condensed chicken broth
1/2 cup water
2 tablespoons brown sugar
2 tablespoons diced pimiento
1/4 teaspoon hot pepper sauce
1/4 cup diagonally sliced green onion
6 cups cooked, sliced potatoes

In skillet, cook bacon until crisp, remove, cool, cut and dice. Pour off all but ¼ cup drippings. Gradually blend chicken broth into flour until smooth and slowly stir into drippings. Add remaining ingredients except potatoes. Cook, stirring until thickened. In 1 1/2 quart shallow baking dish (10"x6"x2") arrange potatoes, pour broth mixture over potatoes. Cover and bake at 400° for 30 minutes or at 350° for 45 minutes. Garnish with bacon and serve. Eggs are easier to separate when very cold, but egg whites beat to greater volume if allowed to warm to room temperature.

GERMAN SPAETZEL DUMPLINGS

1 cup all-purpose flour
1/4 cup milk
2 eggs
1/2 teaspoon ground nutmeg
1 pinch freshly ground white pepper
1/2 teaspoon salt
1 gallon hot water
2 tablespoons butter
2 tablespoons chopped fresh parsley

Mix together flour, salt, white pepper, and nutmeg. Beat eggs well, and add alternately with the milk to the dry ingredients. Mix until smooth. Press dough through spaetzel maker, or a large-holed sieve or metal grater. Drop a few at a time into simmering liquid. Cook 5 to 8 minutes. Drain well. Sauté cooked spaetzel in butter or margarine. Sprinkle chopped fresh parsley on top, and serve.

GARDEN TOMATO SKILLET

4 tablespoons butter or oleo
6 firm tomatoes, red or green
4 tablespoons flour
1 teaspoon salt
1/8 teaspoon pepper
3 tablespoons brown sugar
1 cup heavy cream

Cut tomatoes in thick slices; dip in flour mixed with salt and pepper. Sprinkle with brown sugar. Heat butter in large skillet; add tomatoes and cook slowly. Turn over and sprinkle again with brown sugar. When tomatoes are tender, add cream and cook until bubbles. Arrange tomatoes on a serving platter and pour sauce over them.

BERKS COUNTY CHICKEN POT PIE

2 cups flour, unsifted
1 teaspoon salt
1 egg
3 tablespoons shortening
1/2 teaspoon baking powder
1/4 to 1/2 cup milk

Mix flour, salt, and baking soda; cut in shortening. Add egg and stir in enough milk to make soft dough. Roll out dough (using flour) until thin. Let sit 30 minutes. Cut into 2 inch squares. Cook 1 whole chicken with pinch of saffron, 2 packs Herbox chicken broth, parsley, salt, and pepper until chicken falls off the bones. Remove chicken and put in quartered potatoes (you may have to add some water). Bring to a boil. When mixture starts boiling, drop in one 2 inch square of dough at a time. Then put in shredded chicken and cook until potatoes and dough are done. Add flour or cornstarch to thicken. I also like to add corn to the recipe for a nice chicken, corn, and noodle soup.

SUMMER ZUCHINNI 7 LAYER CASSEROLE

2 pounds of ground beef
3 medium zucchini, cubed
3 tomatoes
2 red or green peppers
1 large onion, chopped
1 bag of frozen corn
Salt
Pepper
Cayenne pepper (this is very important-use to your taste preference)
Parsley

In a very large skillet or frying pan brown the beef and drain any fat off.
Then start layering: Add the zucchini, tomatoes, peppers, onion, corn, salt - pepper and the cayenne pepper. (Do not add parsley yet)
Cover and simmer till veggies are done (approx. 30 minutes)
Just before serving add parsley.

DUTCH SPICED POT ROAST

1-5 or 6 pound beef roast
1 onion, diced
Cloves
Vinegar
Brown flour
Butter

Season meat with spices for 24 hours. (If vinegar is too strong, add a little water.) After meat has sat for 24 hours, brown on both sides, add the juice, and let boil for 2 hours. Thicken with brown flour to make gravy. Add two tablespoons of salt and pepper to taste. Put butter on the beef when finished.

SCHNITZ UND GNEPP

3 pounds smoked ham hock
2 cups apple schnitz (sliced, dried apples)
2 tablespoons brown sugar
Cloves or ground cinnamon to taste

Cover the dried apple slices with water and soak overnight.

When apple slices are ready, put the ham in a pot and add water to cover. Bring to boil, reduce heat, cover, and simmer over medium heat for 2 hours. (Precooked ham will only take about 30 minutes.) When the meat is tender, remove the ham and cut it off the bone. Cut the meat into bite-sized pieces and return it to the pot.

Add the apples and water in which they were soaked. Add the brown sugar and spice and cook 1 hour longer. Add the Gnepp (dumplings) for the last 20 minutes.

To make the Gnepp:
2 cups flour
1/2 teaspoon salt
2 tablespoons butter, melted
1 tablespoon baking powder
1 egg, beaten
1/2 cup milk

Sift the dry ingredients into a large bowl. Stir in the beaten egg and melted butter. Add just enough milk until the batter is stiff.

Drop a spoonful of the dough into the boiling ham and apples. Feel free to add water in order to cook the dumplings.

Cover the kettle tightly and simmer 20 minutes.

Remove the dumplings with a slotted spoon and set aside. Drain off excess water and place the ham and apples onto a platter. Serve with the dumplings on top.

GOOSE STUFFED WITH APPLES

1-7 to 8pound goose
1 sliced onion
2 tablespoons flour
1/2 cup currants
1 1/2 pounds peeled, quartered apples
1 1/2 pints water
Salt
6 peppercorns

Clean and dress the goose, cutting off wings, head, neck, and feet. Trim off all fat, and soak this fat in cold water for 15 minutes. Rub goose with salt inside and outside. Mix the apples well with the currants and stuff into the goose, then sew up. Put the goose in the oven in a covered roasting pan with the water, sliced onion and peppercorns, and roast for 1 hour. Remove the cover then start basting with the drippings every 10 to 15 minutes. If the water boils down, add spoonfuls of it so the fat will not get too brown. It may require from 2 to 3 hours roasting before the goose is well done and crisp. Sprinkle a tablespoonful of cold water over the skin to make it more crisp. Make gravy with flour. Skim off grease if too plentiful.

PORK RIBS AND SAUERKRAUT, BALSER GEEHR

3 pounds salted pork ribs
1 pound sauerkraut
1/4 pound butter
1/2 teaspoon sugar
6 large peeled sliced apples
1/2 bottle white wine
1/4 pound chopped pork
1/4 pound chopped veal
1 egg
1 tablespoon butter
1/4 tablespoon minced onion
Salt
Pepper

Salt the pork for several days then cut into pieces, wash, dry and fry on both sides in hot butter. Put into a pot with sauerkraut on top. (If the sauerkraut is too sour, soak it in water and drain). Add the quarter pound of butter, apples, white wine and sugar, cover and cook slowly for 2 hours. If it gets too dry, pour in some water. For the meat dumplings chop the pork and veal; add a soaked roll of bread, the egg, 1 tablespoonful of butter and onion, mixed. Shape into dumplings and fry well done in the butter in which the fried ribs have been. Serve the sauerkraut in the middle of the platter, the ribs around it and the dumplings piled on top in a heap.

BACON DUMPLINGS

2 cups flour
1 pinch salt
1 1/2 tablespoons baking powder
1 teaspoon dried parsley
Pepper to taste
2 large eggs
Milk
1 tablespoon bacon grease
3-4 tablespoons browned and crumbled bacon

Sift flour, salt and baking powder into large bowl. Gently stir in parsley and pepper. Beat eggs in measuring cup and add enough milk to make 1/2-cup liquid. Add egg mixture and bacon grease to dry ingredients. Stir with fork just until blended. If dough seems too dry, add a little more milk. Add crumbled bacon; stir until combined. To cook, drop spoonfuls onto simmering broth, soup or stew. Cover and cook until done, about 12 minutes

HOMEMADE NOODLE & SAUERKRAUT CASSEROLE

1 cup butter
1 large onion, chopped
1-16 ounces can sauerkraut, undrained
1-16 ounces package wide noodles, cooked

Melt butter. Add chopped onion and sauté until partially cooked. Add sauerkraut. Stir until butter is absorbed in onion and sauerkraut. Stir in cooked noodles.

DUTCH PORK PEPPER, POTTS

2 pounds lean pork
1 1/2 tablespoons vinegar or 1 wineglassful of red wine
1 1/2 tablespoons butter or lard
1/2 onion
3 tablespoons flour
Water or bouillon
1 bay leaf
2 cloves
Salt
Pepper
1/2 cup pig's blood (can be omitted)

Dice the pork into 2 1/2 inch squares. Brown the butter and flour, then add bouillon or water, onion slices, spices and salt, and cook for a few minutes. Put in the meat and cook slowly for 30 minutes. Add the vinegar or red wine and continue to cook slowly until done, about 45 to 60 minutes, or over an hour. Put the mixture in a warm dish and stir the blood into the gravy, strain and pour over the meat.

DUTCH BREADED RABBIT, GRAUL

2 rabbits
1 tablespoon flour
2 eggs, beaten
1 cup milk or cream
1 cup fine bread crumbs
Salt and pepper

Clean, wash, disjoint rabbit in several waters, and parboil for 10 minutes (water boiling when you put in the rabbit). Then take out and dip the pieces in beaten egg, then in bread crumbs; season with salt and pepper and fry in butter and lard, mixed, until brown. Then thicken the gravy with the flour and add the milk or cream and let come to a boil. Pour this over the rabbit, and add also onion sauce.

MENNONITE RABBIT PIE

1 rabbit, cleaned, washed and cut into 2 or 3 pieces
Salt
3 tablespoons butter
2 tablespoons finely chopped onion
2 tablespoons finely chopped parsley
Flour
Dash Tabasco
Pie crust top (make your own or store bought)

Place rabbit in a deep pot and barely cover with water. Cover the pan and simmer rabbit until tender. Add salt to season when partially cooked. Drain and measure the liquid. Set aside. Remove the meat from the bones, keeping it in large pieces. Heat butter in a skillet: add onion and parsley. Cook about 5 minutes, stirring constantly. Measure 1 1/2 tablespoons flour for each cup of reserved rabbit liquid; add flour to onion-parsley mix in skillet and mix well. Add the rabbit liquid slowly, bring to boiling and cook 1 to 2 minutes longer, stirring constantly. Add more salt if needed and the Tabasco. Mix well with rabbit meant and pour into a baking dish. Cover with crust and bake in a 350° oven for 35 minutes.

MENNONITE HASENPFEFFER

8 or 10 slices of dressed rabbit
Flour
1/4 cup fat (such as bacon fat)
1 teaspoon salt
1/4 teaspoon pepper
1/2 teaspoon allspice or cloves
2 medium onions, sliced
2 heads garlic
2 bay leaves
1/2 cup vinegar
1 can stewed tomatoes
1 small can tomato puree
Fresh mushrooms, sliced, optional

Roll pieces of meat in flour and sprinkle with salt and pepper. Fry in fat until a golden brown. Place in a baking pan or casserole dish and add sliced onion, seasoning, vinegar, stewed tomatoes, and tomato puree (add mushrooms if desired). Let simmer or bake at 350° for 1 to 1 1/2 hours.

POTATOES AND SMOKED SAUSAGE

Onions, chopped
Potatoes, sliced
Smoked sausage, sliced
Black pepper
Minced parsley
Chopped garlic

Combine onions, potatoes, and smoked sausage in a cooking pot . . . Add black pepper, parsley, and garlic, if desired, to taste. Just barely cover with water. Bring to boil, and then cook on low heat until potatoes are soft.

CHICKEN WITH KETTLE GRAVY

Place a stewing chicken in a Dutch oven (or 4 to 5 chicken breasts). Cover with water. Salt to taste. Bring to a boil. Cover and reduce heat. Simmer 2 1/2 to 3 1/2 hours until fork tender. Take chicken out and shred meat. Make kettle gravy.

Kettle gravy:

For each cup of broth use 1/4 cup cold water, 2 tablespoons of flour and salt and pepper to taste. Remember that for hot broth use cold water in your gravy. This will reduce lumps. You can use either flour or cornstarch.

DUMPLINGS

1 1/2 cups flour
2 teaspoons baking powder
1/3 teaspoon salt
3 tablespoons shortening
3/4 cup milk

Place all dry ingredients in a bowl. Cut in shortening, until mixture looks like meal. Stir in milk. Drop by spoonfuls onto hot meat or vegetables in boiling stew. (Do not drop directly into liquid). Cook uncovered 10 minutes. Cover; cook about 10 minutes longer or until dumplings are fluffy.

BAKED NOODLES & HAM

6-8 ounces noodles, cooked and drained
3 tablespoons butter
3 tablespoons flour
1 tablespoons mustard
1 1/2 cups milk
2 cups cooked ham, chopped fine or shredded
1 cup celery, chopped fine
Salt and pepper to taste
Bread crumbs
Dots of butter

Combine butter, flour and mustard in saucepan. Add milk gradually, stirring constantly, cook until thick. Add ham, salt, pepper and celery, then stir in the noodles. Pour into greased baking dish. Sprinkle with bread crumbs and dot with butter over the whole mixture. Bake at 350° for 30-45 minutes.

PORK CHOPS

6 plump pork chops
1/4 cup prepared mustard
1 can chicken rice soup
1/2 can water

Spread mustard on both sides of 6 pork chops. Brown in skillet approximately 10 minutes, turning frequently. Add 1 can chicken rice soup, 1/2 can of water and simmer for 25 to 30 minutes. Serve with rice.

Mennonite
Salads/Sides

MENNONITE GERMAN SUMMER SALAD

2 cups raw spinach, finely chopped
1 thinly sliced peeled cucumber
4 green onions, chopped
1/2 cup sliced radishes
2 cups cottage cheese
1 cup sour cream
2 teaspoons fresh or bottled lemon juice
1/2 teaspoon salt
1/4 teaspoon freshly ground pepper
Paprika, to taste
1/2 cup minced fresh parsley

Wash the spinach the day before, then wrap it in a cloth and refrigerate it overnight.
Chop the spinach, add the cucumber, onions and radishes, and then toss lightly. Arrange
in a wooden salad bowl and place a mound of cottage cheese in the middle.
Blend the sour cream with the lemon juice, salt and pepper and pour over the salad.
Sprinkle the paprika in the middle and the parsley all around. Toss when ready to serve.
This is a very good meal in itself, but it's even better when served with thin slices of
lightly buttered black bread.

PENNSYLVANIA DUTCH POTATO SALAD

6 medium potatoes, peeled, cooked, and diced
3 hard boiled eggs, chopped
1/2 cup chopped onion
1/2 cup chopped celery
1 or 2 sprigs parsley, chopped
1 cup mayonnaise
1 tablespoon prepared yellow mustard
2 tablespoons sugar
2 tablespoons apple cider vinegar
1/4 teaspoon celery seed
1/4 teaspoon paprika
1 teaspoon salt
1/4 teaspoon pepper

In a small bowl, mix together mayonnaise and mustard, sugar and seasonings. Stir in
vinegar. Set aside. In a large bowl combine chopped celery, onion and hard boiled eggs.
Stir in diced potatoes. Add mayonnaise mixture to vegetable mixture. Gently stir all
ingredients together. Refrigerate at least 1 hour before serving.

PENNSYLVANIA DUTCH SALAD/HOT BACON DRESSING

3 slices bacon
1 egg, well beaten
1 1/2 tablespoons flour
1 cup milk
1/4 cup sugar
1/2 teaspoon salt
1 quart salad greens

Chop bacon and fry until crisp; add to greens. Add all other ingredients to egg and pour into hot bacon fat. Cook until thickened. Pour at once over salad greens, such as lettuce, endive, dandelion, or spinach. Add a little onion, if desired.

PENNSYLVANIA DUTCH SAUERKRAUT SALAD

1-16 ounces can sauerkraut with caraway seed
1 green pepper, chopped
1 onion, chopped
1 c. celery, chopped
1/4 cup vinegar
1 cup sugar

Drain the sauerkraut and mix with remaining ingredients. Mix with remaining ingredients. Mix sugar and vinegar, bring to a boil and pour over sauerkraut mixture. Place in a covered container and chill overnight.

SOUR CREAM GREEN BEAN SALAD

4 cups cold cooked beans
2 green onions OR 1 small onion
1 tablespoon sugar
1 tablespoon cider or white vinegar
1/2 cup dairy sour cream
Salt and pepper to taste

Schnippel the beans by cutting them on the bias in very long, thin slices. Place in a bowl and add onion. Blend remaining ingredients, pour over beans and toss lightly.

MENNONITE RED CABBAGE AND APPLE SALAD
(ROTKOHL UND APFELSALAT)

4 cups shredded red cabbage
1 cup apples, quartered and sliced thin
1 teaspoon salt
2 tablespoons brown sugar
2 tablespoons vinegar
3 tablespoons butter
1/2 teaspoon mustard
1/2 cup sour cream
Pepper

Melt the butter in a saucepan. Add the cabbage and apple and stir until the butter coats the mixture and there are signs of softening, but the mixture is not really cooked. Add the vinegar, sugar, seasonings and mustard; simmer another 2 minutes, then stir in the sour cream. Serve hot.

CORN CASSEROLE

1-16ounces can cream style corn
1-16ounces can whole kernel corn, drained
1-10ounces package Jiffy Corn Bread Mix
2 eggs
1 cup sour cream
1/2 cup butter, melted
1 cup grated cheese (Cheddar or Swiss)

Mix both cans of corn, corn bread mix, eggs, sour cream and butter. Pour into a 9 x 13 pan or similar size baking dish. Bake at 350° until lightly browned and center is done (approximately 40-45 minutes)
Add cheese and bake an additional 10 minutes or until cheese melts.

RED BEET PICKLES

2 quarts diced or sliced beets, cooked
3 small onions
3 green peppers
2 cups vinegar
3 cups sugar
3 teaspoon salt
1/2 cup grated horse-radish, optional

Dissolve sugar and salt in hot vinegar. Add horse-radish (if used) and bring to a boil. Add beets and chopped onion and simmer 20 minutes. Place in 2 sterilized quart jars or keep in the refrigerator.

RED BEET EGGS

Hard boil eggs for 10 minutes -- no longer or the egg whites will become rubbery. Place in enough red beet juice to cover eggs, and soak for at least 2 days.

You can use the juice from either red beets purchased at the grocery store, or you can pickle your own red beets:

PENNSYLVANIA PICKLED BEETS & EGGS

8 eggs
2-15 ounces cans whole pickled beets, juice reserved
1 onion, chopped
1 cup white sugar
3/4 cup cider vinegar
1/2 teaspoon salt
1 pinch ground black pepper
2 bay leaves
12 whole cloves

Place eggs in saucepan and cover with water. Bring to boil. Cover, remove from heat, and let eggs sit in hot water for 10 to 12 minutes. Remove from hot water, cool, and peel. Place beets, onion, and peeled eggs in a non-reactive glass or plastic container. Set aside. In a medium-size, non-reactive saucepan, combine sugar, 1 cup reserved beet juice, vinegar, salt, pepper, bay leaves, and cloves. Bring to a boil, lower heat, and simmer 5 minutes. Pour hot liquid over beets and eggs. Cover, and refrigerate 48 hours before using.

BUBBAT

This is a Russian Mennonite recipe.

1 yeast cake
1 egg
1 1/2 cups milk
1 tablespoon salt
3 tablespoons granulated sugar
1 pound smoked sausage
3 1/2-4 cups flour (enough to make a stiff dough)

Scald milk, then cool to lukewarm. Add dissolved yeast and sugar. Add beaten egg, salt and enough flour to make a soft dough that can barely be stirred with a spoon.

Let dough rise.

Pour into a greased 14 x 10 x 2 pan. Press 3-inch lengths of sausage into dough at 3-inch intervals. Let rise again. The dough will almost cover sausages.

Bake at 400° for approximately 45 minutes. Serve hot.

MENNONITE SPINACH

4 slices bacon
3 tablespoons flour
1 1/2 cups water or potato water
2 tablespoons sugar
2 tablespoons vinegar
1 egg yolk
1/4 teaspoon dry mustard
1/4 teaspoon salt
Dash of pepper
3 cups chopped cooked spinach
2 hard cooked eggs
Paprika

Dice bacon and cook until crisp. Remove bacon and add four to the bacon fat. Stir until smooth. Mix water, sugar, and vinegar and add this to the flour mixture. Cook together until thick. Remove from heat; stir in the beaten egg yolk and seasonings. Stir and cook about 2 minutes. Combine this sauce and the diced bacon with the chopped spinach. Heat through. Serve garnished with hard cooked eggs which have been sprinkled with paprika.

PEPPER CABBAGE

1 medium size head of cabbage, grated
1 large green bell pepper or 1 red & 1 green small bell peppers, finely chopped
1 cup sugar
3/4 cup cider vinegar
3/4 cup water
1/4 teaspoon salt
1/2 teaspoon black pepper

Mix sugar, vinegar, water and salt & pepper. Blend well. Pour over mixture or grated cabbage and chopped pepper. Mix together thoroughly. Refrigerate several hours before serving for best taste.

MUSTARD BEANS (Sweet & Sour Wax Beans)

2 quarts wax beans (yellow string beans)
1/2 cup sugar
1/2 cup cider vinegar
1/2 cup water
2 tablespoons prepared mustard
1 1/2 teaspoons flour

Cook beans in salt water just until tender - do not overcook. Set aside. Combine other ingredients in a large saucepan and bring to a boil. Cook about a minute. Pour over the cooked beans. Cool, then refrigerate. Serve cold.

SAUERKRAUT SLAW

1-1 pound can sauerkraut
1 cup onion, diced
1 cup celery, diced
3/4 cup sugar
1/4 cup vinegar

Mix together all ingredients. Marinate, covered for 36 hours. (A 2-quart jar is ideal.) Tip occasionally to mix ingredients during marinating period.

BAKED SAUERKRAUT

1 large jar of sauerkraut (your own homemade or store bought)
1/4 pound bacon, cut into small pieces
2 medium onions, diced

Fry bacon and onions together until lightly browned. Rinse sauerkraut and drain. Toss with bacon and onion mixture and bake, covered, at 325° for 1 to 1 1/2 hours.

PENNSYLVANIA GERMAN SAUERKRAUT

2 pounds sauerkraut
1 large or 2 small onions, sliced
2 slices bacon, diced
2 tablespoons butter
1 tablespoon caraway seed
1 tablespoon sugar
1 cup beef bouillon
1/2 cup white wine or beer

Rinse kraut with warm water; drain well. Cook onion and bacon in butter until onion is soft. Add the drained kraut, caraway seeds and sugar. Cover tightly and cook 30 minutes. Add bouillon and wine. Simmer 30 minutes longer. Kraut has a much better flavor if simmered an additional hour.

SAUERKRAUT & APPLES

2 pounds sauerkraut
4 tablespoons bacon grease (for flavor)
2 tablespoons water
3-4 apples (peeled, cored and chopped)
1 slice onion

Cover and cook until sauerkraut and apples are tender. Add a little salt, sugar and sprinkling of caraway seed and 2 finely grated potatoes. Cook a few minutes longer until potatoes are tender.

MENNONITE TOMATO FRITTERS

1 cup all-purpose flour
1 teaspoon sugar
1/4 teaspoon dried basil or 2 tablespoon fresh basil, minced
1 tablespoon parsley, minced
1 egg
Vegetable oil for frying
1 teaspoon baking powder
3/4 teaspoon salt
1-28 ounces can tomatoes, drained
1 tablespoon onions, minced
1/2 teaspoon Worcestershire sauce

In a large bowl, combine flour, baking powder, sugar, basil and salt. Cut tomatoes into 1/2 inch pieces and drain further on a paper towel. Add them to the flour mixture along with onion, parsley and Worcestershire sauce, but do not mix in. In small bowl, beat egg and add it to the flour-tomato mixture. Blend lightly with a fork. Heat oil (about 1/4 inch) in fry pan. Drop the batter by tablespoons into hot oil. Fry until golden brown on both sides. Keep fritters warm in oven until serving.

BAKED LIMA BEANS

2 pounds dried Lima beans
1 pound bacon, cut up
1/2 cup molasses
2 cups ketchup
1 pound brown sugar
1 large onion, chopped

Soak beans overnight. Cook in water until just about half tender; drain. Add other ingredients. Bake at 350° until tender, about 2 to 3 hours.

SWEET POTATO DISH

6 sweet potatoes, washed
3 tablespoons butter
1/2 cup packed brown sugar
1/2 teaspoon salt
1 1/4 cups hot water

Cut pared potatoes in crosswise slices about 1/2 inch thick. Arrange in 2 quart casserole dish and add the mixture of remaining ingredients. Cover tightly and bake 350° for 40 minutes. Uncover and bake 30 minutes longer or until potatoes tender.

DUTCH POTATOES

4 cups mashed potatoes
1 cup sweetened applesauce
1/2 teaspoon curry powder
1/2 teaspoon paprika
1 tablespoon butter

Stir applesauce into mashed potatoes and add curry powder. Mix well and spoon into 1 1/2 quart casserole dish. Spread butter on top. Bake at 375° for 30 minutes or until hot. Broil for about 3 minutes until potatoes begin to brown. Sprinkle top with paprika.

MASHED POTATO FILLING

1 cup diced celery
1 medium onion minced
1 cup butter
2 packages bread cubes
6 eggs beaten
1 quart milk
1 quart mashed potatoes
3 teaspoon salt
2 pinches saffron in 1 cup boiling water
1 teaspoon pepper
1 teaspoon sage leaves, crushed

Cook celery and onion in butter for 15 minutes slowly. Do not brown. Pour over bread cubes and mix. Add the rest of the ingredients, mixing between each addition. Be sure the finished product is very moist; add more milk if necessary. Put into a greased casserole dish.
Bake at 300° for 45 minutes.
Note: You can freeze 1/2 and bake it later.

POTATO FRITTERS

1 cup mashed potatoes
1 egg
4 rounded tablespoon flour
1/2 teaspoon baking powder
1 teaspoon salt
1/4 cup milk

Mix all ingredients together in the above order and fry in skillet with melted shortening 1/4 inch deep. When brown around the edges and cakes are rather well set, turn and brown on the other side.

POTATO KUGEL

4 cups potatoes
1 1/2 teaspoons salt
6 sprigs parsley
1/4 teaspoon pepper
1 large onion
1/3 cup flour
3 eggs
1/4 cup melted butter

Put raw potatoes, parsley and onion through grinder. Rinse in colander
to remove excess potato starch. Drain well. Beat eggs and add seasoning, flour
and melted butter. Mix batter with potatoes and put into 1 1/2 quart casserole. Bake at
350° for 1 hour or until brown. Serve hot.

POTATO PANCAKES

1 cup grated, raw potato
1/4 teaspoon baking powder
1 cup flour
3/4 teaspoon salt
1 egg
1 cup milk
1 small onion, grated

Combine the potato, baking powder, flour and salt. Add to them the beaten eggs. Gradually add the milk and beat until smooth. Lastly, add the onion. Heat oil in a frying pan and drop batter in by tablespoonfuls. When pancakes are full of bubbles, turn and fry other side until brown. This yields about 6 pancakes.

MASHED POTATO CAKES

1 cup left-over mashed potatoes
1 egg
1/3 cup flour
1/2 teaspoon baking powder
1 teaspoon salt
1/4 teaspoon pepper
2 tablespoons milk
3 tablespoons shortening

Beat egg. Mix with mashed potatoes. Add dry ingredients. Add milk and mix well. Batter should be very thick. Melt shortening in frying pan, on medium high heat. Spoon batter into pan by heaping tablespoons to form 3 to 3 1/2 inch wide potato pancakes, about 1/2 inch thick. When well set and browned on edges, turn over to brown on other side.

PEPPER HASH RECIPE

1 head cabbage, shredded
2 green bell peppers, coarsely ground
1/3 cup vinegar
1/3 cup honey or granulated sugar
Salt and pepper to your taste

Shred and chop the vegetables, add the vinegar, honey, salt and pepper. Serve

RICOTTA CHEESE

1 quart milk
2 tablespoons lemon juice

Put the milk in saucepan over low heat and bring to the scalding point 150°. Remove from heat and stir in lemon juice, the milk will curdle. Let the mix sit without refrigeration for 2-12 hours. This is the mellowing process and it seems the more patient you are the mellower the cheese. Put 2 thicknesses of cheesecloth in a strainer and put strainer over a bowl, Pour the cheese mixture in and allow the whey to drain several hours until the curd is dry.

DUTCH SAUSAGE AND GRAVY

2 pounds Dutch pork or smoked beef sausages (or other sausages)
1 cup bouillon or water
1 onion, sliced
1 teaspoon meat extract
2 tablespoons butter or drippings
1 tablespoon flour

Fry the sausage over slow fire in the butter or drippings until brown; take out and put in the onion. Add the flour. When brown, add the bouillon or water and meat extract. Cook for a few minutes. Pour gravy over sausage and serve, with mashed potatoes and sauerkraut or cabbage.

FRIED DUTCH SAUSAGES

2 pounds Dutch pork or smoked beef sausages (or any other sausages)
1 /2 cup flour
1 cup bread crumbs
1 tablespoon butter or drippings
2 eggs

Salt the sausages, dip in a mixture of white of egg, flour and bread crumbs. Fry slowly to a nice brown crisp color. Serve with sauerkraut.

THREE BEAN DRESSING

1 cup green beans, drained
1 cup kidney beans, drained
1 cup yellow beans, drained
1/2 cup chopped green pepper
1/2 cup chopped onion
1/2 cup vinegar
3/4 cup sugar
Season with salt

Mix and refrigerate overnight. Serve chilled

CHICKEN JELLY

2 cups diced chicken
2 cups diced celery
2 tablespoons gelatin
1 1/2 pints meat stock, seasoned hardboiled eggs, pimento, green pepper

Dissolve the gelatin in cold water. Bring the soup stock to point, add the gelatin, take from fire and stir until gelatin is well mixed. Strain. Immerse a mold in cold water and fill to depth of a quarter of an inch. Let chill, then arrange flat slices of hard-boiled egg, pimento, pepper, etc., and cover with another layer of gelatin. Repeat this process once more, and then mix the celery and chicken together and fill the mold. Pour this the remaining gelatin and chill. Serve with lettuce and mayonnaise.

Mennonite

Cookies & Cakes

MENNONITE COOKIES

2 cups sugar
1 cup butter
4 eggs
4 cups flour
1 teaspoon baking soda
1 teaspoon cream of tartar

Combine butter, sugar, and eggs. Mix together & beat for about 2 minutes. Combine dry ingredients. Add to other ingredients and blend well to form cookie dough. Wrap in wax paper and refrigerate overnight before rolling. Cut a portion of the cookie dough and roll to about 1/8" thick on floured pastry cloth. Cut with cookie cutters. Place on cookie sheets. Decorate, then bake at 400° for 8 to 10 minutes. Remove from cookie sheets with spatula and cool on sheets of paper towels spread on a counter top or table. Cool completely before stacking cookies in cans or plastic containers to store.

JAM COOKIES

1 cup sugar
1 cup sour cream
1/2 cup butter
3 eggs
1 teaspoon soda
1 teaspoon baking powder
1/2 lemon juice and rind
Enough flour to make a soft dough
Your favorite flavor of jam

Cream butter and sugar. Add eggs and beat well. Sift dry ingredients and add alternately with sour cream. . Roll out, cut with round cookie cutter and spoon on tart jam. Fold over and seal the edges well. Bake at 375° until golden brown.

BIRDS NEST

4 tablespoons butter
1 1/4 cup sugar
1 cup milk
3 cups flour
3 teaspoons baking powder
2 eggs
1 quart fresh raspberries or blackberries

Cream butter and sugar, add eggs, mix well. Sift flour and baking powder together. Add alternately with milk. Mix well, fold in fresh berries. Place in a greased and floured 9x12 cake pan. Bake at 350°, 40-50 minutes or until toothpick comes out clean. Serve warm with milk, ice cream or whipped cream.

ROLLED OATS CAKE WITH PEANUT BUTTER ICING

1 1/2 cups water
1 1/2 cups flour
1 cup rolled oats (oatmeal)
1 teaspoon salt
1/2 cup (1 stick) margarine
1 1/2 teaspoon baking soda
1 cup white sugar
1/2 teaspoon cinnamon
1 cup brown sugar
1 cup raisins (optional)
2 eggs

Boil water in a medium sized kettle. Remove from heat. Stir in oatmeal. Add margarine and sugars. Stir until all is melted. Add 2 eggs, beat well. Add dry ingredients. Stir well. Add raisins. Pour cake batter into large well-greased pan. Bake 30 minutes at 350°.

Icing: Bring 1 cup sugar and 3/4 cup water to a boil. Remove from heat and cool slightly. Add 3/4 cup peanut butter, stir until smooth. When cake is done, remove from oven and spread icing over hot cake. Return to oven and bake for another 10 minutes.

DUTCH BLACK WALNUT CAKE

2 cups sugar
1 cup butter
5 eggs
1 pint chopped black walnut meats
1 cup milk
3 1/3 cups flour
4 teaspoons baking powder
1/3 teaspoon salt

Cream the butter and then work in the eggs (well-beaten) and sugar. Then sift the flour, baking powder and salt; add the nuts and stir alternately with milk into the first mixture. Oil a large cake pan and pour in the mixture. Bake for 50 minutes in moderate oven (375°). Spread a nut icing over it.

POPCORN CAKE

3 quarts popped corn
2 1/2 cups chopped nuts
2 1/2 cups powdered sugar
3 tablespoons butter

Mix the popped corn and the nuts. Cook together the sugar and butter until it forms a taffy when you test it in water. Mix the syrup well with the popcorn and nuts, using a large spoon. Then butter a bake pan and put in a layer of the mixture, then press together; then another layer and press, until the dish is full. Then you can slice the "cake."

SHOO-FLY CAKE

4 cups flour
1 cup molasses
1 cup margarine
2 cups water
2 cups sugar
2 tablespoons hot water
1 teaspoon salt
1 teaspoon baking soda, dissolved

Mix dry ingredients with margarine. Take out one cup of crumbs to sprinkle over the top of the cake. Add molasses, water and hot water and soda mixture. Mix thoroughly. Pour into well greased and floured pan. Sprinkle crumbs over top. Bake 45 minutes at 350°.

RICH SHORTCAKE

2 cups flour
1 /2 teaspoon salt
4 teaspoons baking powder
1 tablespoon sugar
1 /3 cup shortening
1 egg, beaten
1/ 2 cup milk
Softened butter

Sift flour, salt, baking powder and sugar. Cut in shortening with a pastry cutter or two knives. Stir in egg and milk. Spread half of batter into a greased 9-inch round baking pan. Spread softened butter over mixture, then pat remaining batter over butter. Bake at 450° for 20 minutes.

CINNAMON CAKE

1/2 cup butter
1 cup sugar
2 eggs, separated and beaten
1/2 cup milk
1 1/2 cup cake flour
1 1/2 teaspoon baking powder
1/4 teaspoon salt
2 teaspoons cinnamon

Cream butter and sugar well, then add the beaten egg yolks and beat well. Add the milk. Sift the dry ingredients together and add to mixture. Fold in the stiffly beaten egg whites. Pour into well-greased layer cake pans and bake at 375° for 20 minutes. When cool, ice with favorite frosting.

RED VELVET CAKE

1/2 cup butter
1 1/2 cups sugar
2 eggs
2 teaspoons cocoa
2-1 ounce bottles of Red Food Coloring
1 teaspoon salt
1 teaspoon vanilla
1 cup buttermilk
2 1/2 cups flour
1 1/2 teaspoons baking soda
1 teaspoon vinegar

Cream together butter, sugar, and eggs. Mix together cocoa and food coloring and add to creamed mixture. Add salt, vanilla, buttermilk, and flour, stirring after each ingredient. Mix together baking soda and vinegar and add to mixture. Mix well then pour into greased and floured 9 x 13 baking pan. Bake at 350° for approximately 30 minutes.

Mennonite

Pies

GROUND CHERRY PIE

Ground cherries, also known as husk tomatoes, produce tiny tomato-like fruits in papery husks on low, lanky bushes. This is an old Mennonite recipe with a crumb topping.

1 (9 inch) pie shell
2 1/2 cups pitted Ground cherries
1 tablespoon all-purpose flour
1/2 cup packed brown sugar
2 tablespoons water
3 tablespoons all-purpose flour
3 tablespoons white sugar
2 tablespoons butter

Preheat oven to 425°. Wash Ground cherries and place in unbaked pie shell. Mix brown sugar and 1 tablespoon flour and sprinkle over cherries. Sprinkle water over top. Mix together 3 tablespoons flour and 3 tablespoons sugar. Cut butter in until crumbly. Top cherry mixture with crumbs. Bake at 425° for 15 minutes, reduce temperature to 375° and continue to bake for 25 minutes.

APPLE CRUMB PIE

6 or more apples, peeled, cored & quartered
3/4 cup sugar
1 teaspoon cinnamon
3/4 cup flour
1/3 cup butter or margarine
1 unbaked pie shell

Put apples in pie shell; use more than 6 if desired. Mix 1/4 cup of sugar and cinnamon; sprinkle on top. Put a pot lid on top of pie. Bake at 425° for 10 minutes. Combine 1/2 cup of sugar, flour and butter and use low speed of mixer to make crumbs. Take pot lid off pie; sprinkle with crumbs. Bake at 350° for 35 minutes.

SOUR CREAM APPLE PIE

1 (9 inch) unbaked pie crust (if homemade replace water with apple cider)
Filling:
1 teaspoon cinnamon
6 large MacIntosh apples, peeled, pared, and sliced
1 2/3 cups sour cream (may substitute fat free)
1 large egg
1 cup sugar
1/2 teaspoon salt
2 teaspoons vanilla
1/3 cup flour

Topping:
1/2 cup butter
1/3 cup brown sugar
1/3 cup white sugar
1/2 cup flour
Pinch of salt
1 tablespoon cinnamon
1 cup chopped walnuts

Add 1 teaspoon cinnamon to pie crust. Form crust into pie plate. In a large bowl, gently mix together apple slices, sour cream, egg, 1 cup sugar, 1/2 teaspoon salt, vanilla, and 1/3 cup flour. Spoon mixture into the pie plate and bake at 450° for 10 minutes. Remove from oven and gently stir apple mixture. Reduce heat to 350° and continue baking for 35 more minutes.

Combine topping ingredients. Top apple mixture with topping. Bake for an additional 15 minutes. Serve.

MENNONITE WALNUT PIE

1 cup black walnuts, chopped
3 tablespoons flour
1 1/2 cups sugar
4 eggs
1 1/4 cups water
1 1/2 cups dark corn syrup
2 unbaked pie crusts (homemade or store bought)

Prepare crust for 2 pies, line 2 medium size pie dishes with crusts. Sprinkle the walnuts over the crusts. In a mixing bowl, beat the eggs, then gradually add the sugar. Fold in the flour, corn syrup and water; pour over walnuts in the pie crusts. Bake for 3 minutes at 450°, then reduce temperature to 350° and bake for another 30-40 minutes.

MENNONITE FUNNY PIE

1 unbaked pie shell

Top Part:
1 cup sugar
1 beaten egg
1/4 cup butter or lard (shortening)
1 cup flour
1/2 cup milk
1 teaspoon baking powder
1/2 teaspoon vanilla

Cream together sugar and shortening. Add the combined milk and egg alternately with flour and baking powder. Add flavoring and set aside until lower part is mixed.

Lower Part:
1/2 cup sugar
6 Tablespoon water
4 Tablespoon cocoa
1/4 teaspoon vanilla

Mix together the cocoa, sugar, water, and vanilla. Pour into an unbaked pie shell. Over this pour the "top part". The chocolate will come up around the outside edge which gives a nice crusty edge on the finished product. Bake at 350° for 35 minutes or until firm and a toothpick inserted in center comes out clean.

CORN PIE

6 or 8 ears sweet corn, husked and cut off the cob
dash of sugar
2 teaspoon butter
2 hard-boiled eggs
salt and pepper to taste
2 - 2 1/2 cups milk
Pastry dough for top and bottom crust

Line a 9-inch pie dish with pastry. Place about half the corn in the bottom of the dish. Slice one hard-boiled egg on top of the corn. Add the rest of the corn and slice the other egg on top. Add the sugar, salt and pepper. Cut the butter into pieces on top. Roll out a top crust and cut a hole in the center about the size of a quarter. Place the top crust on the pie, pinch and seal the crust and cut off excess dough. Pour the milk in through the hole, filling until the corn is covered. Bake at 400° for about 1 hour.

MINCE PIE

1 1/4 pounds lean beef stew meat
2-3/4 pounds Granny Smith apples, cored and chopped
1/3 pound beef suet, coarsely ground
3/4 pound dark raisins
1/2 pound dark brown sugar
1/2 cup white vinegar
1/4 cup molasses
1/2 pound currants
1/2 cup apple juice
3/4 teaspoon ground cloves
3/4 teaspoon ground nutmeg
3/4 teaspoon ground allspice
3/4 teaspoon ground cinnamon
1/2 cup bourbon or rum, divided

Place stew meat in 2-quart pot; add water to cover. Bring to boil, cover, and simmer for 1 hour or until tender. Drain and put through meat grinder or processor until coarse.

Place the ground meat and the remaining ingredients, using only 1/4 cup of the liquor, in a 4-quart pot. Stir and bring to a simmer. Cover and cook gently for 1 hour, stirring occasionally. Remove from heat and allow to cool. (Can be made several days ahead and refrigerated or frozen.)

LEMON PIE

2 large lemons
2 cups sugar
4 eggs, well beaten
Pastry for a 2 crust 9" pie

Slice lemons (unpeeled) paper thin. Add sugar, mix well and let stand at least 2 hours. Add beaten eggs and combine well. Roll half the pastry out and place in pie plate. Fill with lemon mixture, cover with top crust. Cut slits and seal edges. Bake in preheated 450° oven for 15 minutes, then reduce heat to 375° and bake 20 minutes longer. Serve at room temperature.

PENNSYLVANIA PEANUT BUTTER CREAM PIE

3/4 cup confectioners' sugar
1/2 cup smooth peanut butter
1 cup granulated sugar, divided
3 cup milk, divided
3 eggs, separated
6 tablespoons cornstarch, divided
3 tablespoons all-purpose flour
1/4 teaspoon salt
2 tablespoons butter
2 teaspoons vanilla extract, divided
1 pie shell, baked
1/4 teaspoon cream of tartar

Beat together the confectioners' sugar and peanut butter until the mixture is crumbly; set aside.

In a large, heavy saucepan, combine 2/3 cup sugar and 2 cups milk; heat to scalding or until bubbles start to form on the bottom. Do not let it boil.

Meanwhile, in a medium bowl, beat the egg yolks to mix; blend in 3 tablespoons cornstarch, flour and salt. Stir to make a paste.

Whisk in the remaining 1 cup cold milk, whisking until the mixture is smooth.

Pour in some of the hot milk mixture, stirring to combine.

Add mixture in bowl to the milk in the saucepan. Cook over medium-low heat, stirring constantly, until the mixture bubbles up in the center.

Add the butter and 1 teaspoon vanilla extract. Remove from heat and let custard cool. Preheat oven to 350° F.

Sprinkle 2/3 of the crumbly peanut butter mixture into the bottom of the baked (and cooled) pastry shell. Pour the cooled custard mixture over the top.

In a large mixer bowl, beat the egg whites, cream of tartar and remaining 1 teaspoon vanilla extract until stiff peaks form.

Gradually, while beating, add the remaining 4 tablespoons sugar and remaining 3 tablespoons cornstarch. Continue beating until whites are very think and glossy. Spread on top of pie. Sprinkle the remaining peanut butter mixture on top.

Bake for 12 to 15 minutes; watching carefully, or until the meringue is golden brown; cool.

SWEET MILK PIE

1 (9 inch) unbaked pie shell
2 1/2 tablespoons all-purpose flour
3/4 cup white sugar
1 cup whole milk
1 cup dark molasses
2 tablespoons butter
1/4 teaspoon ground cinnamon, or to taste

Preheat the oven to 400°. Combine the flour and sugar in the pie shell, and mix together using your fingers. Pour in milk, and again, mix with your fingers to avoid damaging the crust. Dribble in the molasses, and dot with butter. Sprinkle cinnamon over the top. Bake for 10 to 15 minutes in the preheated oven, then turn the oven down to 350°, and bake for an additional 15 minutes, or until the filling is set, and the top is browned.

Mennonite

Other Desserts

CANADIAN MENNONITE PLUM CUSTARD KUCHEN

A tea-biscuit base with neat rows of plums surrounded by custard - this is a delicious dessert.

Crust:
1 1/3 cups flour
1/4 teaspoon baking powder
1 teaspoon salt
2 tablespoons granulated sugar
1/3 cup margarine or shortening
1 egg, beaten
1 cup milk
Plums, pitted and halved (8-10)

Plum Topping:
1/2 cup granulated sugar
1 teaspoon cinnamon

Custard:
1 beaten egg
1/2 cup sour cream
1/2 cup buttermilk or yogurt
1/3 cup sugar

Sift together flour, baking powder, salt, and sugar. Cut in margarine. Beat egg and milk and stir into mixture. Pat the dough over the bottom of a 9-inch cake pan. Arrange pitted plums nicely in rows enough to completely cover the dough. Sprinkle plum topping over the plums. Bake at 400° for 15 minutes.

Meanwhile mix custard ingredients. Take the kuchen out of the oven, drizzle the custard mixture over the plums, and return it to the oven. Reduce heat to 350° and bake for another 30 minutes.

Serve warm.

MENNONITE RELIEF SALE CREAM BUNS

Some Mennonite bakers fill these light buns with sweetened whipped cream; others use the filling that follows. You can fill them with homemade preserves and a final dusting of icing sugar. Without the cream filling, the buns make great dinner rolls.

Buns:
1 1/4 cups milk
1/2 cup granulated sugar
1 tablespoon shortening
1 teaspoon salt
1/2 cup warm water (105° F)
1 tablespoon active dry yeast
1 egg, well beaten
4 1/2 cups all-purpose flour

Cream Filling:
3 cups confectioners' sugar
2 tablespoons milk
2 egg whites
1/2 cup softened shortening
1 teaspoon vanilla extract
Strawberry or other fruit preserves
Confectioners' sugar

For buns: In heavy saucepan, heat milk over medium heat. Add 1/3 cup sugar, the shortening and salt, stirring until sugar is dissolved and shortening melted. Pour into large bowl; let cool until lukewarm.

In small bowl, stir together water and remaining sugar, stirring to dissolve sugar; sprinkle yeast over surface. let stand in warm place for 10 minutes, until puffy. Whisk egg and yeast mixture into milk mixture. Beat in flour 1 cup at a time until smooth dough forms. Turn dough out onto lightly floured surface; knead in remaining flour until soft but not sticky dough forms. Knead for 8 to 10 minutes until dough is smooth and elastic. Form dough into ball; place in lightly oiled bowl, turning dough to coat with oil. Cover loosely with plastic wrap; let rise in warm, draft-free place for 1 1/2 hours or until doubled in size.

Punch dough down; let dough rest for 10 minutes. Divide dough into 24 even-size pieces; shape into buns. Place on greased or parchment-paper lined baking sheets. Cover with damp towel; let rise at room temperature for 1 1/2 hours or until doubled in size. Preheat oven to 375°.

Bake buns for 12 to 15 minutes until well risen and golden brown, and buns sound hollow when tapped on bases. Let buns cool on wire racks.

For cream filling: In large bowl, beat together sugar, milk and egg whites until smooth. Beat in shortening and vanilla extract until smooth and creamy. Split cooled buns, spread with filling and preserves and sift confectioners' sugar over tops

LEBKUCHEN I

1 1/3 cups honey
1/3 cup packed brown sugar
2 cups all-purpose flour
1 teaspoon baking powder
1/2 teaspoon baking soda
1 cup candied mixed fruit
1 tablespoon sesame oil
1/4 teaspoon ground ginger
1/2 teaspoon ground cardamom
2 teaspoons ground cinnamon
1/4 teaspoon ground cloves
1/4 teaspoon ground allspice (optional)
1/4 teaspoon ground nutmeg (optional)
1 1/2 cups all-purpose flour

Preheat oven to 325º. Spray bottom and sides of 10x15 glass pan with non-stick spray. In a 2 cup glass measuring cup, heat the honey and 1/3 cup sugar in a microwave for 1 minute. Pour this mixture into a medium mixing bowl. Sift together the flour, baking powder, and baking soda. Add to the honey mixture. Stir well. Add and mix in by hand the candied fruit, oil, and spices. Add 1 1/2 to 2 cups more flour. Knead dough to mix (dough will be stiff). Spread into pan. Bake for 20 minutes until inserted toothpick comes out clean. Cut into squares. May be frosted with sugar glaze or eaten plain. Best if stored for 2 weeks.

APPLE FRITTERS

1 cup flour
1 1/2 teaspoons baking powder
3 tablespoons powdered sugar
1/4 teaspoon salt
1/3 cup milk
1 egg, well beaten
2 medium sized sour apples, sliced thin
Oil as needed

Combine dry ingredients into a bowl. Beat the egg and add the milk and stir into the the dry ingredients. Mix well. Add the sliced apples. Drop batter by spoonfuls into hot oil and fry.
Lay on paper towels to absorb oil then sprinkle with powdered sugar and serve.

APPLE POT PIE

6 baking apple
4 cups flour
1 teaspoon cinnamon
Water
1/4 pound lard
1/4 teaspoon salt
1/8 pound butter

Make a dough of the lard, flour and salt, adding enough water to moisten and hold together. Roll out like pie dough and cut into 2 inch squares.
Wash and peel apples and cut into eighths. Put alternate layers of apples and dough into kettle, sprinkling each layer of apples generously with sugar and adding a little cinnamon. Have top layer of dough, dot with butter and fill kettle half full of water, cover and cook over a low flame until apples are soft. Serve with milk or cream.

MENNONITE CARAMEL APPLES

6 red baking apples
1 cup brown sugar, firmly packed
1/2 cup water
1 tablespoon cornstarch
1/8 teaspoon salt
1 tablespoon butter
1 cup milk
1/2 teaspoon vanilla extract

Wash and pare apples one-third of the way down from stem end. Leave whole. Remove cores and cook in syrup made of sugar and water. When apples are tender, remove to serving dish. Combine cornstarch and milk to make a smooth paste. Bring cornstarch to the boiling point and then add syrup in which apples were cooked. Cook for 10 minutes or until cornstarch is thoroughly cooked. Remove from heat and add butter, salt and vanilla extract. Pour over apples and serve.

PENNSYLVANIA DUTCH APPLE DUMPLINGS

Pastr:;
1/4 cup allvegetable shortening
1 3/4 cups allpurpose flour
1/2 teaspoon salt
1/4 pound (1 stick) unsalted butter, chilled
4 to 6 tablespoons ice water

For the apples:

4 small tart apples, such as Granny Smith
1 tablespoon raisins
1 tablespoons dark rum
4 teaspoons unsalted butter

For the syrup:

1 cup firmly packed dark brown sugar
1 1/2 cups water
2 tablespoons unsalted butter

To make the pastry, combine the shortening, flour and salt in a food processor fitted with the steel blade. Using on and off pulsing action, combine until the mixture resembles fine meal. Cut the chilled butter into small pieces, and pulse a few times, or until the mixture resembles coarse meal. Sprinkle with 4 tablespoons of the ice water, and pulse a few times. The mixture should hold together when pinched. Add more water, if necessary. (This can also be done using a pastry blender or two knives.) Scrape the pastry onto a floured board, form it into a ball, and wrap it with plastic wrap. Refrigerate at least 30 minutes,

Preheat the oven to 450°. While the pastry is chilling, peel and core the apples. Divide the raisins and rum into the core holes, and place 1 teaspoon of butter in each core hole. Combine the syrup ingredients in a small saucepan, and bring to a boil. Simmer for 3 minutes, and set aside.

Divide the pastry into 4 parts. Form one part into a ball, and place it between two sheets of plastic wrap or wax paper. Flatten with your hands into a "pancake." Roll the pastry into a circle large enough to cover the apple. Place an apple in the center, and bring up the sides to encase it. Pinch the top together, holding the dough with a little water. If the folds seem thick, trim them off and seal the seams with water. Repeat with the remaining apples. Place the apples on a baking sheet, and brush them with the syrup. Place them in the oven and bake for 10 minutes. Reduce the heat to 330°, and brush again with the syrup. Bake an additional 35 minutes, brushing every 10 minutes. Remove from the oven, and allow to cool for 5 minutes. Serve hot or at room temperature.

MENNONITE APPLE STRUDEL

2 1/2 cups flour
1 teaspoon salt
2 tablespoons shortening
2 eggs, slightly beaten
1/2 cup warm water
5 cups sliced apples
1 cup brown sugar, firmly packed
1/2 cup seedless raisins
1/2 cup chopped nuts
5 tablespoons melted butter
1/2 teaspoon cinnamon
Grated rind of 1 lemon

Sift the flour and salt together. Cut in the shortening and add the eggs and water. Knead well, then throw or beat the dough against a board until it blisters. Stand the dough in a warm place under a cloth for 20 minutes.
Cover the kitchen table with a small white cloth and flour it. Put the dough on it and pull it out with your hands very carefully to the thickness of tissue paper. Spread with a mixture of the fruits, sugar, melted butter, cinnamon and lemon rind. Fold in the outer edges of the dough and roll like a jellyroll - about 4 inches wide. Bake in a very hot oven (450°) for 10 minutes, reduce the oven temperature to 400° and bake about 20 minutes longer. Let cool. Cut into slices about 2 inches wide. It should be flaky and moist.

MENNONITE APPLE TURNOVER

4 cups apples
1 tablespoon water
2/3 cup brown sugar
1 tablespoon butter
1/2 teaspoon cinnamon
1 cup sour cream
1/4 teaspoon salt
1 teaspoon baking soda
1 cup flour

Place apples in 9" pie pan. Sprinkle brown sugar, butter, cinnamon, and water over apples make a batter of sour cream, salt, soda, and flour. Spread over apples. Bake in 350° oven until apples are tender and top is brown. Turn upside down on serving plate. Serve hot with cream and sugar or with a homemade ice cream. (Next recipe).

MENNONITE HOMEMADE ICE CREAM
(1/2 gallon)

1 cup sugar
3 eggs
1/8 cup water
2 Junket tablets
Dash of salt
1 pint cream
Milk
2 teaspoons vanilla

Beat eggs until light. Stir in salt, and cream. Warm milk to lukewarm, (test on wrist) add to cream mixture. Add Junket tablets which have been dissolved in water. Pour into ice cream mixer and fill within 1" of top.

PENNSYLVANIA DUTCH BUTTERSCOTCH PUDDING

1 cup brown sugar
2 tablespoons butter
1 tablespoon flour
1 egg yolk
1 cup milk
1 teaspoon vanilla
1/4 teaspoon salt

(Best done in a cast iron skillet.) Boil sugar and butter together until soft. Beat the egg yolk well and add it to the flour, milk, vanilla, and salt. Carefully stir a little at a time into the sugar mixture. Cook, stirring constantly, until thick and bubbly

SPONGE PUDDING

5 to 6 medium apples
1/4 cup butter
2 cups brown sugar
1 teaspoon lemon juice
Grated peel of 1/2 a lemon
2 large egg yolks
1 cup sugar
1/2 cup cold water
1 teaspoon vanilla
1 cup all-purpose flour
1/2 teaspoon salt
1 teaspoon baking powder
2 large egg whites

Pare, core and slice apples. Melt butter in a deep, 9" round cake pan, add brown sugar and stir until well mixed. Top with apples and sprinkle with lemon juice and peel. Beat egg yolks and sugar until light and pale yellow. Mix water and vanilla. Stir flour with salt and baking powder and add to egg mixture alternately with vanilla mixture. Beat egg whites until stiff, fold into batter and pour over apples. Bake at 350° for 45 to 50 minutes. Serve hot or tepid, and it is best not to unmold the pudding.

CRACKER PUDDING
This is a good light dessert after a heavy meal.

2 /3 cup cracker crumbs
1/ 2 cup sugar
1 tablespoon flour
1 /2 cup coconut
1 /2 teaspoon salt
2 1 /2 cups sweet milk

Mix cracker crumbs, sugar, flour, coconut and salt. Mix well. Pour 1 /2 cup milk over this mixture and let soak. Meanwhile, in double boiler, heat the remainder of the milk. Stir soaked cracker mixture into hot milk. Cook 15-20 minutes, stirring constantly. Chill and serve.

MENNONITE STEAMED CARROT PUDDING

1 cup flour
1 teaspoon baking soda
1/4 teaspoon salt
1 teaspoon ground allspice
1 teaspoon ground cinnamon
1/2 to 1 teaspoon ground cloves
1 teaspoon ground nutmeg
2 tablespoons brown sugar
1/2 cup dark seedless raisins
1/2 cup currants
1/2 cup chopped, candied citron
2 cups cooked and mashed carrots
1 cup chopped suet
1 cup molasses
1 egg, beaten
1 teaspoon grated lemon peel

Sift together the first seven ingredients and mix in the brown sugar. Set aside. Break apart the suet, discarding the membrane which coats it and finely chop. Mix the suet and fruits with the dry ingredients. Mix together the carrots, molasses, beaten egg, and lemon peel. Add liquid mixture to dry ingredients and mix until well blended. Pour into a well greased 2 quart mold. Cover mold tightly and steam for 3 to 4 hours.

Steaming Instructions: To steam, place mold on a rack in steamer or deep kettle with tight fitting cover. Pour boiling water into bottom of steam to no more than one-half the height of the mold. Cover steam and bring water to boiling. Keep water boiling vigorously. If necessary, add more boiling water to keep water level at one half the height of the mold during steaming.

MENNONITE BREAD PUDDING
2 eggs, well beaten
1/2 cup granulated sugar
2 cups milk
1/4 teaspoon nutmeg, ground
4 cups day old bread (1/2-inch slices), cubed
1/4 cup raisins
Beat eggs. Add sugar, milk and nutmeg. Butter a 1 1/2-quart baking dish. Put bread cubes into dish and pour egg mixture over the bread. Let the bread cubes become soaked by the mixture. Mix in the raisins. Bake at 350° for 25 minutes.
Serve warm.

CRUSTY PEACH PUDDING

4 cups peaches, sliced
1/2 cups sugar
1/4 cup sugar
1 teaspoon vanilla
1/2 teaspoon cinnamon
1 slightly beaten egg
1 tablespoon butter or margarine
1/2 teaspoon baking powder
1/2 cup flour

Prepare peaches, place in well-greased 1 1/2 quart casserole. Combine 1/2 cup sugar and cinnamon, sprinkle over peaches. Cover with lid or aluminum foil. Bake in moderate oven (375°) for 20 minutes. Meanwhile cream butter, gradually adding 1/4 cup sugar, vanilla and slightly beaten egg. Mix well. Add sifted flour and baking powder, blend with creamed mixture and spread evenly over peaches. Bake 20 to 25 minutes longer at 400° until golden brown. Serve warm or cold, topped with whipped cream, ice cream or plain cold milk.

PENNSYLVANIA DUTCH PEACH COBBLER

1/2 cup (1 stick) margarine
1 cup sugar
1 cup flour
2 1/2 teaspoons baking powder
3/4 cup milk
1 large can sliced cling peaches

Melt margarine in long baking dish. Sift dry ingredients together. Stir in milk and pour over melted margarine (do not stir; it's better to spoon batter over evenly). Lay peach slices on top. Pour peach juice over top. Bake at 350° for 40 minutes until brown.

PEACH CRUMBLE SQUARES

Topping:
1/2 cup brown sugar
3 tablespoons all-purpose flour
Pinch salt
2 tablespoons butter

Base:
1 cup all-purpose flour
1/4 cup icing sugar
1/2 cup softened butter
1/4 teaspoon salt

Filling:
4 cups peeled peaches, sliced 1/2 inch thick
2 teaspoons lemon juice
1/4 cup whipping cream
1 egg yolk

Preheat oven to 425 F.

Butter an 8-inch square pan and line the base and 2 sides with parchment paper. Combine brown sugar, flour for crumble topping and salt in a bowl. Cut in butter until mixture is crumbly. Reserve crumble.

Combine flour, icing sugar and salt to make the base. Cut in butter until mixture is crumbly and press into the prepared pan. Sprinkle half of crumble topping over pastry.

Toss the peaches with lemon juice and fit into the pan in an even layer. Combine the cream and egg yolk and pour evenly over the peaches. Cover with remaining crumble.

Bake for 10 minutes. Reduce the oven temperature to 375° and bake for another 30 to 35 minutes, or until the crumble is golden and the filling is bubbly. Cool before serving.

* To peel peaches: Bring pot of water to boil, add peaches and cook for 30 seconds. Plunge into ice water to stop the cooking and peel with fingers or a paring knife.

CHERRY DESSERT

1 can sour, pitted cherries
1 cup sugar
1 tablespoon butter
2 cups flour
2 teaspoons baking powder
3/4 cup milk
salt

Drain cherries and set aside. Mix sugar and butter and stir well. Add 2 cups flour and 2 teaspoons baking powder, sifted together. Put aside 2 tablespoons for crumb topping. Drain 1 can sour, pitted cherries. 1 c. sugar 1 tablespoon butter Stir well. Add 2 cups flour and 2 teaspoons baking powder, sifted together. Put aside 2 tablespoons for crumb topping. Add 3/4 cup milk and little salt. Stir by hand. Batter will be a little stiff. Layer batter and cherries in greased, floured 11 1/2 x 8 inch oblong or square pan. Bake at 350° for approximately 40 minutes or until lightly browned. Check with toothpick. Serve warm in a bowl, with sugar and milk (as shortcake).

PULLED MOLASSES CANDY

1 cup Molasses
2 cups Brown Sugar
1 cup Water
3 tablespoons cider vinegar
3 tablespoons butter

Combine the first 4 ingredients in a saucepan. Stir hard, cooking over medium heat until mixture boils. Lower heat, cook & stir about 30 minutes. Mixture will thicken. Cooking is done when a small amount (about 1/4 to 1/2 teaspoonful) becomes brittle when dropped in a cup of cool water. Remove from stove. Add butter & stir. Pour into a buttered, shallow pan. Cool long enough so that candy can be handled. With buttered hands, stretch and pull candy until it is a light brown color. Cut in pieces & wrap in wax paper.

Mennonite

Miscellaneous

FRUIT RELISH

20 ripe tomatoes
8 pears
8 peaches
6 large onions
2 red peppers
4 cups sugar
4 cups white vinegar
2 tablespoons salt
2 tablespoons pickling spices (in a bag)

Put all ingredients in a large heavy bottomed pot. Cook over high heat until it begins to thicken then reduce heat. Stir frequently to prevent sticking and burning until very thick (3 to 5 hours in all). To speed up this process, place fruit in a colander overnight to allow some of the juices to drip out first.

HOMEMADE MAYONNAISE

3 cups sugar
2 teaspoons dry mustard
2 cups white vinegar
2 teaspoon salt
4 eggs, well beaten

Thoroughly mix all ingredients. Bring to slow boil in lightly greased skillet at Med/Low heat. Simmer approx. 2-4 minutes, whisking constantly. Remove from stove. Cool to room temperature for a warm dressing, or refrigerate for later use.

MENNONITE BEEF TEA

1 pound lean, tender beef
2 cups water
1 teaspoon salt

Put beef through a food chopper using a coarse blade. Place in top of double boiler and add the water. Cook, covered, over simmering water about 3 1/2 hours. Stir in salt. Strain and keep liquid in a cool place. If too strong, it may be diluted with some boiled water to strength desired.

APPLE BUTTER

Wash apples & cook whole in a large pot with water. When very tender, push through a sieve - now you have apple sauce! No peeling, slicing, or coring needed!

Ingredients for 1 recipe:
8 cups of apple sauce
4 cups of sugar (2 white - 2 brown)
1/2 cup of vinegar or cider
2 teaspoon cinnamon
1 teaspoon ground cloves

Put all into a crock pot & cook on high for approximately 7 hours. Make sure the steam can escape when it starts boiling. You can store in the fridge or freeze for later use. Canning would also be a good option. Fill jelly jars with hot sauce & process in a boiling water bath.

SWEET PICKLE RINGS

3 quarts cucumber rings
1 tablespoon salt
2 cups sugar
1 pint vinegar
3 dozen whole cloves

Sprinkle salt over sliced cucumbers. Mix lightly. Let stand 1 1/2 hours. Press out the juice, but do not bruise. Add the sugar, vinegar and cloves, and enough water to cover. Heat to the boiling point and can. This makes 5 pints.

SOFT PRETZELS

2 1/2 cups warm water
1 teaspoon course pretzel salt
4-5 cups flour
1/2 cup brown sugar
1 package dry yeast

Mix water, brown sugar, and salt together. Dissolve yeast and add the flour to make moderately stiff dough. Knead well until very smooth. Set aside, cover and let rise about 1 hour. Punch down the dough and divide into about 18 to 20 pieces. Let the pieces rest for a few minutes. Roll out into long ropes about 24 inches long. Twist into a pretzel shape. By the time you have all the pretzels shaped, you can start placing the pretzels in a simmering water solution of 1 tablespoon baking soda per 1 gallon of water. Simmer them for about 1 minute, take out and drain on towels. Place on lightly greased baking sheets and cover the pretzels with course pretzel salt. Bake in a hot oven at 400° for about 10 minutes. Serve immediately.

ONION PRETZELS

1 pound large pretzels broken up
3/4 cup (1 1/2 sticks) butter
1-2 packages of Mrs. Grass' Onion Soup Mix (or your favorite)

Break up pretzels into bite-sized pieces. Put into a large baking pan.
Then melt the butter and pour over the pretzels. Mix well!
Stir in the onion soup dry mix. Mix well till coated.
Bake at 200° for about 2 hours. Stir every 1/2 hour.

Made in the USA
Charleston, SC
01 May 2010